FUNDAMENTALS OF STRATEGIC WEAPONS
Offense and *Defense* Systems

The Gods must weep . . .

. . . War is Father of All

FUNDAMENTALS OF STRATEGIC WEAPONS
Offense and *Defense* Systems

James N. Constant

1981

MARTINUS NIJHOFF PUBLISHERS

THE HAGUE / BOSTON / LONDON

Distributors:

for the United States and Canada

Kluwer Boston, Inc.
190 Old Derby Street
Hingham, MA 02043
USA

for all other countries

Kluwer Academic Publishers Group
Distribution Center
P.O. Box 322
3300 AH Dordrecht
The Netherlands

Library of Congress Catalog Card Number: 80-83263

ISBN 978-94-015-0157-6 ISBN 978-94-015-0649-6 (eBook)
DOI 10.1007/978-94-015-0649-6

Softcover reprint of the hardcover 1st edition 1981

Preface

The purpose of this book is to provide graduate students, professional engineers, military officers, and weapons-systems planners with a comprehensive grounding in the technology, evolution, functions, costs, impacts on society, utility, and limitations of modern strategic weapons systems. Since the subject is often left to the specialists, this work should introduce the general reader to the fundamentals of such systems in an informed manner.

Nowadays the intense interaction of means and ends symbolized by strategic weapons has stimulated a changing discipline in which new missile systems and the intricate logic of nuclear force and counterforce hold the stage alongside the truths of conflict, alliances, fears, games, and subtle gains and losses. Many readers with new personal interest or public responsibility in this complex field will require an overall guide to it.

This book will not prepare the reader to become an expert in the vast subject of strategic weapons systems. It will, however, enable him to understand, evaluate, and form reasonable opinions about these systems, their capabilities and effectiveness. The subject is dealt with more from the viewpoint of the user (investor) rather than the architect (systems engineer) and builder (design engineer). While the user will be concerned with both political as well as technical options which may be available to solve a problem, the systems and design engineers are concerned with analyzing and building technological weapons devices once their requirements are generally known. Basic to the book's purpose is the underlying effort to expose the technical data used in decisions involving the selection, design, and operation of weapons systems.

The book originated in notes and articles which I have found helpful in analyzing and evaluating the performance and effectiveness of a variety of weapons systems. I am greatly indebted to the numerous authors whose works are referenced at the close of the chapters. My own work—necessarily condensed—rests on the solid foundation of these references. I wish also to thank those authors and their publishers who have so generously allowed me to reproduce or adapt illustrations from their publications.

Since it is presented from the user's viewpoint, this work reduces the many theoretical niceties to practical terms. Any responsibility for oversight or error is mine alone. Suggestions and comments, whether constructive or critical, are invited from readers. It is hoped that these comments will keep possible future revisions to the work abreast of the developing technologies.

Claremont, California
January, 1980

James Nickolas Constant

Table of Contents (Part Two)

PREFACE V

INTRODUCTION XVII

CHAPTER 1. STRATEGIC SYSTEMS AND WAR GAMES

1.1. U.S. Strategic Systems 4
1.1.1. Intelligence and Warning 5
1.1.1.1. Ground Radars 5
1.1.1.2. Airborne Radars 9
1.1.1.3. Satellites 9
1.1.2. Defense 10
1.1.2.1. Anti-Ballistic Missile (ABM) . . . 11
1.1.2.2. Anti-Satellite 15
1.1.2.3. Anti-Aircraft 15
1.1.3. Offense 16
1.1.3.1. International Ballistic Missiles (ICBMs) 16
1.1.3.2. Sea-Launched Ballistic Missiles (SLBMs) 20
1.1.3.3. Strategic Bombers and Cruise Missiles 22
1.2. The Game 24
1.2.1. Geometric Technique 29
1.2.2. Monte-Carlo Approach 31
1.2.3. Probabilistic Technique 33
1.2.3.1. Single-Shot Kill Probability . . . 33
1.2.3.2. Lethal Area Approach 34
1.2.3.3. Vulnerable Area Approach . . . 36
1.2.3.4. Cumulative Kill Probability . . . 37
1.3. Game Strategies 38
1.4. Systems Evaluation 41
1.4.1. Strategic Policy 41
1.4.2. Mission Requirements 42
1.4.3. Systems Engineering 43
1.4.3.1. Systems Analysis 43
1.4.3.2. Systems Development 45

1.4.4. System Performance 45
1.4.5. System Cost 48
1.4.6. Value Model 49
1.4.6.1. Factors Normalization 53
1.4.6.2. Weightings Assignments 55
1.4.7. Value Sensitivity Assessment 56
References 58

CHAPTER 2. ABM DEFENSE

2.1. Safeguard System 71
2.1.1. Perimeter Acquisition Radar (PAR) . . . 72
2.1.2. Missile-Site Radar (MSR) 75
2.1.3. Spartan Interceptor 78
2.1.4. Sprint Interceptor 79
2.1.5. Data-Processing System (DPS) 80
2.1.6. Costs 82
2.2. Radar Operations 83
2.2.1. The General Radar Equation 84
2.2.2. Probability of Detection and False Alarm . 88
2.2.3. Target Detection 91
2.2.4. Tracking Operation 92
2.2.5. Beacon and Telemetry Operations . . . 93
2.2.6. Jamming Environment 94
2.2.7. Chaff Environment 95
2.2.7.1. Chaff Glint 98
2.2.8. Nuclear Environment 99
2.2.9. Radar Waveforms 101
2.2.10. Resolution 103
2.2.11. Accuracy 106
2.3. Computer Operations 110
2.3.1. Communications 113
2.3.2. Search, Target Selection, and Tracking . . 115
2.3.2.1. Search 115
2.3.2.2. Target Selection 116
2.3.2.3. Target Tracking 116
2.3.3. Radar Control 117
2.3.4. Interceptor Guidance 118
2.3.5. Signal Processing 118
2.3.6. Testing 119
2.3.7. Local Display and Input/Output 120
2.3.8. Human Operator 120
2.4. Interceptor Operations 121

2.4.1. Parabolic Trajectory 122
2.4.2. Navigational Triangle 124
2.4.2.1. Pursuit Course 127
2.4.2.2. Lead Pursuit Course 127
2.4.2.3. Lead Collision Course 128
2.4.3. Guidance 128
2.4.3.1. Preset Guidance 129
2.4.3.2. Command Guidance 131
2.4.3.3. Terminal Guidance 132
2.4.4. Kill Mechanisms 133
2.4.4.1. Conventional Warheads 133
2.4.4.2. Nuclear Warheads 134
References 137

CHAPTER 3. ABM SYSTEM DESIGN AND PERFORMANCE

3.1. System Requirements 143
3.1.1. Data Accuracy 144
3.1.2. Target Characteristics 145
3.1.3. Target Trajectory and Dynamics 146
3.1.4. Tracking Geometry 146
3.1.5. Subsystem Location 147
3.1.6. Array Characteristics 147
3.2. The Radar Design Problem 154
3.2.1. Radar Performance Factors 155
3.2.1.1. Detection 155
3.2.1.2. Resolution 156
3.2.1.3. Measurement Accuracy 158
3.2.1.4. Reliability Considerations 159
3.2.1.5. Operation in Adverse Environment 159
3.2.2. Array Advantages and Disadvantages . . 160
3.2.3. Conclusions 161
3.3. Error Analysis 162
3.3.1. Error Components 163
3.3.2. Total Target Error 165
3.4. Smoothing of Position and Velocity Data . . . 166
3.5. Noise Reduction 168
3.5.1. General Filter Equation 168
3.5.2. Midpoint Filter 169
3.5.3. Curvefitting error 171
3.5.4. Computation Time 172
3.6. Error Model 172

3.6.1. Error Inputs 173
3.6.1.1. Stabilization System Errors 173
3.6.1.2. Target-Dependent Errors 173
3.6.1.3. Radar-Dependent Errors 176
3.6.1.4. Flexure Errors 176
3.6.1.5. Platform Errors 176
3.6.1.6. Position and Velocity Uncertainty . 177
3.6.2. Parametric Error Curves 177
3.7. Design Procedure 179
3.7.1. Mission Objectives and Radar Functions . 179
3.7.2. Design Selection 180
3.7.3. System-Design Analysis 184
3.7.3.1. Facility 184
3.7.3.2. Prime Electronic Equipment . . . 185
3.7.3.3. Design Features 189
3.7.4. Performance Analysis 189
3.7.4.1. Search 190
3.7.4.2. Tracking 192
3.7.4.3. Resolution 194
3.7.4.4. Accuracy 194
3.7.4.5. Smoothing 199
3.7.4.6. Target Capacity 202
3.7.4.7. Noise Jamming Environment . . 203
References 204

CHAPTER 4. ABM SYSTEM SURVIVABILITY ANALYSIS

4.1. Survivability with respect to ECM 209
4.1.1. ECM Threat 210
4.1.1.1. ECM Location 210
4.1.1.2. ECM Antenna Characteristics . . 210
4.1.1.3. ECM Techniques 211
4.1.1.4. ECM Power, Frequency, and Frequency Stability 211
4.1.2. ECM Vulnerabilities 212
4.1.2.1. Confusion ECM 212
4.1.2.2. Deception ECM 213
4.1.2.3. Probability of Survival 214
4.1.3. ECM Analysis 215
4.1.3.1. Confusion ECM 215
4.1.3.2. Deception ECM 217
4.2. Survivability with respect to Nuclear Bomb Effects 222

4.2.1. Nuclear Threat 222
4.2.1.1. Weapon Parameters 222
4.2.1.2. Weapon Effects 223
4.2.1.3. Weapon Targeting 226
4.2.2. Nuclear Vulnerabilities 226
4.2.3. Nuclear Effects Analysis 227
4.2.3.1. Bonus Point Target 228
4.2.3.2. Direct Point Target 229
4.2.3.3. Replicated-Dispersed Targets . . 231
4.2.3.4. Line Target 231
4.2.3.5. Combination of Simple Targets . . 233
4.3. Survivability with respect to Chemical, Biological Radiological Warfare, and Sabotage (CBR&S) . . 236
4.3.1. CBR&S Threat 236
4.3.1.1. Lethal Biological Weapons . . . 236
4.3.1.2. Incapacitating Biological Weapons . 238
4.3.1.3. Lethal Chemical Weapons 240
4.3.1.4. Incapacitating Chemical Weapons . 243
4.3.1.5. Anti-Plant Agents 244
4.3.1.6. Radiological Weapons 244
4.3.1.7. Sabotage 245
4.3.2. CBR&S Vulnerabilities 246
4.3.3. CBR&S Effects Analysis 248
4.3.4. International Agreements Evolution . . . 250
4.3.4.1. Present International Agreements . 251
4.3.4.2. U.S. Policy 252
4.4. Summary 253
References 254

CHAPTER 5. RADAR COUNTER-COUNTER-MEASURES

5.1. Effects of Jamming 259
5.1.1. Confusion ECM 259
5.1.2. Deception ECM 260
5.2. ECCM Design 260
5.2.1. General Objectives 261
5.2.2. Design Philosophy 262
5.2.2.1. Maximize Signal-to-Noise Ratio . . 262
5.2.2.2. Maximize Discrimination Altitude . 265
5.3. ECCM Techniques 266
5.3.1. Noise Avoidance 267
5.3.1.1. Frequency Agility 267

5.3.1.2. Frequency Diversity 268
5.3.1.3. Spatial Diversity 268
5.3.1.4. Range Diversity 268
5.3.2. Noise Suppression 269
5.3.2.1. Noise Cancellation 269
5.3.2.2. Noise Silencing 270
5.3.2.3. Hard Limiting 271
5.3.2.4. Constant False-Alarm Rate (CFAR) 271
5.3.2.5. Automatic Gain Control (AGC) . . 272
5.3.2.6. Logarithmic Receivers 273
5.3.2.7. Forward Biasing 274
5.3.2.8. Fast Time Constant 274
5.3.2.9. Sensitivity Time Control 275
5.3.2.10. Interference Supression 275
5.3.2.11. Synchronization and Blanking . . 276
5.3.3. Signal Enhancement 276
5.3.3.1. Pulse Compression 276
5.3.3.2. Frequency Modulated Pulse . . . 277
5.3.3.3. Predetection Integration 277
5.3.3.4. Postdetection or Video Integration . 278
5.3.3.5. Receiver Range and Angle-Tracking Bandwidths 279
5.3.4. Signal Discrimination 279
5.3.4.1. Doppler 279
5.3.4.2. Pulsewidth 280
5.3.4.3. Pulse-Repetition Frequency (PRF) . 280
5.3.4.4. Polarization 281
5.3.4.5. Balanced Mixers 281
5.3.5. Radar Parameters Optimization 282
5.3.5.1. Transmitter Power 282
5.3.5.2. Antenna Gain 283
5.3.5.3. Duration of Observation Time . . 284
5.3.5.4. Antenna Sidelobe Suppression . . 284
5.3.5.5. Ohmic Losses 285
5.3.5.6. Propagation Losses 286
5.3.5.7. Siting 286
5.3.5.8. Receiver Noise Figure 287
5.3.5.9. Matched Filter 288
5.3.6. Computer Parameters Optimization . . . 288
5.3.6.1. Memory Extension (or Time Compression) 289
5.3.6.2. Data Smoothing 293
5.3.7. Jamming Parameters Suppression 296

5.3.7.1. Jammer Power 296
5.3.7.2. Jammer Antenna Gain 297
5.3.7.3. Jammer Range 297
5.3.7.4. Chaff Cloud Volume 298
5.3.7.5. Chaff Dipole Cross-Section . . . 298
5.3.7.6. Total Number of Chaff Dipoles . . 300
5.3.8. Non-Catastrophic Operation 300
5.3.8.1. Wide Dynamic Range 301
5.3.8.2. Rapid Reacquisition 301
5.3.8.3. Passive Angle Tracking 302
5.3.8.4. Triangulation 302
5.3.9. Trained Operators 302
5.3.9.1. Operator-Aided Detection . . . 303
5.3.9.2. Operator Over-Ride 304
References 305

CHAPTER 6. BOMBER DEFENSE

6.1. U.S. Bomber Defenses 313
6.1.1. Early Warning Radar 314
6.1.2. Airborne Warning and Control System (AWACS) 315
6.1.3. Interceptor Aircraft 315
6.1.4. Surface-to-Air Missiles (SAMs) 317
6.1.5. Projected Costs 318
6.2. Russian Bomber Defense 318
6.3. Multipurpose Weapons 320
References 321

CHAPTER 7. SURVEILLANCE AND RECONNAISANCE

7.1. Advanced Surveillance 326
7.1.1. Over-the-Horizon (OTH) Radar 329
7.1.1.1. Reflections from the Atmosphere . 330
7.1.2. Satellite Reconnaisance 332
7.1.2.1. Sensors 341
7.1.2.1.1. Spectral Windows . . . 345
7.1.2.1.2. Photography 347
7.1.2.1.3. Television (TV) 349
7.1.2.1.4. Infrared (IR) 349
7.1.2.1.5. Radar 350
7.1.2.2. Resolution and Contrast 351

7.1.2.3. Coverage 355
7.1.2.4. Sensor Equations 359
7.1.2.4.1. Ground Signals Spectral Distribution 359
7.1.2.4.2. Receiver Sensitivity . . . 362
7.1.2.4.3. Background and Contrast 363
7.1.2.5. Communications 366
7.1.2.5.1. Capacity 366
7.1.2.5.2. Actual Data Rate . . . 367
7.1.2.5.3. Efficiency 368
7.1.2.6. Data Processing 370
7.1.2.6.1. Contrast Statistics . . . 370
7.1.2.6.2. Computer Transfer Function 371
7.1.2.7. Data Display 373
7.1.3. Satellite Early Warning 375
7.1.4. Satellite Inspection 375
7.1.4.1. Rendezvous Techniques 376
7.1.4.1.1. Minimum Thrust . . . 376
7.1.4.1.2. Catch-Up Maneuvering . 376
7.1.4.1.3. Shortest Time 377
7.1.4.1.4. Rendezvous Turn . . . 377
7.1.4.2. Propulsion Requirements 377
7.1.4.3. Launch Window Requirements . . 379
7.2. Treaty Verification 380
References 383

CHAPTER 8. COMMAND CONTROL AND COMMUNICATIONS

8.1. Warning 391
8.1.1. Signal Intelligence 391
8.1.2. Airborne and Shipborne Warning Systems 392
8.2. Command and Control 394
8.2.1. Airborne Command and Control 395
8.2.2. Measures of Performance 396
8.3. Communications 397
8.3.1. Defense Satellite Communication System (DSCS) 397
8.3.2. Data Processing 398
8.3.3. Security 400
References 401

APPENDICES. ELECTROMAGNETIC PROPAGATION EFFECTS

A1. Attenuation of RF Waves by Absorption . . . 409
A2. Attenuation of RF Waves by Precipitation . . . 411
A3. Refraction of RF Waves by the Ionosphere . . . 412

Introduction

The ability one nation has to destroy another is no longer questioned. Of the utmost concern, however, is the price a nation must pay to obtain this awesome capability. Modern strategic weapons systems are costly in terms of a nation's resources, priorities, and willingness to pay. Consequently, it is important to understand the peaceful impacts of obtaining such systems as well as their performance in the event of hostilities.

The two functions of modern strategic weapons systems are offensive and defensive. A nation's missile forces, airborne and seaborne, inflict the direct damage upon the opponent while anti-missile forces take the opponent's blows. Thus, simply stated, strategic weapons systems are the modern analogs of the sword and shield. Importantly however, while the sword and shield impacts were less perceived than understood, modern strategic weapons now place heavy demands and burdens on society. Apart from the moral questions involved, the implementation of modern strategic weapons systems has impacts which are well perceived but least understood.

As they are now structured, strategic forces consist of six elements: land-based intercontinental ballistic missiles (ICBMs); submarine-launched ballistic missiles (SLBMs); manned bombers and their associated refueling tankers; anti-ballistic missile defenses (ABMs); surveillance and reconnaissance systems (S&R); and command, control and communications systems (CC&C). The first three components make up the strategic offensive capability and are known as the triad. The last three components also form a triad but are less well-known; ABM and air defenses—the strategic defensive forces—play a secondary though not inexpensive role in the strategic postures of both the U.S.A. and the U.S.S.R. Since large-scale ABM defenses of urban and industrial centers have been preempted by the overwhelming capabilities of offensive missiles and cost-effectiveness considerations, such defensive forces as do exist are confined to protecting fixed land-based forces and top national commands in Washington and in Moscow, respectively, and perhaps to defending against light or accidental attacks. S&R networks and CC&C systems are essential to the effective functioning of all the other forces. Because these systems are numerous, diverse in purpose, and usually highly classified they cannot be examined closely. This book examines the triad of ABMs, S&Rs and CC&Cs while a companion book presents the triad of ICBMs, SLBMs and bombers.

The material is developed in a logical sequence of key topics which have been chosen to fulfill a number of objectives; (1) to present each subject in its sequence to other subjects, (2) to indicate the inter-relationships between similar or related subjects, (3) to arrive at conclusions expeditiously, and (4) to maintain a continuity of development in a necessarily short treatment.

Chapter 1 summarizes U.S. strategic systems and sets forth the game criteria for evaluating the performance of defense systems. A detailed description of the characteristics of an anti-ballistic missile (ABM) defense system is given in Chapter 2. Chapter 3 is devoted to some key design aspects and performance of ABM systems while Chapter 4 is a discussion of ABM system survivability with respect to various types of threats. Radar counter-countermeasures are discussed in Chapter 5. Chapter 6 presents elements of bomber defense. Surveillance and Reconnaissance Systems (S&R) are discussed in Chapter 7. Finally, Command, Control and Communications Systems (CC&C) are discussed in Chapter 8. Appendices are provided giving data on electromagnetic propagation effects affecting the performance of radars.

The references following each chapter are of two types: first, each section is referenced to indicate a major source and, second, general references are also listed although these may not be specifically keyed to the text. It is assumed that not every reader will be satisfied with the material presented on a given topic. The references, therefore, and their own excellent bibliographies, may be used to obtain additional information.

It is hoped that this book will aid the interested reader in obtaining an independent understanding of strategic weapons systems and some of the basic issues regarding their uses and impacts.

Part Two

Defense Systems

1

Strategic Systems and War Games

In order to describe the U.S. strategic defense apparatus it is necessary first to evaluate the Russian threat. This of course is the function of U.S. intelligence and no doubt intensive effort is directed to this end. However, while the public literature provides much information about U.S. systems it provides little about the Russian ones. For this reason, in this chapter we use the U.S. offense capability to obtain some perspective about what the strategic defense problem is all about. Following chapters will be devoted to the U.S. strategic defense systems.

The U.S. deterrent consists of three main components: land-based intercontinental ballistic missiles (ICBMs), Polaris and Trident submarines carrying Polaris, Poseidon and Trident sea-launched ballistic missiles (SLBMs) and bombers carrying gravity bombs and guided missiles. Each of these components alone is actually capable of delivering far more warheads than is needed for deterrence and each is currently defended against surprise destruction in a quite different way. ICBMs are in hard silos and are numerous. SLBMs are hidden in the seas. Bombers are at various levels of alert status and are geographically dispersed. At present, both the U.S. and Russian deterrent capabilities are no longer a matter of numbers. By assuring missile force availability, penetration and effectiveness, the credibility of these nuclear deterrents can be maintained through qualitative improvements as well as through a simple increase in numbers. Such improvements to the nuclear deterrents are being continuously made and these have the advantage that they cost less than the proliferation of numbers. In this manner, the USA has tried to maintain its deterrent capability through such improvements as multiple independently-targeted reentry vehicles (MIRVs) and the accuracy of missile-carrying platforms and the missiles themselves rather than through an increase in the number of weapons. It does not appear that the Russians have taken the same approach and this may well reflect the difference in basic technological capabilities between the two opponents.

1.1. U.S. Strategic Systems

References 1–36, 41–44, 49, 50, 54–63, 67, 68, 74, 75, 97–108

At present the USA has a wide numerical lead over the USSR in independently-targetable strategic nuclear warheads (MIRVs). Already, by mid-1975 the USA had emplaced about 8,500 such strategic weapons deliverable by land-based missiles (ICBMs), sea-based missiles (SLBMs) and bombers, compared with the Russian total of perhaps 2,800. Since 1971 the USA has been producing strategic nuclear weapons at the rate of about three per day. By the November 1974 Vladivostok understanding between the USA and the USSR the USA could have as many as 21,000 strategic nuclear weapons deployed by 1985.

Not included in the strategic arsenals is the vast stockpiling of tactical, or comparatively short-range, nuclear weapons by both sides. It is estimated that the USA alone has approximately 22,000 tactical nuclear weapons distributed around the world. This total includes about 7,000 on land in Europe, 1,700 on land in Asia, 2,500 aboard U.S. Navy combat ships and the remainder, 10,800, assigned to bases in the USA. The total number of tactical nuclear weapons deployed by the USSR is not public knowledge, but it is believed the Russian forces in Europe have some 3,500 tactical nuclear weapons at their command, giving the U.S. forces in Europe a roughly 2:1 advantage in this category.

Tactical nuclear weapons were first introduced in Europe by the USA in 1954, three years before the Russians began their deployment. At present the U.S. tactical nuclear weapons in Europe include four different kinds of surface-to-surface missiles (Lance, Sergeant, Honest John and Pershing), two sizes of nuclear artillery shells (155 millimeter and 203 millimeter, nuclear surface-to-air missiles (Nike-Hercules), nuclear air-to-surface missiles (Walleye) and nuclear demolition munitions and nuclear depth bombs as well as large numbers of nuclear air-to-surface bombs capable of being delivered by more than 500 fighter-bombers. The U.S. tactical nuclear forces are stationed in a number of European countries but are most heavily concentrated in West Germany. France also maintains some of its tactical forces in West Germany.

To support the continually on-going improvement process of its strategic nuclear deterrent forces the USA has initiated a variety of strategic warfare projects. These fall in the three main areas of Intelligence and Warning, Defense and Offense Systems.

1.1.1. Intelligence and Warning

References 12, 16, 26–36, 70, 71

Intelligence and Warning systems include ground radars, airborne radars and satellite sensors which provide the early detection of attacking missiles and the first alert to an impending attack. Since the warning time in the case of an ICBM attack is generally taken as being about thirty minutes, and since fractional orbiting missiles (FOBs) and sea-launched ballistic missiles (SLBMs) can reduce this time to below fifteen minutes, it is extremely important to maintain the longest possible time period as a minimum. This is particularly needed in the case of the strategic bomber force to enable its launching. Thus, the Intelligence and Warning Systems are physically situated to enable the early detection of attacking missiles. Table 1.1 provides a listing of U.S. strategic warfare projects for Intelligence and Warning and gives the project status, milestones, funding and contractors involved.

1.1.1.1. *Ground Radars*

The launch-phase detection of ICBMs is accomplished by the 466L Electronic Intelligence System (ELINT), while midcourse ICBM detection is accomplished by the 474L B-MEWS System. The 440L system is an over-the-horizon (OTH) radar system and can be used to detect FOBs. The older SAGE system has been modified to detect submarine-launched ballistic missiles. FOBs detection is also made by the 446L ELINT and 496L Spacetrack systems. The detection of aircraft is made by the SAGE-BUIC system which is also complemented by the airborne AWACS system. Satellite monitoring is made by the 496L and SPASUR systems. To some degree, the 440L OTH radars, the 474L Ballistic Missile Early Warning system (B-MEWS) and the 496L Spacetrack system have shown that they can also perform, in addition to their early warning functions, the weapons-development surveillance function needed for arms control. In either role, however, the ground-based L-systems are not optimally located, suffer seriously from radar am-

Table 1.1. *U.S. strategic warfare projects*

INTELLIGENCE AND WARNING			
	MISSIONS, TECHNICAL GOALS	STATUS, MILESTONES	FUNDING, CONTRACTORS
STRATEGIC WARNING AND DETECTION	Ground-based radar: midcourse ICBM detection by 474L B-Mews; launch-phase ICBM detection by 466L Elint, 440L OTH; SLBM detection by modified Sage; FOBS detection by 440L OTH, 446L Elint, 496L Spacetrack; aircraft detection by Sage-Buic (complemented by AWACS; satellite traffic monitoring by 496L. *USAF-ESD, Navy-NRL.*	AN/FPS-85 operational; backscatter OTH operational. Studies of integration of Sentinel with 496L and 474L. 474N (SLBMW) may be complemented by Sam-D. Continuing studies of satellite systems to replace overseas ground-based sensors.	Through FY '68, $1.2b; FY '69, $100m. Avco (FPS-26), Bendix (FPS-85), RCA-MSRD (FPS-95). FY '73 $3.2b; FY '75 $3.7b; FY '78 $4.7b est.
SAMOS 	Complete recon system evolved from Discoverer program for photographic and other electromagnetic surveillance of earth from 100-300-mi polar orbits. Photo data returned in recoverable capsules. Advanced Samos equipped with ir and uv sensors, tv link readout. Weights: early Samos, 4000+lb; advanced Samos, 8000 lb. Boosters: Atlas-Agena; Titan 3B- Agena D (advanced Samos). *USAF-SAMSO.*	First launch '60; advanced Samos operational '66. Effective network requires 9 satellites in orbit, 12 launches per year. High-definition readout lacks real-time capability. Much of Samos' nonphotographic capability may go into multipurpose recon early warning satellites; MEL may perform photographic and other Samos missions.	Through FY '68, $2.2b; FY '69, $340m; est. FY '70, $300m. LMSC (prime).

ICM ALARM SATELLITE 	Detection of ICBM trails with ir and uv sensors as little as 2 min after launch. Random orbits; piggyback launches on various USAF boosters. *USAF-SAMSO.*	Operational since '63. Used as sensor development testbed. Will be replaced eventually by multipurpose satellite.	Through FY '68. \$550m; FY '69, \$35m; est. FY '70, \$35m. LMSC (prime).
VELA (NUCLEAR DETECTION SATELLITE) 	Carries radiation sensors for detection of nuclear explosions in space. Later versions detect nuclear explosions in atmosphere; may develop capability to detect underground explosions. 500-lb polyhedrons launched in pairs to fly 180 deg apart in 50,000-60,000-mi polar and equatorial orbits. 5 pairs now flying. Boosters: Atlas-Agena, Titan 3C, *ARPA*, *USAF-SAMSO.*	First pair launched in '63 to near-circular 55,000-mi orbit, second pair in '64. Third pair in '65 was first to keep station, carry sensors for detecting atmospheric explosions. EM pulse, ir, optical, uv, x-ray, gamma-ray, and particle detectors. '67 launch added advanced sensors for very-long-range detection of space explosions.	Through FY '68. \$175m; FY '69, \$40m; est. FY '70, \$40m. TRW Systems (prime); Aerospace Corp. (payload).
INTEGRATED MULTIPURPOSE WARNING-RECON SATELLITE 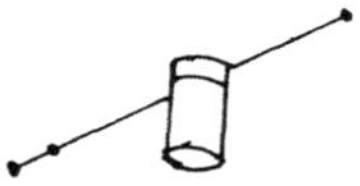	Multipurpose operation in synchronous orbit combining ICBM alarm with functions of Vela, weather satellite, portion of Samos in real time. To detect launch and possibly reentry points of land- and sea-based ICBMs; nuclear explosions in space, atmosphere, and underground; troop movements; assess post-attack damage. Real-time readout, perhaps adaptive data processing. Booster: Titan 3C. *DDR&E, USAF-SAMSO.*	System may employ variety of orbits and spacecraft. ETR launch into elliptical orbit (Aug. '68) suggests Molniya-type surveillance being explored besides low circular orbits. Synchronous orbits with satellite-satellite relay in millimeter bands also may be studied. Integration with 496L and Sentinel under consideration.	Through FY '68, \$125.5m; FY '69, \$94.3m; est. FY '70. \$30m. Aerojet-General, GE-MSD, TRW Systems.

Table 1.1. (*continued*)

INTELLIGENCE AND WARNING			
	MISSIONS, TECHNICAL GOALS	STATUS, MILESTONES	FUNDING, CONTRACTORS
AWACS (AIRBORNE WARNING AND CONTROL SYSTEM) 	Airborne surveillance and strike-recon-intercept vectoring of fighters to stop low-level penetrators. Must interface with Sage-Buic for strategic defense and 407L Tacs for tactical applications. Program could divide into strategic and interim tactical systems. *USAF-ESD, ASD.*	Overland radar technology (ORT) testing completed, including studies of preferred pulse repetition frequencies, antenna configurations, signal processing. Airframes feature tail or aft-fuselage-mounted rotodomes. Twelve aircraft now operational.	Through FY '68, \$35m; FY '69, \$75m; est. FY '70, \$30m; GE-LMED, Hughes-Fullerton. Westinghouse, Raytheon (overland radar); AIL (systems test): Boeing, McDonnell Douglas (airframe); Mitre (system integration).
OCEAN SURVEILLANCE SATELLITE 	Worldwide surveillance of sea state for surface ship routing; monitoring of all surface ocean traffic to prevent sneak attack. Also may attempt detection of submarine trails. *Navy Oceanographic Office, NRL.*	Feasibility studies; radar sensors and power supplies are limiting factors to date. Signal processing to distinguish ship and ocean noises requires improved sea state measurements. Flight of some sensors possibly by '75; could be operational by late '70s.	Through FY '69, \$0.5m; est, FY '70, \$0.3m, Philco-Ford (space-craft); GE-MSD.

SOURCE: See reference 16.

biguities and are slow. Thus, the space-borne systems are emerging as the main devices for solving the problems of strategic intelligence and warning.

1.1.1.2. Airborne Radars

The detection of low-level aircraft penetrators is accomplished by the Airborne Warning and Control System (AWACS), which provides airborne surveillance and directs interceptors to stop the intruders. The system interfaces and complements the SAGE-BUIC ground radar system and is used in conjunction with the 407L Tactical System (TACS) for tactical applications. AWACS is also being considered in connection with the SAGE-BUIC role to detect SLBMs.

The use of aircraft in the intelligence and early-warning role has its own merits and is complimentary to the capabilities afforded by the ground-based radars and satellites used in this role. Thus, airborne radar can be optimally located and eliminates the ambiguity problem of VLF-operating OTH radars and, being at altitude, has a sufficient long-range capability although not operating in the OTH mode. In relation to satellites, airborne radars provide excellent resolving capabilities against low-level aircraft penetrators.

1.1.1.3. Satellites

Satellite systems being used are the SAMOS, ICBM Alarm Satellite, VELA, Integrated Multipurpose Warning Reconnaissance Satellite and the Ocean Surveillance Satellite. These systems have been specifically engineered for the intelligence and warning function.

SAMOS is a reconnaissance system which has evolved from the Discoverer program for photographic and other electromagnetic surveillance of the earth from 150-500 km polar orbits. The photographic data is returned to the earth in recoverable capsules. The advanced SAMOS satellite is equipped with IR and UV sensors and a TV link. The detection of ICBM trails with IR and UV sensors as little as two minutes following their launch is accomplished by the ICBM Alarm Satellite which is placed in random orbits. The Nuclear Detection Satellite (VELA) carries radiation sensors for detection of nuclear explosions in space. Advanced VELA has the capability to detect nuclear explosions in the atmosphere and underground as well. The Integrated Multipurpose Warning Reconnaissance Satellite is used for multipurpose operation in synchronous orbit. It combines functions of ICBM alarm with functions of VELA, the Weather Satellite and portions of SAMOS in real time. It detects launch and possibly reentry points of ICBMs and SLBMs, nuclear explosions in space, the atmosphere and underground, troop movements, and it can assess post-attack damage. It has real-time readout and perhaps data processing. Finally, the Ocean Surveillance Satellite performs the worldwide surveillance of the sea-state for surface-ship routing and it monitors all surface ocean traffic to prevent a sneak attack using surface ships. It may also attempt the detection of submarine trails.

Reconnaissance from space is still limited by sensor resolutions and downlink communications channel bandwidths. As a consequence, satellite reconnaissance missions will continue to use recoverable payloads. It is interesting to note that the return beam vidicon technology of NASA's earth resources program of high resolution from space does not yet approach the spatial resolution needed for military missions. However, satellites can provide missile early launch warning, FOBs deceleration detection and orbital space surveillance through hot and cold body tracking using IR sensors. Satellite orbits include circular ones at 500 to 800 km as well as some elliptical and geosynchronous ones.

1.1.2. Defense

References 4–9, 15–17, 37–40, 51, 72, 73, 99, 108

Defense systems include those against ballistic missiles, satellites and aircraft. These systems are designed to intercept the Russian offensive forces. Since the defense must contend with the near simultaneous arrival of a large number of missiles and aircraft it must possess both the capacity and reaction time to accommodate this heavy traffic in order to minimize damage to the civilian population, industry and deterrent forces. The latter of course must survive the attack and be available in sufficient numbers for retaliation. At present, the cost of defense exceeds the cost of offense by a factor 5-10, i.e., it takes 5-10 times more resources to implement a defense system against an offensive force of given cost. Thus, a breakthrough in defense technology may deny the opponent first-strike capability or give him this capability uncontested if not implemented. Table 1.2 provides a listing of U.S. strategic warfare projects for defense and gives the project status, milestones, funding and contractors involved.

1.1.2.1. Anti-Ballistic Missile (ABM)
References 15, 16, 40, 72, 73, 99

The Safeguard ABM system is a light defense system against ICBMs and SLBMs. It employs the Spartan missile for exoatmospheric intercepts at 1000 km range in the area defense mode of operation and the high-g Sprint missile for intercepts within the atmosphere at 60 km range in the terminal defense mode of operation. The Multifunction Array Radar (MAR) is used to acquire, discriminate and track many targets while the Missile Site Radar (MSR) is used to track and guide interceptor missiles. The Perimeter Acquisition Radar (PAR) is used for distant low reflectivity targets, ultra-high-speed data processing and decision-making.

The Hibex is a very high acceleration, perhaps 100 g, two-stage missile which is used to defend hard-point targets such as ICBM silos. The missile is a potential replacement for the Sprint missile and is capable of intercepting at only 3-10 km range using a simple phased-array radar.

The Sabmis is a sea-launched anti-ballistic missile and is used for the early intercept of hostile ballistic missiles from surface ships and submarines before they separate into multiple warheads or release their payloads.

The development of missile defense technology beyond the Sentinel and Safeguard ABM systems is under the Advanced Ballistic Reentry Systems (ABRES) program sponsored by the U.S. Air Force and the Defender program sponsored by ARPA. This latter program includes continued work on the Nike-X as well as the Hibex and Prestage-Upstage advanced projects. The cone-shaped High Impulse Booster Experiment (Hibex) missile is about 5 m long and is launched from a 15 m × 3 m silo. Its single engine uses a high-impulse propellant molded with a filament-wound glass fiber cage. Although Hibex has a lower burnout velocity than Sprint it accelerates faster up to 100-g and could intercept and destroy hostile warheads at altitudes as low as 3 km above the ground.

Table 1.2. *U.S. strategic warfare projects*

		DEFENSE	
	MISSIONS, TECHNICAL GOALS	STATUS, MILESTONES	FUNDING, CONTRACTORS
SENTINEL (ANTI-ICBM SYSTEM) 	"Thin-line" anti-Chinese system, is scaled- down version of Nike-X. Modular defense against ICBMs, SLBMs with Spartan missile for exoatmospheric intercepts at 600+ mi in area defense; high-g Sprint missile for in-atmosphere intercepts at 35 mi in terminal defense; Multifunction Array Radar (MAR) to acquire, discriminate, track many targets; Missile Site Radar (MSR) to track, guide interceptor missiles; Perimeter Acquisition Radar (PAR) for distant low-reflectivity targets; ultra-high-speed data processing for control, decisionmaking *Army-SSM, SENS-COM, SENSEA*.	Nike-Zeus development began in '55, program reoriented as Nike-X in '63 with emphasis on phased-array radar able to track many targets without saturation, terminal phase intercept to take advantage of atmospheric filtering of light decoys, chaff, debris. Extensive R&D has led to successful ICBM intercepts; missile, radar, and computer hardware development; knowledge of reentry phenomena. First Sprint launched '65; first Sentinel battery was to be deployed in '71; full system to handle limited number of attacking missiles and penaids was to be operational by '75.	Through FY '68, $3.52; FY '69, $964m; FY '70, $1.75b. Western Electric (prime); McDonnell Douglas (Spartan); Martin-Orlando (Sprint); GE, Raytheon (MAR); Bell Labs, Raytheon (MSR); Univac (data processor).
HIBEX-UPSTAGE (HARD-SITE MISSILE DEFENSE) 	Feasibility studies of very-high-acceleration (100-g), 2-stage missile to defend hard points such as ICBM silos. Upstage second stage with external burning for attitude control; filament-wound Hibex first stage. Potential replacement	Development of silo-launched Hibex begun '63; unguided flight tests with successful maneuvering in '65-'66, used double-fast burning propellant. Prestage development of Upstage subsystems begun '65; tests com-	Through FY '68, $100m; FY '69, $30m; est. FY '70, $30m. Boeing (prime, vehicle); Hercules (propulsion); McDonnell Douglas (Prestage); Sperry Rand (Hapdar).

	for Sprint. Intercept at only 2-6 mi, permitting use of simple phased-array radar. Hard-point demonstration radar (Hapdar) under development as complementary effort. *ARPA, AMC.*	pleted '67. Upstage development begun '67; testing completed '71, then probably Hibex-Upstage mating. Continuing Hapdar tests include multitarget acquisition and tracking.	
SABMIS (SEA-LAUNCHED ANTIBALLISTIC MISSILE) 	Early intercept of hostile ballistic missiles from surface ships and submarines before warhead separates into multiple warheads or releases decoys and chaff. Atmospheric intercepts also conceivable. *USN-ONR.*	Sabmis among strongest of Advanced Sea-Based Deterrent (ASBD) concepts, second in priority only to advanced Poseidon. Feasibility studies show practicality of mounting missiles and phased-array radars on ships.	Through FY '68, $3.4m; FY '69, $6m; est. FY '70, $25m; Hughes (prime); LMSC (missiles), Newport News Shipbuilding (shipboard installation).
ANTISATELLITE SYSTEMS 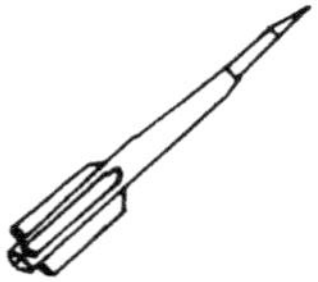	Inspect or destroy hostile satellites and FOBS using manned vehicle (possibly MOL), Zeus and Thor-Burner 2 missiles (from Kwajalein and Johnson Island, respectively). "Window-limited" intercepts based on Spadats orbital predictions have been expanded by using the highly maneuverable Burner 2 upper stage. Ship- and submarine-launched missiles also being considered for global deployment covering all possible orbits. *USAF-SAMSO, AMC.*	Space intercept with Delta and Agena upper stages (Programs 437 and 505) now incorporated in Spin (Program 922). Limited-range, off-the-shelf Nike-Zeus (Program 505) operational since '64; of more psychological than real value. Thor with Burner 2 tested '68. Sentinel system also has intrinsic anti-satellite capability in Spartan missile with guided upper stage (e.g., Upstage); alternate systems include Minuteman 1, Burner 2.	Through FY '68, $255m; FY '69, $40m; est. FY '70, $40m. McDonnell Douglas (Thor, Delta, 437 guidance), Western Electric (launch and engineering support, guidance), Hughes, LTV, Northrop (studies).

Table 1.2. (*continued*)

DEFENSE			
	MISSIONS, TECHNICAL GOALS	STATUS, MILESTONES	FUNDING, CONTRACTORS
YF-12A, SP-71 (ADVANCED STRATEGIC AIRCRAFT) 	Air Force interceptor and strategic recon versions of CIA-developed A-11 with Mach 3 cruise and 100,000-ft-altitude capability. Titanium airframe. YF-12A carries long-range search radar, inertial navigation, onboard computer for operation independent of ground control in degraded environment; with F-106X, could be part of interceptor fleet vectored by AWACS. Larger SR-71A carries multisensor recon package to detect military buildup and for targeting, post-attack damage assessment. *USAF-ASD.*	A-11 development begun '59; first flight '61; 12-13 operational A-11s delivered to CIA by '63. First YF-12A delivered Dec. '62; 3 test models built. First SR-71A flight '64, operational in Jan. '66; 17 planes built. Production line now expected to close down unless decision is made in favor of F-12/F-106X interceptor fleet.	A-11: est. $200m through FY '62 (CIA). YF-12A: through FY '68, $258m; FY '69, none; est FY '70, $50m, SR-71A; through FY '68, $815m; FY '69, $5m; est. FY '70, none. Lockheed-California (aircraft); P&W (engines).

SOURCE: See reference 16.

1.1.2.2. Anti-Satellite
Reference 109

Anti-satellite systems are used to inspect or destroy hostile satellites and FOBs. Both direct ascent techniques and orbital platforms have been studied as satellite interceptors. Two current programs for space intercept (SPIN) are the Army-sponsored program 505 and the Air Force-sponsored program 922. The 505 system is based on Kwajelein Island in the Pacific and is built around the surplus Nike-Zeus missiles and appears to be dormant. The 922 program is based on Johnston Island and combines the Thor missile with an upper stage which expands the normally window-limited ascent capabilities of earlier satellite interceptors.

The simple method of destruction of a satellite by x-ray energy offers the largest kill radius. An alternative method is that of the hyper-velocity kill which requires terminal homing. In this method clusters of energetic particles of mass are directed towards the target in a shotgun blast pattern and this causes its destruction.

1.1.2.3. Anti-Aircraft
References 17, 51

Early advanced strategic aircraft were the YF-12A and the SR-71, used as reconnaissance and interceptor platforms. Both aircraft have titanium air frames and were outgrowths of the CIA-developed A-11 which has a Mach 3 at 30,000 m cruise altitude. The YF-12A carries long-range search radar, inertial navigation and an on-board computer for operation independent of ground control and in a degraded environment. Together with the F-106X interceptor, the YF-12A could be part of an interceptor fleet vectored by the AWACS. The much larger SR-71A carries a multisensor reconnaissance package to detect military buildup and to assist targeting of other weapons. It can also be used in post-attack damage assessment.

Current strategic aircraft are the F-14 and F-15. The F-14 is a Navy plane that has a 160 km radar and uses the Phoenix air-to-air missile while the F-15 is an Air Force plane with an 80 km radar and additional fuel tanks for increasing its range.

1.1.3. Offense

References 10, 11, 16, 41–48, 52–63, 104

Offensive systems include ICBMs, SLBMs and bombers and these are designed to withstand a first-strike launch by the opponent and, following their survival, to launch against, penetrate and effectively inflict unacceptable damage upon the opponent. Since an attack on all three components cannot be made simultaneously (simultaneous launch by an opponent precludes simultaneous arrival times at the various offensive systems being attacked and staggering the launches to insure simultaneous arrival times can both be detected in sufficient time), it appears quite likely that the combined offensive forces being attacked can survive the first strike. Furthermore, any one of the three offensive components is capable of inflicting unacceptable damage upon the opponent. Table 1.3 provides a listing of U.S. strategic warfare projects for Offense and gives the project status, milestones, funding and contractors involved. Offensive forces include the Minuteman I, II, and III land-based ICBMs, the Polaris, Poseidon and Trident SLBMs, SRAM and advanced strategic missiles.

1.1.3.1. Intercontinental Ballistic Missiles (ICBMs)

References 10, 16, 101, 102

The Minuteman missile is the main nuclear strike force with 1,000 missiles deployed. Some 450 are Minuteman IIs and 550 are Minuteman IIIs with MIRVed warheads. All Minuteman I missiles have now been retired and the 450 Minuteman II missiles will eventually be retired in favor of Minuteman II. An intriguing possibility for the missiles being retired is their possible conversion and use as ABM interceptors for the mid-course interception of Russian missiles. The Minuteman II has a larger second-stage motor than the Minuteman I and has improved guidance, 13,000 km range and terminal guidance. The Minuteman III incorporates a Post-Boost Control System (PBCS) which deploys three Mk 12 MIRVs. This force is augmented by fifty-four Titan II liquid propellant ICBMs which will eventually be phased out.

Advanced strategic missiles include both land-based and sea-based ballistic missiles to replace the Minuteman and Poseidon in the late 1970s and 1980s. Land-based concepts include new

Table 1.3. *U.S. strategic warfare projects*

OFFENSE

	MISSIONS, TECHNICAL GOALS	STATUS, MILESTONES	FUNDING, CONTRACTORS
MINUTEMAN 	Operational main nuclear strike force, now at full 1000-missile strength MM2 has larger 2nd-stage motor, improved guidance, range to 8000 mi, terminal homing. MM 3 under development, will further improve payload/range, penetration capabilities. Augmented by 54 Titan 2 liquid-propellant ICBMs, which will be phased out as MM 3 becomes operational. *USAF-SAMSO.*	First MM 1 launched '61; MM 2 has improved CEP (reportedly 1.5 mi), eight times higher kill probability, TVC instead of MM 1's gimbaled nozzles for trajectory control. Problems with MM 2 guidance and control led to reprogramming of funds from MM 3, slowing MM 3 deployment. MM 3 incorporates PBCS (post-boost control system) to alter ballistic trajectory, Mark 12 warhead with 3 MIRV First MM 3 flight Aug. '68. About 600 MM 3s and 400 MM 2s planned for deployment by '73; eventually entire missile force will be MM 3. Advanced missile to follow, possibly MM 4.	Through FY '68, $9.06b; FY '69 $1b; FY '70, $900m. Boeing (missile); Avco, GE (reentry bodies); Autonetics (guidance and control, PBCS); Bell (PBCS).
POLARIS, POSEIDON 	Missile portion of Fleet Ballistic Missile (FBM) system. Poseidon has twice Polaris' weight and payload, range of 3000+ mi, terminal guidance, MIRV (eventually up to 10 warheads) To be launched from reamed-out tubes on 31 Polaris subs; another 10 subs to retain Polaris A-3. All subs carry 16 missiles.	Polaris A-1, A-2 to be phased out by '70. Poseidon development began in Jan. '65, stepped up in '67 for deployment by '70. First-stage motor fired July '67, underwater pop-up launch, Nov '67; first full flight test, Aug. '68.	Through FY '68; $7b; FY '69; $665m; est. FY '70, $475m. LMSC (prime); Hercules, Thiokol (engines); Westinghouse (launch system); GE (guidance); Nortronics (checkout).
FB-111A (INTERIM STRATEGIC BOMBER)	Strategic bomber version of F-111A to replace aging B-52C through F series. Armed with nuclear-tipped Sram	First prototype flight, July '67, first production flight, July '68; operational '69, extensive deployment by '71. Pro-	Through FY '68, $762m; FY '69, $587m; est. FY '70, $450m. GD-Ft. Worth (aircraft): P&W

Table 1.3. (*continued*)

OFFENSE

	MISSIONS, TECHNICAL GOALS	STATUS, MILESTONES	FUNDING, CONTRACTORS
	standoff missile to roll back enemy defenses, hit primary targets. Basic variable-geometry F-111A with added avionics (mostly penaids), larger wing, up to six 600-gal external fuel tanks for 4000-mi unrefueled range. Mach 2.5 at altitude; Mach 1.2 at very low altitudes for penetration. *USAF-ASD.*	gram likely to be cut back from 253 to 90 aircraft, with funds being transferred to Amsa. Several crashes in '68 have placed entire F-111 program in doubt. FB-111A uses Mark 2B integrated avionics, including long-range inertial navigation, terrain-following radar, ECM gear, armaments control system. Two afterburning turbofans of 19,000-lb thrust each on reheat; engines for operational aircraft may have higher thrust.	(engines); Autonetics (Mark 2B avionics).
SRAM (SHORT-RANGE ATTACK MISSILE) 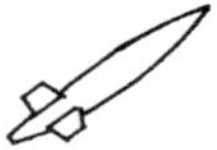	Standoff missile for attack against well-defended air defense and primary targets from FB-111A and possibly B-52. Lightweight missile with small radar cross section, range of 100+ mi. Restartable solid motor, inertial guidance, and terminal homing likely. *USAF-ASD.*	In development for deployment in '69. FB-111A will carry up to 6 Srams; B-52G and -H will get avionics required for Sram launch. Flight tests from B-52 to begin early '69. Amsa could carry 32 Srams internally. Variety of interchangeable terminal guidance heads likely, including radar command, tv map-matching, antiradiation homing, and combinations thereof.	Through FY '68, $120m; FY '69, $102m; est. FY '70, $115m. Boeing (prime); LPC (propulsion); General Precision (guidance).
ADVANCED STRATEGIC MISSILES	Studies for land-based (ICBM-X, ICM, WS-120A) and sea-based (SLMS/ULMS) missiles to succeed Minuteman and Poiseidon in late '70s and '80s. Land-based concepts include new hard-rock	Air Force studies began '63; Improved Capability Missile (ICM) defined '67. Strat-X studies in '67–'68 concluded new missiles should be deployable in superhard silos rather than mobile. Contract	Through FY '68, $70m; FY '69, $70m; est. FY '70, $115m. Avco, GE, Goodyear (studies), LMSC, Martin, TRW Systems.

	silos compatible with MM 3, larger booster with greater throw weight, improved motors, "self-aligning boost and reentry" (Sabre) guidance. Minuteman 4 could result from this effort. Navy's Surface/Undersea Long-Range Missile System (SLMS/ULMS) is part of ASBD (Advanced Seaborne Deterrent) effort, covers studies of deep-diving ballistic submarines, bottom-mobile weapon systems. All concepts use Abres technology for advanced warhead. *USAF-SAMSO, USN-ONR.*	definition phase for new missile (possibly MM 4) to begin '70. Navy's advanced development work began '65; production unlikely before late '70s. All systems contain advanced precision guidance and terminal homing with map-matching techniques for zero CEP. Indent SLBMs operational '80.	
AMSA (ADVANCED MANNED STRATEGIC AIRCRAFT) 	Follow-on to B-52 and FB-111A to penetrate air defenses of '70s and '80s. Would have variable-sweep wing, 10,000-mi unrefueled range, carry 32 Sram nuclear-tipped standoff missiles plus bombs or large nuclear and conventional internal bomb load (100,000 lb), extensive active and passive penaids. Gross weight about 370,000 lb. Supersonic at very low altitudes; Mach 2.5 cruise at altitude. Large undercarriage, self-contained avionics permit dispersed deployment to small fields and highways. *USAF-ASD.*	Extensive mission and airframe studies continue, plus R&D on propulsion and avionics. Project stalled by DOD concentration on ICBMs and SLBMs as prime strategic force. FB-111, developed as interim Amsa, could be cut to provide funds for Amsa development. Soviet's deemphasis of strategic bombers (they have 155 to our 680) has given program low priority, but reports of mobile missile deployment could make low-altitude, penaid-laden Amsa attractive. Dispersed deployment improves second-strike capability. B-1 bomber cancelled '78.	Through FY '68, $118m; FY '69, $30m; FY '70, $30m. Boeing, GD-Ft. Worth, NAR-LA (aircraft, missions); GE, P&W (engines); Autonetics, IBM Federal Systems (avionics).

SOURCE: See reference 16.

hard-rock silos compatible with Minuteman III, a larger booster with greater payload weight, improved motors and improved (SABRE) guidance. Sea-based systems include newer, faster and deeper-diving Trident submarines as well as bottom mobile weapons systems.

An important aspect of the offensive force is the Command-and-Control (C&C) network which is needed as a war-prevention network in peacetime and to provide force management and control in time of conflict. The latter function is particularly critical for a credible deterrent since if it does not work the ICBMs, SLBM submarines and SAC bombers may be focusing their firepower on empty silos and empty airstrips. C&C is also needed to coordinate between the offensive forces and the ABM forces to stop defending missiles from shooting friendly ICBMs after they are launched or to keep offensive missiles outside corridors being defended against attacking warheads. To improve the survivability of ICBMs, the use of ABMs and active and passive hardening have been studied. Both former choices are expensive however so that presently superhard silos are being built to protect the Minuteman III force. These superhard silos are capable of withstanding 200 kg/cm^2 of overpressure or about ten times more overpressure than the present silos. In addition, guidance systems are being continually improved. The Minuteman III can send a 900 kg payload a distance of 8,500 km with a circular error probability (CEP) of 400 meters. The next-generation missiles will be capable of throwing 3,200 kg payloads to 10,000 km ranges with a CEP of below 300 m. However, increasing the size and throw-weight of missiles could make each missile a more lucrative target for the enemy so that a larger number of smaller missiles offers flexibility in targeting and escalation and as the smaller missiles are more dispensible they also tend to exhaust the attacker's arsenal and to minimize the damage done by his large-yield weapons.

1.1.3.2. Sea-Launched Ballistic Missiles (SLBMs)
References 10, 16, 48–49, 103

The Polaris, Poseidon and Trident missiles are the missile portions of SLBM systems. The Poseidon has twice the range of Polaris and is twice as heavy. Its range is over 5,000 km and it has terminal guidance. Each Poseidon missile carries some ten

MIRV warheads and is launched from widened tubes on thirty-one of the forty-one Polaris submarines. The remaining ten Polaris submarines carry the Polaris A-3 missile with three warheads. Both the Poseidon and Polaris missiles use the Polaris submarine, each of which carries sixteen missiles. The Trident I missile which carries twenty-four MIRV warheads has the same size as the Poseidon missile and was deployed in Polaris submarines in 1973. The Trident II missile will be carried by a new and larger submarine which is not yet operational. Trident I has a range of 6,400 km, while Trident II reaches the range of 10,000 km.

The USA currently has 41 submarines equipped with SLBMs. Of these, 31 boats have been equipped with the longer-range Poseidon missiles and the remaining 10 boats carry the Polaris missile. All 41 boats are nuclear-powered and are capable of staying on their stations for prolonged periods of time. Each of the 41 Polaris submarines has sixteen tubes, thus giving the capability to launch 656 missiles. Of these, 496 are Poseidon warheads and the remaining 160 are Polaris A-3 warheads with a range of 4,600 km. Each Poseidon missile carries some ten MIRVs with yields of somewhat less than 100 KT each, while each Polaris missile carries three MIRVs. The U.S. Navy has under contract three of the ten Trident submarines it plans to build. The vessels which will be the backbone of the U.S. sea-based nuclear deterrent forces in the 1980s and beyond, will carry twenty-four long-range missiles. The nuclear-powered Tridents will be huge submarines displacing 18,700 tons compared to the 8,250 tons displaced by the current Polaris submarines.

The Soviets have led the world in submarine production since the end of World War II, building more than 600 submarines compared with 150 for the USA. The Soviets have emphasized nuclear power, building about twelve submarines per year, ten of them nuclear, in contrast with the U.S. average of three or four per year. The Soviet submarines have been for some years armed with cruise missiles capable of submerged launch against surface ships or land targets. The USA is still developing such.

In 1978, the Soviet Navy's active fleet deployed 43 nuclear powered and 24 diesel powered attack submarines armed with anti-ship cruise missiles in addition to their conventional torpedoes. The Soviet navy also operates 38 nuclear powered and 154 diesel powered torpedo attack submarines giving a total of

259 attack submarines. Against these attack submarines, the total U.S. navy surface combatant fleet in 1978, including frigates or larger, was 165 vessels. However, the U.S. Navy considers the Russian's lead in numbers to be offset in two key areas: anti-submarine sensors and techniques for quieting submarines in operation.

The Soviets currently maintain from 40-60 ships, including a dozen submarines, year round in the Mediteranean. The Soviet Indian Ocean fleet is comprised of one submarine, four to five destroyers, three minesweepers, two amphibious ships and a dozen support ships including oilers. Most of the Indian Ocean fleet cruises or is anchored near the Gulf of Aden.

Diesel submarines are gradually giving way to nuclear powered submarines, gun destroyers and support ships are being added to give the Soviet fleet a much greater operational range. Prior to 1975, the Soviet Pacific fleet exercised almost exclusively in the Sea of Japan. Now, the Soviet fleet is well established in the Western Pacific and Indian Oceans. The Russian Pacific fleet, which had only 95 support ships in 1960, now consists of an armada with more than 240 support vessels, 125 submarines—50 of which are nuclear powered, 10 cruisers, 30 destroyers, 50 destroyer escorts and 300 smaller fighting ships such as fast patrol boats, amphibious ships and mine sweepers. One important task of the Soviet Pacific fleet is to restrict the operational range of the U.S. SLBM carrying submarines.

1.1.3.3. Strategic Bombers and Cruise Missiles

References 50-63, 104

Strategic bombers include the FB-111A and AMSA. The strategic bomber force serves as a backup force especially in the event that the opponent obtains a breakthrough in his ABM technology and, as a result, can minimize his damage from attacking missiles.

The FB-111A is an interim strategic bomber which replaces the B-52C through F series. It is armed with the nuclear SRAM standoff missile and has a 6,500 km unrefueled range. It reaches Mach 2.5 at altitude and maintains Mach 1.2 at the very low altitudes near the ground needed for penetration. The Advanced Manned Strategic Aircraft (AMSA) will replace the B-52 and FB-111A in the late 1970s and 1980s. It will have a variable swept wing, 16,000 km unrefueled range, will carry thirty-two SRAM missiles, other weapons, penetration aids and will be supersonic at low near-ground altitudes. In the early 1970s AMSA evolved as the B-1 bomber whose production phase was cancelled by the U.S. Congress in 1978. However, several B-1 prototype aircraft were built and tested and the B-1 remains as one option for replacing the aging B-52 and FB-111A bomber fleet.

The Short-Range Attack Missile (SRAM) is a standoff missile for attack against well-defended air defenses and is launched from both the FB-111A and B-52 aircraft. It is a lightweight missile with a small radar cross-section and has a range of over 160 km. The U.S. Air Force has entered into full production and deployment of the SRAM missile, which is a nuclear warhead supersonic missile and gives the attacking airplane a standoff capability which avoids terminal anti-aircraft defenses or gives the capability to suppress the defenses and thereby penetrate to the target. The SRAM standoff missile is the primary armament for the 104 B-52Gs and -Hs, each of which carries twelve missiles in underwing racks. The FB-111A carries six SRAMs and AMSA carries thirty-two SRAM missiles. In order to increase AMSA's chances of penetration its radar cross-section will be minimized by the elimination of sharp frontal corners and by optimal canting of the windshield. The IR signature will be reduced through camouflaging of the engine exhaust by means of a flood nozzle. About two-thirds of this bomber's electronics will be ECM gear.

One of the more obvious shortcomings of the agreements reached as a result of the strategic arms limitations talks (SALT I and II) between the USA and the USSR was the omission of any restriction on the development, construction or deployment of modern long-range bombing aircraft. As a consequence this omission could lead to a major revival of the strategic arms race in this most expensive of all weaponry. The USA now plans for an entirely new strategic bomber, the B-1, which already has been tested. So far the Air Force has built and test-flown several prototype units. If the bomber had been approved, a total expenditure of $11 billion over the next ten years would have been spent for 244 new bombers. The B-1 force, if it is ever built, could carry as many as 5,280 nuclear weapons, 1,140 more than could be carried by a similarly projected fleet of modernized B-52 bombers. The additional 1,140 weapons would increase the total number of deliverable nuclear warheads in the 1980s to some 18,600 (2,500 in ICBMs, 6,500 in SLBMs and 9,600 in aircraft). This projected total would nearly triple the number of deliverable nuclear weapons now possessed by the USA, which currently has some 531 long-range strategic bombers compared to the USSR, which has only 140. B-1 improvements over the B-52 include a faster reaction, increased resistance to nuclear effects, shorter escape times, longer range, greater payload, higher speeds at both high and low altitudes, reduced IR signatures, decreased radar cross-section and greatly increased electronic countermeasures.

An entirely new dimension to the strategic triad of land- and sea-based missiles and bombers is the cruise missile—a small pilotless jet air-plane that has a range of about 3,700 km and can hit the target within 30 m. It now replaces the B-1 strategic bomber, which has been canceled, in U.S. planning for strategic offense. Thus, the U.S. Air Force plan to build 244 B-1 bombers as a replacement for the aging B-52 bomber fleet has been canceled in favor of cruise missiles whose tests have already been successfully completed and which can be expected to enter service by the early 1980s. Since the cruise missile in either its strategic or tactical versions is basically a small jet aircraft with a low radar cross-section, it is a conventional weapon to defend against. Thus, while the SALT I and II treaties limit defenses against ballistic missiles they do not limit bomber defenses and these therefore will be expected soon to be upgraded in their capabilities against cruise missiles. For example, the upgraded SAGE system against bombers and the AWACS airborne radar against low-altitude aircraft will now find new roles to counter the cruise-missile threat. Apart from its escalating the offensive arms buildup of opponents it appears that the cruise missile will reestablish defensive armaments at the strategic level.

1.2. The Game

References 41–44, 64–66

In the analysis of a modern weapon system, the crucial question is whether or not the system is capable of surviving a tactical encounter of the type it was designed to meet. Usually, this question is extremely difficult to answer because of the great number of factors that enter the picture. To intelligently and accurately analyze the situation, a tactical model is introduced. This model is derived from a systematic examination of both the attacking and defensive weapons systems and is basically a logical manifestation of their relative deployment, tactics, performance, countermeasures, counter-countermeasures, etc. This type of construction permits a war-game approach to weapon system evaluation.

Game theory is a new tool which is employed to determine the nature and outcome of actions in a play before the play is begun, and is based on the premise that given a set of rules in a game between adversaries, its outcome can be predicted. The task therefore of game theory is to (1) discover the rules of the game, and (2) apply these rules to the prediction of the game outcome. Often, it becomes necessary to assume a spectrum of rules for the game and then to obtain solutions for the game outcome based on each set of rules.

In the subject of real-life war games, a cardinal or prime objective of each adversary in the game is to destroy his opponent while at the same time minimizing side effects and retaliation possibilities. A secondary objective is to accomplish the primary objective at the least expenditure of resources and compromise of convenience.

Two factors influence each adversary's play or non-play in a game: his ability and motivation to play the game. Both factors, because of the many uncertainties, must be established with some degree of confidence before a play or non-play decision is made. Uncertainties involving these two factors, therefore, restrain each adversary at each play.

The war game begins with the appearance of attack-type weapons, actions, and manifestations by each adversary. This is followed next by the appearance of defense-type weapons. Each side alternates moves and escalates its position and advantage in accordance with restrictions imposed by its ability and motivation. The game moves from a passive phase to an active phase which is characterized by a rapid rate of play. The rate at which the game proceeds is determined by the uncertainties, real or imagined, in the estimates by each side of its own or its opponent's capability. The game ends when one or both adversaries accept the futility of any further play. In the present chapter, and indeed in the present text, we will discuss those measures and countermeasures which sustain the ability to play the game, leaving the subject of motivation to other qualified sources.

Consider an attack-system which is directed to an un-

Table 1.4. *First-play attack system performance characteristics*

A. HARDWARE AVAILABILITY	
	1. HARDWARE OPERATIONAL ASSURANCE
	2. COMMAND AND CONTROL
B. HARDWARE EFFECTIVENESS	
	1. WARHEAD
	2. DELIVERY OR GUIDANCE

defended target. The target generally consists of individual point- or area-type targets typically found in an opponent's territorial domain. The attacker must insure primarily the availability and effectiveness of his weapon system. In this first play, the attacker must provide the basic attack characteristics summarized in Table 1.4.

Consider the undefended target reaction to this first play. Assuming it has the ability and motivation, it must provide for both its defense and counterattack. The former is required to minimize the effects of attack while the latter is needed to win the game. In this round of the first play, the defender both checks the attacker and escalates the play decision.

In the second play, the attacker checks the defender by providing his own defense-system and also further escalates the play decision by providing a hardware penetration capability in his attack-system. Thus in the second play, the attacker must provide the basic attack characteristics summarized in Table 1.5.

Comparison of Tables 1.4 and 1.5 indicates the trend of escalation between the first and second play attack-systems. The latter system includes provisions which insure the attack-system survivability, penetration, and real-time confirmation of damage against the now defended and counterattacking target.

In this manner, the game continues with each play escalating the availability, penetration and effectiveness of both attack and defense systems. These measures and countermeasures will be discussed in more detail during the course of this text.

The rate at which plays proceed in the game appears quite analogous to the reaction rate of a propellant or fuel. When the rate of play is very slow, actions happen in a controlled manner; when the rate of play becomes rapid, the game, plays, and actions then become uncontrollable. To the extent that each

Table 1.5. *Second-play attack system performance characteristics*

A. HARDWARE AVAILABILITY
   1. HARDWARE OPERATIONAL ASSURANCE
   2. COMMAND AND CONTROL
   3. HARDWARE SURVIVABILITY
B. HARDWARE PENETRATION
   1. OPERATIONAL FLEXIBILITY
   2. PENETRATION AIDS
C. HARDWARE EFFECTIVENESS
   1. WARHEAD
   2. DELIVERY OR GUIDANCE
   3. REAL TIME DAMAGE ASSESSMENT

opponent in the game can be described as a servo system, the game itself can be mathematized and the response can be obtained to each action.

The particular configuration of the attack and defense systems existing at the commencement of the uncontrollable phase in the game defines the initial state of affairs for this portion of the game.

The rules of the game are determined primarily by restrictions and limitations imposed upon each adversary by his own ability and motivation to play. Thus, in examining the choices of implementing attack and defense systems and their improvements at each play in the game, each adversary is limited by his own resources, reaction time, targeting or defending plan. Resources are normally allocated to meet the most pressing requirements in a given play, consistent with acceptable reaction times and coverage of targets. In a first play, resources are generally committed to implementing the basic attack and defense weapons systems, while in the later plays resources are directed to the evolutionary availability of weapons systems and their increased sophistication and flexibility. As the play continues in its controllable phase, both attack and defense systems grow in their sophistication and flexibility to engage more and more targets in less and less time. If the strain of resource consumption, response-time achievement, and target-area coverage can be borne by the adversaries, either by their own effort or through mutual agreement, a temporary standoff ensues. The game resumes at some point and proceeds to its critical phase as the rate of play becomes rapid and actions become increasingly uncontrollable for each side. At this time overt hostilities begin and the accompanying physical damage is added to the list of inconveniences previously sustained by each adversary in his allocation of resources during the previous phase in the game.

The outcome of the game is determined primarily by the initial state of affairs at the commencement of the uncontrollable phase in the game as well as upon the rules of the game. It becomes therefore an urgent task for both adversaries to closely study the nature and outcome of plays in both phases of the game.

The classic truth about any weapons system is that its kill probability P_k is the product of several other probabilities; that it is available, that it can penetrate, and that it is effective. Mathematically:

$$P_k = P_A P_P P_E \tag{1.1}$$

where

P_k = probability of kill,
P_A = probability of availability,
P_P = probability of penetration, and
P_E = probability of effectiveness.

Each of the probabilities P_A, P_P, P_E in turn are themselves the products of several other probabilities. Thus P_A may be expressed as the product of the probabilities of launch and guidance, P_P the probability of detection, discrimination and interception and, P_E the probability of fuse action, detonation of the warhead, and the probability of damage resulting from the detonation.

As equation (1.1) is a factorial statement of a large number of factors with each factor being less than one, the probability of kill P_k must always be smaller, and often is substantially smaller, than the smallest factor in its product. The obvious aim of the offense is to increase P_k while the exact opposite is sought by the defense. The probability of kill of a given weapon system defines its performance. The system performance is but one of three factors which determine the value of the system, the other two being the cost and the social and political stress.

Several techniques have proved helpful in the attempt to

mathemetize the game and thereby to obtain a rigorous result of its outcome. One technique is the geometric technique, which is a simple technique based on the well-known laws of the movement of objects in space. A second technique is the so-called Monte Carlo technique, which is based on the well-known laws of random numbers as they may become applied to the operation of a weapon system. This technique is more effective because of its flexibility to include aspects of the play other than spatial, which is the only aspect considered in the geometric technique. The third technique is the so-called probabilistic technique which assigns a probability to each event or aspect of the play and is based on the well-known theory of probability. This technique involves a rather extended mathematical procedure but is no less valuable in attempting to solve the attack-defense problem.

Basically, each of the aforementioned techniques seek to discover both the rules of the game and its outcome through the study of the impact each play or action has upon the probabilities of kill or survival of each adversary in the game. These techniques will now be discussed.

1.2.1. Geometric Technique
Reference 64

Figure 1.1 illustrates a profile of the geometrical technique when applied to the terminal engagement of incoming missiles by a local defense system. The first step in this technique is to establish a protective surface S above the defended area which defines the volume within which an exploding warhead

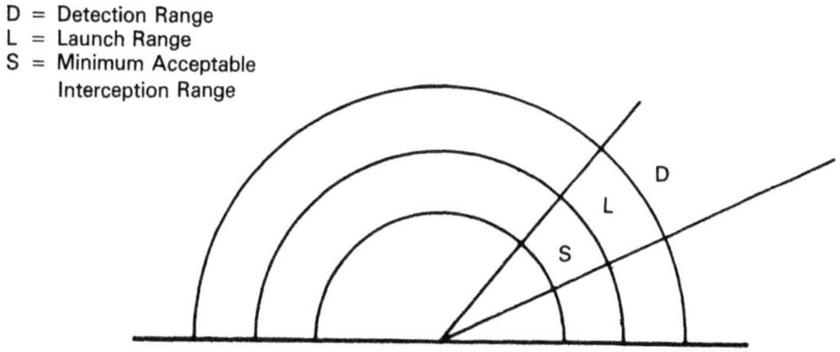

Fig. 1.1. *Profile of the geometric technique for terminal defense*

achieves a certain level of damage, usually expressed as the probability of kill. The surface *S* can be established in a straightforward manner by using tables or graphs which give the effects of a nuclear burst against the target and its characteristics. Such effects can be used in the determination of the protective surface *S* for a given warhead and target.

Given the protective surface *S* above a target, some other parameters can now be determined. To accomplish an interception of the incoming missile at or prior to its arrival at the surface *S*, the defense must launch its interceptors when the missiles are at or prior to the surface *L*. To make a launch decision at or prior to the launch surface *L*, the defender must decide whether or not to fire at the earliest possible time. This is accomplished following missile detection at the surface *D*. Shown in Fig. 1.1 are the incoming missile reentry angles, and times between missile arrivals at the surfaces *D*, *L*, and *S*.

The *D* or decision altitude, which has been set by the firing doctrine, is a measure of the discrimination and target-tracking capabilities of the defense. The *D* surface signifies that launching of defending missiles before threatening missiles reach this altitude would be wasteful of defending missiles.

The *L* or launch altitude, which is computed as a function of the intercept point and the time required to intercept, is used to compute the surface of last penetration. Once the incoming missile has penetrated this surface, no defending site could launch its missiles to intercept the incoming missile and prevent its penetration of the protective surface *S*.

Geometrically speaking, the surface *S* provides protective coverage for people, industry, resources and deterrent forces, i.e., the constituents of the local target. Most likely, however, the total target area extends beyond the protection offered by the surface *S* so that these other areas of the total target area remain undefended. The effectiveness of the surface *S* can therefore be measured by (1) the fraction of the total target area it protects and (2) the fraction of the local target area it protects. In this regard, an area defense system tends to protect the total target but with local unprotected gaps while a local defense system tends to protect a localized small target area with no gaps.

In the geometric technique, the game consists of plays which have as their objective to push the surface *S* lower and nearer the target for the offense and higher and farther away for the defense. To begin with, the undefended target does not possess any protective surface whatsoever. The introduction of a defense first establishes the surface S. Finally, in the alternation of succeeding plays, the surface S moves in or out depending upon the success or failure in each play.

In addition to its determination by the warhead and defended target characteristics, the surface S is pushed even higher and farther away by such criteria as the firing doctrine requirements, avoiding fratricide, avoiding bad cross angles between incoming and defending missiles, avoiding radar blackouts due to incoming precursor or defending missile bursts, incoming missile maneuvering, and penetration aids.

1.2.2. Monte Carlo Approach

Reference 64

If n missiles are launched against a target, then the probability of killing the target should increase as the number of missiles is increased. In the case of a defended target, the probability of killing the target would remain relatively low if the number of attacking missiles did not exceed the defense system traffic-handling capacity but would increase substantially as this capacity is exceeded. A further increase in the number of attacking missiles would buy little increase in the probability of killing the target.

In the Monte Carlo approach, the game is played by selecting a particular value for the number of attacking missiles n and then determining, by random sampling, whether this particular number of missiles has killed the target. By playing enough games at each missile level n, we can establish a probability of kill for that particular number of missiles and be thus able to produce a point on the curve. By doing this for every missile level n, we can obtain the curve which depicts the probability of killing the target as a function of n, the number of attacking missiles. Such a curve for a defended target is shown in Fig. 1.2. Also shown in the figure is a curve for an undefended target.

The curve D shown in Fig. 1.2 is a continuous curve and is obtained by smoothing the data from a discrete number of n-missile attacks. The curve is sometimes called the S-curve and is well known in statistics as the expression for the integral of a distribution.

Referring to the figure, we desire to find the point on the

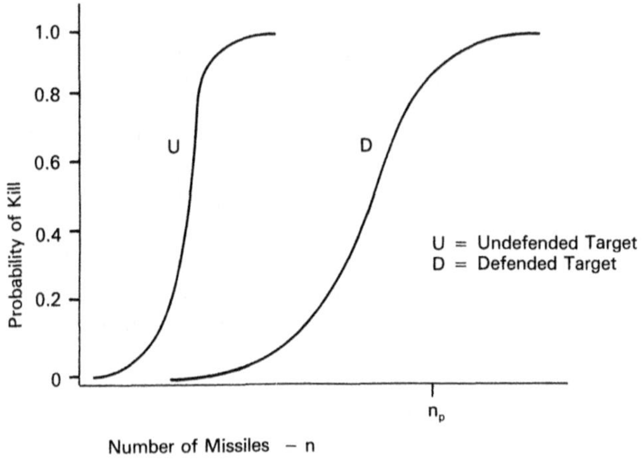

Fig. 1.2. *Probability of kill as a function of the number of missiles launched against a target*

curve where, if the number of missiles n is further increased, the probability of kill is very little improved and if n goes below this number, the probability of kill drops off quite a bit. This point occurs at the knee of the *S*-curve and designates the number of missiles required in a salvo or simultaneous attack to achieve a high probability of kill, additional missiles in the salvo would contribute little to the achievement of this probability of killing the target. The number of missiles n_p required to achieve the high probability of kill in the vicinity of the knee is called the price of admission for the offense, i.e., the number of missiles which are needed in a simultaneous attack to kill the target with a high probability. Any further increase in the number of missiles must be weighed against the importance of the target.

The procedure for implementing the Monte Carlo approach is as follows:

1. Select a number of missiles n to salvo attack the defended target.
2. Play a number of games for each missile attack, i.e., vary aim points, trajectories, warhead yields, and penetration aids.
3. Perform a smoothing operation through the discrete num-

ber of points obtained for each n-missile attack to obtain the S-curve and compute the price of admission n_p.
4. Perform a summary of all possible attack conditions.

Basically, in war games, there are two types of simulation: (1) event-oriented, and (2) clock-time-oriented. Both types accomplish the same purpose but in a slightly different manner. Event-oriented simulation is the easier of the two, since the chain of events determines the end result, which needs only be observed. In time-oriented simulation, each clock pulse determines success or failure. In general, time-oriented simulations are too time-consuming on computers but are best suited for determination of simultaneous events in the play; i.e., the interceptor and warhead must be at the same place at the same time. Defense multi-data-processing, radar tracking and high target traffic problems can best be implemented in time-oriented simulation. In the event-oriented simulation, the game is centered about the basic defense functions of target detection, discrimination, tracking, site assignment, intercept point, interceptor salvo, launch interceptors, and interception computations.

After the price of admission has been determined, the summation takes place with respect to all the threats, as is also the case for the geometric approach. We can then determine the performance of a particular offense or defense system or strategem.

1.2.3. Probabilistic Technique

References 65–66

Every event or aspect of play in the tactical encounter is subject to uncertainties. A method, therefore, can be developed to produce the probability of killing the target. This method, by which probabilities are assigned to each event in the tactical encounter, results in a long mathematical procedure of finding the kill probability and is referred to as the probabilistic technique of assessing the performance of a weapon system.

1.2.3.1. Single-Shot Kill Probability (SSKP)

The single-shot kill probability is a measure of the chance of destroying a target with a single missile. It involves a number of conditional probabilities of events prior to the event of destroying the target. Thus, the single-shot kill probability involves the probability p_1 of launching the missile, the probability p_g of guiding the missile to the target, the probability p_f of fuse action, the probability p_w of detonation of the warhead, and the probability p_d of damage resulting from the detonation:

$$P_s = \int_A p_1 \cdot p_g \cdot p_f \cdot p_w \cdot p_d \cdot dA \tag{1.2}$$

where the product of probabilities is integrated over the area A, which defines the lethal area centered about the target which is considered to be a point. When the target is a large area target, the area A defines the so-called vulnerable area. The most critical conditional probability is p_g, the probability of guidance. It is evident that if p_g is so low in magnitude that it is not likely that the missile will hit the target within a concentric circle of radius equal to or less than the lethal radius, the probability of killing a target will be small. Three recourses are possible: increase the accuracy of guidance, increase the lethal radius of the warhead or increase both. This decision usually has far-reaching consequences on the design of the missile; hence the warhead-guidance tradeoff is usually made first. Consequently, the probabilistic technique provides expressions for the single-shot kill probability that interrelate guidance accuracy (CEP) with lethal, or vulnerable, area for a particular warhead and particular target.

1.2.3.2. Lethal Area Approach

In this approach the target is assumed to be a point target or at best a small-size target over which the damage density can be assumed constant. The missile crosses the lethal plane at some distance r from the center of the target. If we combine the probability functions p_f, p_w and p_d, as designated in the previous section, to obtain the overall probability of damage for the warhead and fuse, which will now be denoted p_D, then, if enough energy hits the target, the damage function may be expressed as:

$$p_D = p_f p_w p_d = k\, e^{-(r/a)^2} \tag{1.3}$$

where k is a constant which is determined by the conditional probabilities p_f, p_w and p_d and a is the radius of the circle of damage.

As the ratio r/a increases, the conditional probability p_D decreases exponentially. The radius a is called the lethal radius. It should be noted at this point that the level of damage has not been specified. The damage will depend on the specific physical characteristics of both the warhead and the target. Hence, in specifying the lethal area, the particular warhead and particular target must be specified. It is customary to identify the lethal areas as the area within which the damage is sufficient to destroy the target, although it may also be preferable to identify a damage area for a specific level of damage.

If the flight errors are entirely random and mutually independent, the probability of guidance p_g, i.e., that the trajectory will intersect the lethal area is given by:

$$p_g = \frac{1}{\sigma^2} e^{-(1/2)(r/\sigma)^2} \tag{1.4}$$

where σ is the standard deviation of the missile dispersion, as measured in the target plane with respect to the center of the lethal area, i.e.,

$$\sigma = \sigma_r = (\sigma_x + \sigma_y)^{1/2}.$$

If the probability of launching the missile p_l is unity then equations (1.3) and (1.4) can be used in the defining equation (1.2) to compute the single-shot kill probability:

$$P_s = \frac{k}{\sigma^2} \int_0^\infty e^{-(1/2)(r/\sigma)^2} e^{-(r/a)^2} r \, dr \tag{1.5}$$

It has been assumed in the derivation of equation (1.5) that the conditional probabilities given by equations (1.3) and (1.4) are both independent of the angular orientation of the missile and target. In practice angular distributions must be considered for a more exact solution. The single-shot kill probability can be obtained through integration of equation (1.5):

$$P_s = \frac{ka^2}{a^2 + 2\sigma^2} \tag{1.6}$$

which relates the accuracy of missile flight to the damage capability of the warhead.

When the target is offset from the aim point or when a systematic error or bias is attributed to the guidance, the aim point is displaced by a distance d from the center of the target. In such circumstances, the single-shot kill probability is given by:

$$P_s = \frac{ka^2}{a^2 + 2\sigma^2} e^{-d^2/a^2+2\sigma^2} \tag{1.7}$$

which reduces to the previous equation for $d = 0$, which implies that no bias exists for the missile guidance and that the aim point and target center can be co-located.

1.2.3.3. Vulnerable Area Approach

The single-shot kill probability discussed in the previous section was based on a definition for the lethal area which assumes an average damage density to be constant over the lethal area about a small, essentially a point, target. In many instances the target area is large and contains many such point targets. In such cases the single-shot kill probability may be considered as being the product of the single-shot kill probabilities of individual targets within the target area:

$$P_s = \prod_i \frac{ka_i^2}{a_i^2 + 2\sigma^2} \exp -\sum_i d_i^2/a_i^2 + 2\sigma^2 \tag{1.8}$$

where the subscripts enumerate the N individual point targets in the target area. The lethal radius for the ith target is a_i while d_i represents the separation between the ith target and the target to which the missile is directed. Also included in d_i may be the bias error which may be present. Both the product and summation which appear in equation (1.8) extend over the N individual point targets in the target area. The variance due to the missile guidance is of course $2\sigma^2$.

For the particular case in which all targets have the same lethal radius, equation (1.8) can be simplified as follows:

$$P_s = \left[\frac{ka^2}{a^2 + 2\sigma^2}\right]^N e^{-d^2/a^2+2\sigma^2} \tag{1.9}$$

where a is the lethal radius of individual targets and

$$d^2 = \sum_i d_i^2 \tag{1.10}$$

The sum in equation (1.10) has the dimensions of an area and actually is a measure of the total target vulnerability. If the vulnerable area of the target is defined by equation (1.10), it can be seen that the single-shot kill probability will decrease for a fixed vulnerable area as the lethal radius decreases. Thus, smaller kill probabilities are produced when using smaller warheads against harder targets.

Three cases can be further distinguished in equation (1.9), according as to whether the lethal radius a vastly exceeds, in about equal to, or is vastly less than the variance due to missile guidance $2^{1/2}\sigma$. Thus, equation (1.9) can be further simplified for each case:

$$a^2 \gg 2\sigma^2 \qquad P_s = k^N \, e^{-d^2/a^2} \tag{1.11}$$

$$a^2 = 2\sigma^2 \qquad P_s = \frac{k^N}{2^N} \, e^{-d^2/4\sigma^2} \tag{1.12}$$

$$a^2 \ll 2\sigma^2 \qquad P_s = \left[\frac{ka^2}{2\sigma^2}\right]^N \, e^{-d^2/2\sigma^2} \tag{1.13}$$

1.2.3.4. Cumulative Kill Probability

The probability that the target is killed in n mutually independent shots may be found providing that the single-shot kill probability P_{ks} is considered to be the same for all the missiles. Thus, the n shot kill probability is given as:

$$P_{kn} = 1 - (1 - P_{ks})^n \tag{1.14}$$

If the magnitude of the single-shot kill probability P_{ks} is small, then equation (1.14) can be written as follows:

$$P_{kn} = n \cdot P_{ks} \tag{1.15}$$

However, the effects of cumulative damage cannot be disregarded so that the single-shot kill probability P_{ks} cannot be considered to be the same for all the missiles. Thus, in practice:

$$P_{kn} = 1 - \prod_{n}(1 - P_{i,ks}) \tag{1.16}$$

where the product $\prod_{n}$ extends over the number of missiles fired and $P_{i,ks}$ is the conditional probability of kill associated with the ith missile.

The probability of survival of the target, following the ith missile, $Q_{i,ks}$ equals $1 - P_{i,ks}$, and decreases as the number of missiles increases and is determined primarily by the size of the warhead and hardness of the target. It can be seen that if n missiles are fired at the target, the gain or increment in the kill probability is less following the first, or first few, missiles.

1.3. Game Strategies

There are two pure strategies for attack against a ballistic missile defense. The first is an all-warhead attack in which the missile force carries a large number of small warheads. These warheads are separated at some instant between the time the missiles leave the atmosphere and the time of reentry. The warheads from one missile can all be directed against the same large target such as would be a city; these multiple reentry vehicles (MRVs) are purely a penetration aid. Alternatively, each of the reentry vehicles can be given an independent boost to a different target, thus making them into multiple independently-targeted reentry vehicles (MIRVs), which are not penetration aids as much as they are a counterforce weapon, especially if the reentry vehicles have a high accuracy. In this case each MIRV can be used to destroy an enemy missile silo. The Minuteman III for example carries three MIRVs while the Poseidon missile carries ten MIRVs. The Titan II liquid fuel rocket could carry twenty or more nuclear weapons. If these were employed simply as MRVs, the fifty-four Titans alone could provide 1,080 reentry vehicles for the Russian defense to deal with. An additional 7,540 could be delivered by the 1,000 Minutemen and forty-one Polaris/Poseidon submarines. Thus, in a coordinated attack, the Russian defense would have to contend with the near simultaneous arrival of perhaps 8,620 reentry vehicles, excluding the bombers, SRAMs and short-range tactical missiles.

The fact is that nuclear warheads can be delivered at a fraction of the cost necessary to intercept them. This is exemplified by the fact that the total investment in the entire Minuteman force is about $7 billion whereas the expected total investment for a heavy defense system is estimated to be about $50 billion. The Safeguard system fully deployed is expected to cost about $5 billion. Spartan interceptors each cost $1–2 million. The 1974 SALT II ABM Treaty limits deployment of ABM systems to one site in each country. However, the USA has not selected a preferred site and Safeguard has not been deployed but is confined mostly to the testing of prototype hardware at various testing locations. Thus, an all-warhead attack favors the offense insofar as costs are concerned. The attacker can make a further relative saving by concentrating his attack so that most of the enemy interceptors, all bought and paid for, have nothing to shoot at. The all-warhead strategy is a high-reliability penetration strategy which also minimizes costs. This may not always be the case, however, since total investment costs may be more or less equal if both opponents have already deployed both missile and ABM forces.

The second pure strategy for attack against an ABM defense is to precede the actual attack with an all-decoy attack or to mix real decoys with the warheads. This strategy reduces the number of missiles available to carry the real warheads. However, if the ABM defense is an area defense it must fire against these threatening objects at very long range before they reenter the atmosphere where, because of their lightness, they would be slowed down considerably faster than the heavier warheads. Several thousand such decoys could readily exhaust an area defense system. The attack with real warheads would then follow.

A hybrid strategy for attack against a ballistic missile defense is the employment of various penetration aids mixed with the real warheads. This strategy again reduces the payload which is available to carry the warheads. However, if the number of missiles which succeed in penetrating the defense exceeds the number which would have penetrated had a pure strategy been employed, then a smaller missile force using penetration aids can accomplish the same damage as can a somewhat larger missile force which employs real warheads alone. This strategy therefore represents a possible additional saving for the attacker. In fact the only disadvantage of a pure strategy is that should the enemy defense prove relatively easy to penetrate using penetration aids the offense in fact is obliged to pay a higher than necessary price for admission.

There are two pure strategies for defense against a ballistic missile attack; area defense and terminal defense. The first strategy—area defense—is to intercept incoming missiles as far away as possible from their intended targets, possibly even during their launch phases or during early climbout. This strategy reduces the number of objects the defense system must contend with, i.e., detect, discriminate and intercept. Objects carried by individual missiles would have to be spaced far apart following their separations from the missile boosters and would have to be placed on threatening trajectories in order to avoid multiple kills by a single defensive burst. Such requirements involve precision achievements by the missile deployment system. In theory the area defense system would require the use of no greater number of interceptors as there are missiles in the attacking force. In fact, a single nuclear burst from a defensive interceptor could possibly intercept an attacking missile salvo.

The second strategy—terminal defense—intercepts incoming missiles as close as possible to their intended targets. This strategy employs the atmosphere to separate the lighter-weight objects from the real warheads. Objects accompanying the RVs would have to be sufficiently heavy and credible to reach the defense's final decision altitude. Thus the attacker is forced to use an increased portion of his payload in carrying penetration aids or alternatively to use real warheads, i.e., to employ a pure strategy for attack.

Several other possible defense strategies come under the heading of advanced basing techniques. These approaches apply particularly to the protection of the land-based missile force and include super-hardening, proliferation, a shell game in which there are more silos than missiles and land-mobile missiles. Silo-hardening reduces the kill radius, thus forcing the attacker to provide better accuracy, higher warhead yield or a combination of both in order to destroy the silo. Proliferation of silos is intended to draw the attacker's firepower that otherwise could be focused on the missile force, urban and industrial areas. The shell game employs a network of silos which may or may not contain missiles and in which the missiles are continually rotated at random between silos. Road-mobile and rail-mobile strategies are intended to force the attacker to continuously keep track of the mobile missiles and upgrade his own guidance systems.

1.4. Systems Evaluation

References 3, 7–9, 21, 45–47, 51, 68, 86, 97, 98

The question of value of a given strategic weapon system involves questions of system performance, costs and the ensuing social and political stresses. These must be derived from a careful analysis of the national strategic policy and statements of the mission objectives for the system. These in turn must be obtained by the broadest possible consensus of the electorate and through illuminating public and Congressional debates. In the past the social and political aspects have not been appropriately included in the deliberations leading to the selection and buildup of strategic forces. The fact is, in view of the pressing priorities to obtain the weapons capability, many such systems were built and deployed without regard to costs as well. As a consequence, strategic systems have been built by institutional fiat, with result the expense of burgeoning expenditures and social and political stresses. A more mature approach to the problem of balancing the desirable and undesirable aspects of weapons systems is to consider the social and political impacts as well as the costs and performance of weapons systems. There is no advantage to be gained in acquiring a strategic arsenal if its social and political impact is to change a country into an ideological form against which the arsenal is supposed to defend.

1.4.1. Strategic Policy

References 5, 6, 11, 43, 44, 74, 75, 92, 98, 101, 106, 107

Solutions to the problem of national security theoretically can assume one of two forms; pure political and pure technological. Nevertheless these two forms are not independently viable and, as a consequence, a more realistic form of strategic policy evolves as a combination of both. This is a necessary result which obtains in a conflict which manifests, in their ascending order of mutual impact upon the adversaries, in debates, games and fights. Thus, pure technological approaches to the problem of national security cannot be isolated from the pertinent social and political influences and, as a result, can always be questioned in terms of the equally legitimate demands of society and the electorate in particular.

The utility of a technological approach to national security is twofold: first, it enforces the national will in peacetime and, second, it provides a means to conduct a fight if one should break out. It must be recognized however that a pure technological approach forces and fuels an armaments race and in this respect tends to waste the resources of both adversaries as well as implementing their policies. Allocation of resources to strategic armaments of course always has and will compete with the possible use of these resources in more humane and civilized projects. Thus the main question which must be addressed by each adversary is his willingness to assume the burdens of armaments escalations and possibly a devastating war to protect his society or way of life, i.e., is his society worth protecting? To be credible this question must be answered by those mostly being affected—the members of society and the electorate—and not by those who may profit—the armaments industries and their politicians. Never before in the history of human affairs are so many people being affected by the decisions of so few. A credible and enforceable strategic policy therefore requires the widest supporting base both by the participating constituents and, equally important, by world-wide public opinion. Thus social and political questions undoubtedly influence the pure technological approach to national security. A nation's strategic policy is therefore one which defines the national consensus, and international as well, and backs this up with the best technical approaches—not the reverse of this. Once the national objectives have been decided openly and unambiguously, of course, the political problem becomes one of stating the mission requirements for each of the individual strategic weapons systems. It is at this point that the armaments technologist becomes involved in the task of defining the best technical approach in terms of feasible hardware to meet the mission requirements.

1.4.2. Mission Requirements

The mission requirements of a given strategic weapon system are usually stated in terms of the results which must be achieved by the system. Thus a missile force is required to produce a certain level of damage inflicted upon the opponent, for example, one-third to one-fifth of the opponent's population and one-half to three-quarters of the opponent's industrial capacity may be required to be destroyed. On the other hand, a defense system may be required to prevent such levels from being obtained by the opponent's missile force, perhaps by tolerating only a much smaller level of damage in terms of destruction of population and industrial capacity. It should be noted that these requirements are quite subjective in nature and are given as the levels above which the opponent would cease to be a viable society. Such purely subjective statements of the mission requirements may well be academic since the burdens of armaments impacts may affect both opponents in many ways other than their physical annihilation. For example, a society can be destroyed in its substance whether or not it survives a physical destruction in an actual nuclear exchange. This of course can be accomplished by the social and political stresses of the armaments burden.

1.4.3. Systems Engineering

References 3, 65, 66, 70, 76–96

The broad objective of systems engineering is to provide the optimum weapon system which meets or exceeds the mission requirements which have been quite generally specified for the weapon. The functions which are usually classed as systems engineering fall into two broad categories: systems analysis and systems development. In systems analysis the tasks are to (1) define the problem, (2) examine all possible solutions and (3) select the optimum solution. In system development the tasks are to (1) detail the design, (2) develop the equipment, (3) test and evaluate the equipment, (4) provide consulting for the production, installation, testing and maintenance of equipment, (5) conduct operations analyses and (6) conduct equipment upgrading programs following force installations.

1.4.3.1. Systems Analysis

The first three phases of engineering a system are in defining the problem, examining all possible solutions and selecting the optimum solution. To begin with, the problem must be defined and the objective which is desired must be clearly understood. In this regard a careful analysis of the mission objectives must be made in order to define the system requirements. The factors which appear to be of universal interest and the focal points for defining the performance of a strategic weapon system are: system availability, system penetration and system effectiveness. Based on the system objectives a specification of the system performance in terms of its availability, penetration and effectiveness can be made. A result of this activity is a listing of system requirements each accompanied by a subjective, perhaps judgemental, statement as to the degree of importance of obtaining a particular requirement. Thus, for a given weapon system a preliminary set of requirements may contain the following general statements:

1. The system must be available on demand.
2. The system must penetrate a given defense.
3. The system must inflict a given level of damage on the opponent.
4. A given cost is considered to be the maximum acceptable cost for the system.

Such preliminary and general system requirements can be further detailed to fully specify the system performance and cost. Notably excluded in the present discussion are general system requirements in terms other than purely technical approaches, i.e., in terms of social and political stresses to be inflicted upon the opponent as well as domestic impacts of a similar nature. It should be noted that willingness to pay is an important factor in this phase of the systems analysis.

Once the problem has been defined through a set of system requirements the second step in system analysis is to examine all possible approaches to obtaining a solution. In this activity the system requirements are translated into mathematical formulations and models which permit the logical assessment of a variety of systems which can be proposed to meet the system requirements. Each possibility is carefully analyzed to determine whether it can satisfy the requirements. The various proposed solutions are evaluated on a relative basis. In addition to mathematical and model formulations being used to evaluate systems, certain limited testings of critical components may be employed to check their feasibility, or new developments may be indicated to achieve high-payoff solutions.

The third step in the systems analysis effort is to select the system which provides the optimum solution, i.e., which best meets or exceeds the requirements. Mathematical and model analyses continue to be helpful in determining the best system approach as do limited tests and new developments. It should be recognized, however, that such activity may be limited, especially if the problem is too complex in nature or its parameters are too ill-defined. Various approaches are therefore needed, including variational analysis, which best compliment each other. Also, a certain amount of system design must take place when the various alternative systems are examined and the optimum system is to be selected. Most of the design in these states is of a preliminary nature and is aimed at establishing relative feasibility between systems. The design is specified by means of a functional block diagram of the optimum system.

1.4.3.2. System Development

Once the optimum system has been selected a more detailed design must be made and the block diagram first established during systems analysis must be considerably detailed. In fact, a complete design of the equipment is made with the objective of producing the system.

The second step in systems development is the testing and evaluation of the system. This is accomplished through a series of tests of prototype equipment from which appropriate data are obtained which serve to further perfect the final system design. Final equipment specifications and design drawings are made available in this step.

Following systems testing, the system developmental effort is directed to consult in the production, installation, operation and maintenance of the system and furthermore to conduct continuing operational analyses and upgrading programs.

1.4.4. System Performance

References 7, 21, 68

The performance of a weapon system can be classified into three main categories: system availability, penetration and effectiveness. Table 1.6 enumerates the three main categories and gives a further detailed breakdown of system performance characteristics for the special case of a ballistic missile system. Similar tables can be constructed for other weapons systems as well.

The system performance characteristics given in Table 1.6 are not all-inclusive and are given to illustrate the sizeable number of items that must be considered in defining the system performance. Thus, in the overall definition of the system performance many characteristics contribute their small parts to the system total performance. Each contribution may or may not be equally important to the other contributions and as a consequence, although measurable, they must be assigned weightings which are indicative of their relative importance. These weightings may not in themselves be measurable quantities but rather subjective assignments. In this manner the system performance can be defined as the sum of performances of the individual system characteristics properly weighted. For example:

$$P = \sum_i w_i p_i \tag{1.17}$$

where

P = system performance,
p_i = performance of individual characteristics, and
w_i = weightings ascribing relative importance of contributions by the individual characteristics.

Thus, any one of the three major characteristics of Table 1.6 can be expressed in terms of the individual contributions of characteristics p_i and its weight w_i signifying relative importance.

Equation (1.17) simply states the relative fact that the system performance is the sum of the individual system attributes properly weighted. This may not be the case in a given application but will be used to illustrate the present discussion. In a more general context, in view of the mutual influences which exist in many cases between the various characteristics in Table 1.6, it may be more appropriate if the definition of system performance is made in some other manner, perhaps in the form of a function which accounts for the presence of mutual

Table 1.6. *Weapon system performance characteristics for a ballistic missile system*

AVAILABILITY	PENETRATION	EFFECTIVENESS
RELIABILITY	TARGETING FLEXIBILITY	SIZE
MAINTAINABILITY	TRAJECTORY FLEXIBILITY	RANGE
HARDWARE OPERATIONAL	PENETRATION AIDS	WARHEAD
ASSURANCE	RV HARDENING	YIELD
FLEXIBLE C&C NETWORK	ECM, ECCM	WEIGHT
CABLE	DECOYS	NO RVs/MISSILE
GROUND RADIO	MULTIPLE RVs	ACCURACY
AIRBORNE	MANEUVERABLE RVs	INSTRUMENT ERROR
SATELLITE	DEFENSE KNOCKOUT	MIDCOURSE GUIDANCE
SURVIVABILITY		TERMINAL GUIDANCE
BLAST HARDENING		PENETRATION
ACTIVE DEFENSE		FLEXIBILITY
DECEPTIVE BASING		
MOBILE SYSTEMS		

influences. In practice, equation (1.17) is one of many ways of defining the performance of a weapon system.

1.4.5. System Cost

References 84, 88–96

The cost of deploying a new missile weapons system generally increases with the number and complexity of unit missiles, range, payload and the system accuracy. Table 1.7 enumerates ten major sources of costs involved in the deployment of such systems.

Table 1.7. *Sources of cost in the deployment of a weapons system*

RESEARCH	PROCUREMENT
DEVELOPMENT	INSTALLATION
DESIGN	TRAINING
FABRICATION	OPERATION
TESTING	MAINTENANCE

Following the definition of requirements for the system the initial costs are directed in the areas of research, development, design, fabrication and testing of prototypes and test models, following which heavy procurement and installation costs are made along with costs of training personnel. Once the weapon system has been installed the main costs are for its operation and maintenance.

In the overall consideration of system costs the first step is to consider the three major performance characteristics of the system: its availability, penetration and effectiveness as outlined in Table 1.6. This leads to the conclusion that in the overall definition of the system cost, many individual system characteristics will contribute their small parts to the total system cost. Each contribution may or may not be equally important to the other contributions and as a consequence, although measurable, they must be assigned weightings which are indicative of their relative importance. These weightings may not in themselves be measurable quantities but rather subjective assignments. Thus, the system cost must be defined in a similar manner as was the system performance:

$$C = \sum_i u_i c_i \tag{1.18}$$

where

C = system cost,
c_i = cost of a given system characteristic; system characteristics are given in Table 1.6 while sources of cost are given in Table 1.7, and
u_i = weightings ascribing relative importance of costs between system characteristics.

In equation (1.18) c_i is the cost of a given system characteristic and u_i is its weight signifying relative importance. Equations (1.17) and (1.18) are identical except for the definitions of the parameters appearing in these equations. The point to be made is that both the performance and cost of a weapons system are at best subjective determinations since only subjective criteria may be available for assigning the relative importance of the various characteristics that make up these systems.

1.4.6. Value Model
Reference 86

One of the principle problems of the systems engineer involved in the preliminary design phases of a military or space system development is the selection of the "optimum" design from a number of proposed system designs. In general, at an early stage in the life of a system there will be a number of design approaches, each having somewhat different technical performance and costs. The value of a system for a given application will depend upon the relative importance of the various performance factors as well as upon cost factors. In the past, the assessment and comparison of proposed designs was often made by the systems engineer mostly on the basis of the urgency to field a system, with result that R&D costs far overshadowed other costs, including those for system procurement. As the program matures, the R&D costs for developing state-of-the-art solutions become more amenable to control and precedent since major capital investments have been made so that they no longer assume the significant portion of total program costs. On the other hand, system procurement costs which were previously adversely weighted as to their relative importance in establishing the system value now become important.

For this class of systems where the broad questions of strategy are no longer a factor and where the required performance of the system can be stated explicitly to be based on the state-of-the-art, selection and evaluation of a design can be made on a simple objective basis. The lowest-cost system that provides the required performance is selected. Here cost of a weapon system is considered in the broad sense to include R&D, procurement, training and operating and maintenance costs associated both with hardware and software aspects of the system. In many military and space systems, however, the value of particular performance characteristics are difficult to assess either because of their strategic implication, performance before cost or the uncertainty in the state-of-the-art, for example, in the environments of the sea or space. The objective of the present systems approach is to provide a simple mathematical model formulation of all factors which affect system design and whose performance can be reliably bounded or stated explicitly, based on the state-of-the-art. The model will include all the factors, both in the areas of performance and cost, on which an optimized system can be determined. This formulation will also illustrate by way of example the variation in value of a particular system as the weights to the different performance and cost factors are varied.

Value is defined as the sum of all measurable performance and cost factors or parameters that apply to the system, each factor normalized and weighted in accordance with its importance to the particular system application. Such a definition of system value serves as a computable numeric yardstick to aid in system valuation. Identifying, evaluating and mathematically relating these parameters are the steps necessary for objective system evaluation. A careful statement of the objectives and mission requirements of the system is therefore necessary and indicates the relative importance of each performance and cost factor. These factors can be normalized using appropriate functions or graphical techniques, based on knowledge of the state-of-the-art, to provide normalized performance and cost coefficients. The sum of the weighted products of the normalized coefficients is then the numerical value of the system. A value equation can be written in the following form:

$$V = \sum_i w_i x_{ij} \tag{1.19}$$

where

V = value of weapon system,
x_{ij} = normalized system factor,
w_i = weighting,
i, j = running indices.

The normalized system factors are such that:

$$0 \leq x_{ij} \leq 1$$

and the weightings are assigned so that:

$$\sum_i w_i = 1$$

The running index i covers all measurable performance and cost factors.

Typical factors or attributes which may be expected to affect the value of a weapon system may be inferred from inspection of Tables 1.6 and 1.7. The four principal factors are listed in Table 1.8.

The listing of performance and cost factors given in Table 1.8 is by way of example for a missile system. More factors may be added to the list or a given factor may be expanded into sub-factors, for example, as indicated in Tables 1.6 and 1.7. Thus, the system-availability factor can be expanded into reli-

Table 1.8. *Four principal weapon system factors (attributes)*

x_{1j}	AVAILABILITY
x_{2j}	PENETRATION
x_{3j}	EFFECTIVENESS
x_{4j}	COST

ability, maintainability, hardware operational assurance, C&C network flexibility and survivability. In particular, the cost factor can be expanded into those sub-factors given in Table 1.7. In practice, an exhaustive list of all factors and sub-factors is included in the list of Table 1.8.

The running index j covers all the possible subfactors under a given factor. Thus, x_{11} indicates reliability, x_{12} indicates maintainability, x_{13} indicates hardware operational assurance, and so forth.

Equation (1.19) states the simple fact that the value of a weapon system is the linear sum of its performance and cost attributes. The weightings describe the relative importance of the various factors. Once these factors have been identified and compiled into a list they must be normalized for use in (1.19) which is the value model equation. The determination of the normalization functions is based on the interpretation one gives to the mission and system requirements.

Before proceeding any further, it will be well to consider the limitations of this and any other scheme for the comparative evaluation of a complex system. It must be recognized that no approach will unerringly select the optimum system unless a very specific set of performance and cost data is available. The data would have to cover the detailed knowledge of the system performance and costs and the exact weights which are to be assigned to the various factors. Clearly, if such reliable data is available, the selection of the optimum system is a simple problem, since both what is wanted and what is available are already known.

Generally, there are many uncertainties, particularly in the value which is to be assigned to aspects of the system performance and the assessment of either the mission requirements or of strategic or national objectives, the importance of these determinations influencing the price in performance and cost one is willing to pay. These uncertainties cannot be removed by the manipulation of variables in equations. Selection of a particular technique of evaluation will depend on the nature of the problem and the most convenient or agreed-upon method of assessing the uncertain data. Where data is in a probabilistic form a probability approach is indicated. The method being outlined in this discussion is felt to be convenient when many aspects of performance are important and must be taken into account with appropriate weighting factors.

In the approach described herein the weighting factors will

usually incorporate the major subjective judgements. It is in these weighting factors that the tradeoffs between performance desired and the price which is to be paid are incorporated.

1.4.6.1. Factors Normalization

The procedure for normalizing factors will be illustrated by a simplified example. Consider now a weapon system which is *desired* to be available on demand with a 90 percent availability, is *desired* to penetrate a given defense with a 90 percent penetration, is *required* to inflict a given level of damage on the opponent with a 90 percent effectiveness and is *desired* to cost no more than 90 percent of a given cost. Those systems which result in less performance or more cost will be penalized, while systems which exceed performance and have less cost will be favored—although perhaps at a diminished rate excepting the cost since the added performance in no way significantly increases the system value once a certain capability has been achieved.

Figures 1.3–1.6 illustrate the normalization functions which can be utilized in normalizing the various performance and cost factors. These conversion graphs have been derived from the statement of the problem and the interpretation of system requirements given for this simplified example.

Figure 1.3 depicts the system availability normalization function in which values of blast damage below a certain level *a* are penalized at a high rate at the lesser damage levels while being favored at a lesser rate at the higher damage levels. The slopes of both lines in the figure have been determined by the fact that the system must be available on demand, especially following an attack by the opponent's missile force, i.e., the weapon system must withstand a certain level of blast damage and yet be capable of being available. Those systems which cannot withstand blast damage are penalized while those which do withstand the blast are favored. Furthermore, it is important to penalize the former systems than reward the latter. Of course, in this simplified example the system availability has been implied as that due to blast damage alone. As mentioned previously many factors combine to define the total system availability and these should all be considered and normalized in practice.

Figure 1.4 depicts the system penetration normalization

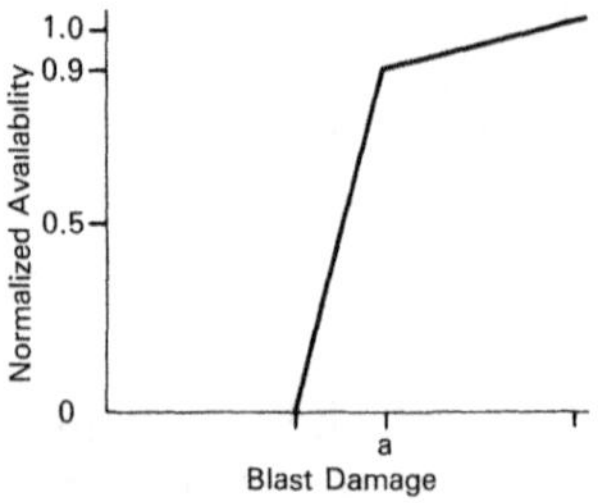

Fig. 1.3. *Availability normalization function*

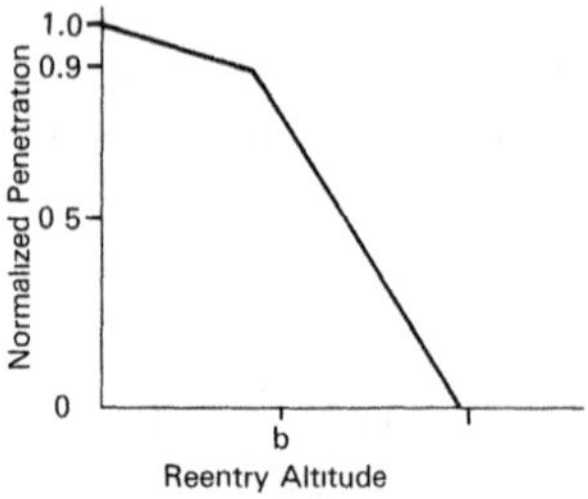

Fig. 1.4. *Penetration normalization function*

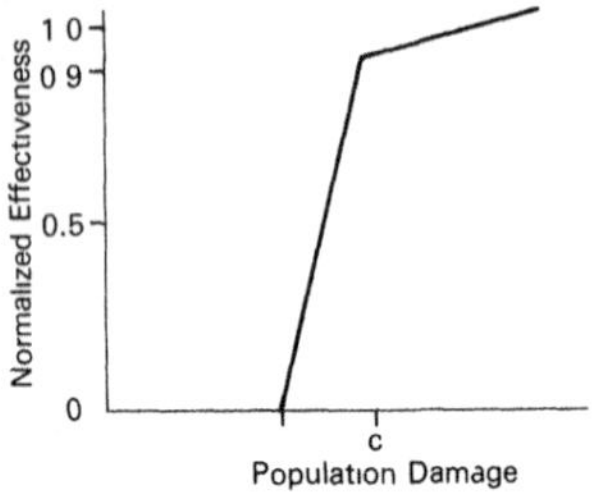

Fig. 1.5. *Effectiveness normalization function*

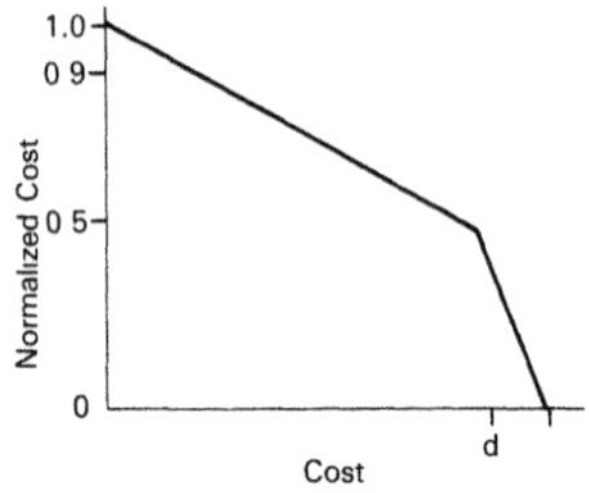

Fig. 1.6. *Cost normalization function*

function in which values of reentry altitude above a certain altitude *b* are penalized at a high rate at the higher altitudes while being favored at a lesser rate at the lower altitudes. Penetration, of course, is successful, provided the reentry vehicles penetrate to a certain low altitude beyond which the defense cannot react. The slopes of the two lines have been determined by the fact that the system must penetrate, especially to some low altitude. Those systems which cannot penetrate to this altitude are penalized while those which do penetrate are favored. Furthermore it is more important to penalize the former systems than it is to favor the latter. In this simplified example the system penetration has been implied to be its reentry altitude alone. Since many other factors combine to define the total system penetration these should all be considered and normalized in practice.

In a similar fashion, Fig. 1.5 depicts the system effectiveness normalization function in which population damage below a certain level *c* is penalized at a high rate while being favored at a lesser rate at the higher levels. The slopes of both lines in the figure have been determined by the fact that the system must inflict a certain level of damage on the opponent. Those systems that cannot inflict damage on the opponent are penalized while those which do are favored. Furthermore it is more important to heavily penalize the former systems than to heavily reward the latter. Again the example has been simplified.

In Fig. 1.6 the system cost normalization function is depicted. A system which just attains the level *d* in costs is normalized to 0.5. Those systems which exceed the level are severely penalized while those systems which cost less are increasingly favored. Again many factors determine cost and these all should be considered and normalized in practice.

Using the graphs of Figs. 1.3–1.6, the normalized factors of system availability, penetration, effectiveness and cost may be determined. As mentioned, these graphs must be constructed from a careful analysis of the statement of the problem and the interpretation of system requirements. In the present discussion a simple example has been used to illustrate the procedure. In practice many individual performance and cost factors must be considered and normalized. Examples of such factors for a missile system have been listed in Tables 1.6 and 1.7.

1.4.6.2. Weightings Assignments

Before solving the value equation (1.19) it will be necessary to determine appropriate weightings w_i to be applied. In the case of the foregoing example the system availability, penetration and cost will each be weighted 20 percent, i.e., $w_1 = w_2 = w_4 = 0.2$ and the system effectiveness will be weighted 40 percent, i.e., $w_3 = 0.4$. These weights have been derived on a subjective basis from the statement of the system requirements which *desire* certain levels of system availability, penetration and cost but *require* a certain level of system effectiveness. The variations analysis is concerned primarily with alternative settings for the weightings.

Consider now two proposed weapons systems. The first will have a system availability of 90 percent, system penetration 90 percent, system effectiveness 90 percent and system cost 50 percent, i.e., this system exactly meets the requirements for the system. The second system will also have a system availability of 90 percent, system penetration 90 percent, but for this system the system effectiveness has been increased to 100 percent but at the expense of added costs which reduce the normalized cost factor to 25 percent. The value of both systems can be computed using equation (1.19) and is found to be 0.82 for the first system and 0.81 for the second system. Therefore, on the basis of the weightings selected, the second system is one of lesser value even though it has achieved outstanding effectiveness, albeit at added cost. It is easy to see that this result will reverse if a different set of weightings is chosen.

1.4.7. Value Sensitivity Assessment

Reference 86

The sensitivity of the value of a given system to variations in the weightings is also a matter of interest. Many procedures may be used to study the impact on the value of variations in weightings and this would certainly be indicated in cases of nearness of results, as was shown to be the case for the two systems in the example of the last section. Thus the system value analysis should feature the computation of the system value which is based on a most likely set of weights which must be determined during the analysis and following the examination of the system requirements.

The weighting assignments for the system availability, penetration and cost in our previous example were set at 0.2 while the system effectiveness weighting was set at 0.4. Thus the former factors being *desired* rated less than the latter which was *required*, i.e., the interpretation of these statements indicates that the emphasis should be put on the system effectiveness. In this manner the selected weightings are quite likely since other conceivable sets of weightings are decidedly unlikely. No basic rules exist to confirm the relative likelihood of different sets of values for the weightings. Thus, had the weightings been set at 0.21 for the system availability, penetration and cost at 0.37 for the system effectiveness, no discriminant is possible to distinguish between the likelihood of the two sets. However, an unlikely set of weightings, for example, would be the assignment of 0.25 for each of the four factors. This last assignment is contrary to the statement of requirements for the system, which specifically points out that the system effectiveness level is *required* whereas all other factors are but *desired*.

The value sensitivity assessment of the system to variations in the settings of the weightings can be obtained by taking small increments of the weightings dw_i about their chosen likely values w_i. Thus the increment or decrement of the system value, as the case may be, can be expressed as a function of the individual incremental changes dw_i in the weightings w_i:

$$dV(w_i) = \sum_i dw_i x_i \tag{1.20}$$

If the weightings are set at 0.21 for the system availability, penetration and cost and 0.37 for the system effectiveness, it can be found using either equations (1.19) or (1.20) that the value of the first system becomes 0.816 while the value of the second system becomes 0.8005. Thus both systems reduce their overall value using the second set of weightings but the first system reduces at a lesser rate so that the systems now are separated by a higher differential value using the new settings for the weightings.

Again it is pointed out that the simple example of the present discussion is intended to illustrate the procedure for the assessment of the value of a system under a variation of the weightings w_i. In practice a large number of factors and weightings must be analyzed and this often requires the use of a computer to make the tedious but rather straightforward computations.

Summarizing the systems-approach technique described above for determining the system value involves several steps. First, the normalizing functions must be ascertained from the significant technical performance and cost factors involved in the problem. Second, weightings are assigned to each factor and the values of proposed systems are computed. Finally, the sensitivity of the value of a particular system to variations in the choice of weightings may be studied.

Judgement plays an important role in the determination of what factors are important, in the normalizing function and in the assignment of the weightings. Thus, the use of our proposed numerical yardstick to evaluate advanced and complex designs of weapons systems on the basis of the system value will have to be made, not as a substitute for judgement, but in order to provide a systematic and organized method of assessing the many often interrelated factors and subfactors and their tradeoffs with system performance and cost.

References

1. Eds., "The SALT-free Weapons", Science and the Citizen, *Scientific American*, May, 1975.
2. A. H. Quanbeck and B. M. Blechman, *Strategic Forces: Issues for the Mid-Seventies*, The Brookings Institution, 1775 Massachusetts Ave., N.W., Washington, D.C. 1973. The Institute publishes studies in the form of monograms. This study summarizes the present and projected state of strategic forces.
3. D. Fink, "Strategic Warfare", *Science and Technology*, October, 1968. Explains how engineering a strategic force requires continual interactions between offense and defense to find the level of strength that deters attack without triggering further rounds of escalation.
4. S. Tilson, "The Antiballistic Missile", *IEEE Spectrum*, August, 1969. Presents a report on a debate on the ABM controversy by G. Rathjens who argues against and D. Brennan who argues for the ABM.
5. H. York, "Military Technology and National Security", *Scientific American*, August, 1969. Analyzes the ABM debate in the context of a larger dilemma: the futility of searching for technological solutions to what is essentially a political problem.
6. A. Frye and J. Foster, "Defense Policy and World Environment", *Astronautics and Aeronautics*, August, 1970. Authors present their respective viewpoints on defense policy.
7. UPI, "Army Missile Intercepted by ABM System", news item, 31 August, 1970.
8. P. Friel, "The Merits of a Limited Ballistic Missile Defense", *Astronautics and Aeronautics*, October, 1970. Discusses an area ABM system and covers the possible response of the Soviet Union to the deployment of Safeguard.
9. R. L. Garwin and H. A. Bethe, "Anti-Ballistic Missile Systems", *Scientific American*, March, 1968. The authors argue that offensive tactics and cheap penetration aids could nullify the effectiveness of a light ABM system and any other visualized so far.
10. H. York, "Multiple Warhead Missiles", *Scientific American*, November, 1973. MIRVs increase the number of strategic nuclear weapons and now threaten the stability of the nuclear balance of power. Their history shows why they present special problems of arms control.
11. H. Scoville, Jr., "The Limitations of Offensive Weapons", *Scientific American*, January, 1971. It is argued that the best

possible outcome that can be expected to emerge from the current (1971) strategic arms limitations talks (SALT) would be a freeze on the existing offensive forces of both sides.
12. T. Greenwood, "Reconnaissance and Arms Control", *Scientific American*, February, 1973. Reconnaissance satellites are the chief means relied on by the USA and the USSR to verify each others compliance with the SALT I accords. What bearing will they and related systems have on SALT II?
13. Ed., *The Strategic Survey*, Inst. of Strategic Services, 18 Adam Street, London W.C.2. Contains an excellent survey of strategic offense and defense systems.
14. Ed., *The Military Balance*, Inst. of Strategic Services, 18 Adam Street, London W.C.2. Contains an excellent survey of strategic offense and defense systems.
15. C. M. Johnson, "Ballistic Missile Defense Radars", *IEEE Spectrum*, March, 1970. Describes the Safeguard ABM system and its major components, the PAR and MSR radars, the Spartan and Sprint interceptors, and the data-processing system.
16. Ed., "Strategic Warfare", *Space/Aeronautics*, January, 1969. Provides a perspective summary of U.S. strategic warfare projects covering intelligence and warning systems, defense and strike systems.
17. R. English and D. Bolef, "Defense against Bomber Attack", *Scientific American*, August, 1973. Among the weapons not covered by the strategic arms limitations talks are anti-aircraft systems. The Department of Defense has now proposed that the nation's air defenses be modernized.
18. J. Batchelor et al., eds., *The Illustrated Encyclopaedia of 20th Century Weapons and Warfare*, Columbia House, 1400 N. Fruitridge Ave, Terre Haute, Indiana, 47811, 1978.
19. R. Pretty, ed., *Jane's Weapons Systems 1977–78*, ISBN 531-03284-1, Franklin Watts Inc., 730 Fifth Ave, New York, NY 10019, 1978. Covers missiles of every type; aircraft armament, radars, sonars, torpedoes, EW equipment, including active, passive, deception and noise jammers, monitoring and DF systems. Gives individual description of every missile, drone, RPV, fire-control system, armed vehicles, underwater equipment, reconnaissance equipment, etc.
20. K. Gatland, *Missiles and Rockets*, Macmillan, New York, 1978. Gives eighty color pictures of world's ballistic missiles, chiefly military, including cutaway and exploded views, with engines, carriers, silos, etc. The Russian Delta-class nuclear submarine "seems to have become operational in January 1973" with a missile range of 4,000 nautical miles, not enough to reach the USA from the Barentz Sea.
21. "Red A-Subs cruise closer to U.S.", Associated Press, 9 June 1975.
22. "The SALT-free Weapons", Science and the Citizen, *Scientific*

American, May, 1975. It is estimated that by the middle of 1975 the USA will have emplaced approximately 8,500 strategic weapons (deliverable by land-based missiles, sea-based missiles and bombers), compared with the Russian total of perhaps 2,800. By the 1974 Vladivostok undertanding, the USA could have as many as 21,000 strategic nuclear weapons deployed by 1985. What is ignored is the vast stockpiling of "tactical", or comparatively short-range, nuclear weapons, by both sides. The USA alone has 22,000 tactical weapons and it is believed the Russian forces in Europe have some 3,000–3,500 tactical nuclear weapons.

23. A. Gavshon, Which is the Strongest Nuclear Power—U.S. or Russia, Associated Press, 30 July, 1977. Provides numerical strengths of various weapons forces of opponents.
24. "Far Stronger", editorial, *Wall Street Journal*, 29 December, 1976. Provides numerical strengths of various weapons forces of opponents.
25. "Missile Scoreboard", *Microwave Systems News*, September, 1977.
26. "The silent war", *Countermeasures*, March, 1975. Few basic articles exist on Electronic Warfare (EW). The general tendency for the EW community is to mask its efforts in secrecy, even to the point of not discussing information requiring no security classification. Consequently, individuals entering industry, workers in peripheral areas, and non-technical managers are relatively uninformed about EW even though they serve as program and hardware development managers. This article provides a general understanding and evaluation of the EW role, along with a glossary of associated terms.
27. "Communications Satellite Systems", Session 32, WESCON, San Fransisco, 16–19 September, 1975.
28. F. Lightfoot et al., "An Experimental TDMA Network for Airborne Warning and Control Systems Interoperability Demonstrations: Part I", *Microwave Systems News*, December/January, 1976. With its TDMA (Time Division Multiple Access) communication system AWACS has demonstrated improved operational capability in such areas as surveillance, control, detection of penetrating aircraft, distribution of radar tracking information to surface C^2 centers, and control of friendly interceptors.
29. "Defense Satellite Communications in the 1980s", *Countermeasures*, December/January, 1976. The defense satellite communications system (DSCS) is a worldwide military command-and-control communications network in use by the USA. Because of an unsuccessful attempt in May 1975 to deploy two additional satellites, new direction and urgency has taken over the program. Other improvements in the launch weight capability of the Titan IIIC launch vehicles, redesign of the satellite payload, improvements in the capability and effectiveness of the electronics, and

the advent of the space shuttle as a primary launch vehicle have added new dimensions to the DSCS.

30. "Airborne Navonics", *Countermeasures*, February, 1976. Airborne EW, a concept which began in World War II, has matured into an essential element of fleet defense and air control operations because of the added capabilities afforded by electronics. The E-2C aircraft, a veritable storehouse of computers, sensors, displays, and communications equipment, incorporates the latest airborne Navonics (naval electronics) technology to complete its mission as an airborne EW&C platform for the U.S. Navy fleet operations throughout the world.
31. W. Shockley, "The E-4 Airborne Command Post", *Countermeasures*, July, 1976. Airborne command-and-control systems are used for both strategic and tactical applications. In strategic situations, these systems are used by top-level command personnel to monitor worldwide military operations and allow integrated command of all friendly forces. In tactical applications, airborne command-and-control systems are used by local area commanders to monitor battlefield operations and control the ground and airborne forces under their command.
32. L. Pashall, "Command Control and Technology", *Countermeasures*, July, 1976. Lt. General L. Paschall, U.S. Air Force, is Director, Defense Communications Agency (DCA). As Director, General Paschall is responsible for management and direction of the worldwide Defense Communication System. He is also responsible for system engineering and technical support to the National Military Command System, and for provision of technical support to the worldwide military command-and-control standard automatic data-processing systems. In his capacity as Manager, National Communications System, he is responsible for providing effective direction to the worldwide National Communications System, which includes the communications facilities of the various Federal agencies. The Director, DCA, is also Chairman, Military Communications–Electronics Board, providing a liaison point for joint and international communications matters. In this article General Paschall gives an overview of the Defense Communications System (DCS).
33. B. Walsh, "An Eagle in the Sky", *Countermeasures*, July, 1976. An article on AWACS, an all-altitude surveillance, command, control, and communications airborne warning and control system.
34. B. Walsh, "C^3—The Big Picture", *Countermeasures*, March, 1977.
35. T. Bibbens and J. Lake, "COMINT/SIGINT", *Countermeasures*, July, 1976. Signals intelligence (SIGINT) and communications intelligence (COMINT) are two vital elements in the structure of Intelligence, Electronic Warfare (EW) and Command, Control and Communications (C^3). SIGINT encompasses both COMINT

and ELINT (electronic intelligence). ELINT is associated with the detection, identification and location of radars and similar emitters of weapons systems. COMINT is associated with communication, telemetry and other information links.

36. J. Fawcette, "C^3: Key Challenges Face Military Planners", *Electronic Warfare,* June, 1978. Spiraling costs, technological complexities loom as major obstacles at both the strategic and tactical levels.
37. P. Nahin, "The Laser BMD and other Radient Energy Weapons: Some Thoughts", *IEEE Transactions on Aerospace and Electronic Systems,* March, 1977. Describes ABM Treaty and the particular language in it that presently excludes the deployment of a laser BMD. Eight pressure points are then presented that may cause both the Soviets and Americans to reevaluate the payoff of such ACD agreements to their overall national security, even to the extent of abrogating the Outer Space Treaty and SALT I. Next, an extended presentation of the current state of affairs in the American laser research program is given to show the magnitude of the U.S. commitment to developing radiant energy weapons technology. The paper concludes with a brief excursion into the psychological and strategic systems impact of laser weapons.
38. T. Tiede, "$5 billion defense site abandoned, left to decay", NEA news item, June 1, 1977.
39. "Extraterrestrial Eavesdropping Seen Possible via Radar Leakage", *Microwave Systems News,* June, 1978. Leakage from U.S. and Soviet BMEWS-type radars could enable extraterrestrials to detect the presence of civilization on Earth at interstellar distances.
40. "PAVE PAWS Tests Scrutinized", *Microwave Systems News,* June, 1978. Pave Paws is a detection and early-warning system to guard against a ballistic missile (SLBM) attack on the continental U.S. It will also track satellites in orbit.
41. "Shell Game", Science and the Citizen, *Scientific American,* March, 1977. Describes the Air Force plan to replace the present force of 1,000 Minuteman ICBMs with a more powerful and more accurate missile X which would not be confined to an identifiable launching silo but could be moved around in a random pattern.
42. K. Bacon, "Movable Deterrent", *Wall Street Journal,* 14 March, 1978. Plan to bury missiles in vast tunnel system sparks defense debate. "MX" would alter location to elude enemy strike: Cost: up to $40 billion.
43. F. Kaplan, "Enhanced Radiation Weapons", *Scientific American,* May, 1978. Although President Carter has deferred production of the neutron bomb, it is still an alternative of U.S. policy. It remains a weapon of doubtful utility that could result in an all-out nuclear exchange.

44. S. Drell and F. von Hippel, "Limited Nuclear War", *Scientific American,* November 1976, and in Letters, February, 1977. The USA may be committing itself to preparing for a war limited to attacks on military bases, with relatively few civilian casualties. Would the casualties really be few, and could the war stay limited?
45. K. Tsipis, "The Accuracy of Strategic Missiles", *Scientific American,* July, 1975. The USA has initiated the development of ultra-accurate missiles capable of destroying the land-based missile force of the USSR. The technology behind the new "counter-force" strategy is reviewed
46. W. Yengst, and reply by K. Tsipis, Letters, *Scientific American,* October, 1975.
47. W. Panofsky, Letters, *Scientific American,* December, 1975.
48. "65 Attack Nuclear Subs now in fleet—30 on rails", UP news item, 25 May, 1977.
49. J. Moore, ed., *Jane's Fighting Ships 1977–78,* ISBN 531-03277-9, Franklin Watts Inc., 730 Fifth Ave, New York, N.Y. 10019.
50. J. Taylor, ed., *Jane's All the World's Aircraft 1977–78,* ISBN 531-03278-7, Franklin Watts Inc., 730 Fifth Ave, New York, N.Y. 10019.
51. P. Nahin, "Can Land-Based Strategic Bombers Survive an SLBM Attack?", *IEEE Transactions on Aerospace and Electronic Systems,* March, 1977. A recent study by the Brookings Institution on the B-1 bomber contains an erroneous analysis of its survivability in an SLBM attack. This issue is discussed and an alternative analysis is presented.
52. R. Hartman, "B-52 Update", *Countermeasures,* March, 1978. The B-52 update is a long-range program begun in 1969 and expected to continue into the 1980s. Cancellation of the B-1 has shortened this timetable.
53. "B-52 Avionics to be Completely Updated", *Microwave System News,* February, 1978.
54. "The Cruise Missile", *Wall Street Journal,* 19 June, 1975.
55. "Buzz Bombs", Science and the Citizen, *Scientific American,* 1975.
56. K. Bacon, "Cruise Missile's Future", *Wall Street Journal,* Vol. XCVO, No. 1. Cheap and effective weapon upsets the traditionalists and faces SALT hurdle. Better than the B-1 Bomber?
57. "Air Launched Cruise Missile in Flight Tests", *Countermeasures,* July, 1976. The first ALCM prototype was launched at 15,000 feet and flew for ten minutes and forty seconds some 70 miles.
58. B. Walsh, "The Dawn of a New Era—Cruise Missiles", *Countermeasures,* January, 1977. The USA recently completed advanced development of a new intermediate range cruise missile that will have a positive impact on the way the military conducts its naval and land attack missions of the future. These pilotless aircraft, capable of flying only a few feet above water or any terrain, have

gone through exhaustive design and evaluation by the Air Force and Navy. This month, the Defense Systems Acquisition Review Council is scheduled to decide the future of cruise missiles. This article summarizes the work completed by the two services, and reviews some of the important issues being considered by the Defense Department if future development is to continue.

59. K. Tsipis, "Cruise Missiles", *Scientific American,* February, 1977. This new category of inexpensive, highly-accurate weapons presents a difficult but not insuperable problem to arms-control negotiators.
60. J. Deutch et al., and reply by K. Tsipis, Letters, *Scientific American,* August, 1977.
61. Special Report, "Cruise Missile Technology", *Microwave Systems News,* September, 1977.
J. Fawcette, "Map-Reading Missile Takes on Soviets and B-1". A cruise missile is merely a small, pilotless jet plane, but several advances in technology have made this simple missile a weapon the Soviet Union fears more than our ICBMs and multi-million dollar, electronic-laden fighters and bombers.
R. Davis, "Millimeter Waves: A solution that has finally found a problem". MM wave technology has languished in the lab for about twenty years now, seeking applications that never materialized. The need for accurate delivery of ordnances, under all weather and illumination conditions, has brought about renewed interest in this field by the military. An enormous new market is developing.
62. "Capital Capsules", *Electronic Design,* June 21, 1978. The Air launched Cruise Missile (ALCM) won't enter service until June 1980 and won't be fully operational until September, 1981.
63. "Air Force is contemplating future bombers", *Electronic Design,* July 5, 1978.
64. J. P. Dix, "Game Theoretic Applications", *IEEE Spectrum,* April, 1968. Provides a discussion of gaming based on the defense problem and develops the geometrical and Monte Carlo approaches for obtaining the solutions to this problem.
65. R. B. Dow, *Fundamentals of Advanced Missiles,* John Wiley and Sons, 1958. This is a useful engineering reference covering many subjects on a first principle basis and is recommended for reference to the material presented in this chapter. In particular, Chapter 4 gives the elementary theory of probability and discusses much of the theory of probability of killing a target. Chapter 9 presents aspects of systems engineering.
66. G. Merril, ed., *Principles of Guided Missile Design,* D. Van Nostrand, 1955. This reference is a comprehensive treatment of the principles of missile guidance and considers the fundamentals of guiding a controlled missile reliably to its target. Attention is

directed to Chapter 8 for a discussion of target damage definitions. Chapters 15 and 20 discuss important aspects of systems engineering.

67. D. Archer, *Defense and Foreign Affairs Handbook 1978,* ISBN 531-03275-2, Franklin Watts Inc., 730 Fifth Ave, New York, N.Y. 10019. Gives defense and political information for every country in the world. Includes key defense manufacturers and personnel; complete cabinet listings; economic data; a Who's Who in politics and defense; glossary of defense and political acronyms; balance of payment, power, and SAM tables.
68. "Soviets step up sub tests", Associated Press, 12 December, 1977.
69. G. Clancey, "New Breakthroughs in ASW Signal Processing", *Countermeasures,* October/November, 1975.
70. "Airborne Counter Threat to Submarines", *Countermeasures,* September, 1975. An overview of the long-range patrol aircraft whose mission is to locate and attack enemy submarines.
71. "Modern Shipboard EW", *Countermeasures,* February, 1976. In modern warfare, a complex electromagnetic environment poses many threats to the survivability of naval vessels in a tactical theater of operation. New sophisticated electronic equipments provide threat early-warning protection for the U.S. Navy's fleet of surface and subsurface platforms.
72. A. Arnot, "SM-2 Passes Naval Tests", *Countermeasures,* June, 1977. Recent at-sea firings of the U.S. Navy's new Standard Missile-2 (SM-2) demonstrate a new shipboard defense capability against airborne threats.
73. R. Hartman, "SAM-D is a Patriot", *Countermeasures,* July, 1977. Being developed to replace both Nike Hercules and Hawk systems, the highly mobile, all-weather Patriot will be the field army's air defense cornerstone against medium- to high-altitude targets in the sophisticated land-warfare environment predicted for the 1980s and beyond.
74. G. Myrdal, "Political Factors in Economic Assistance", *Scientific American,* April, 1972. It is argued that the global downward trend in aid from the rich countries to the poor countries can be reversed only by stressing the moral argument.
75. E. Rostow, "For an Adequate Defense", *Wall Street Journal,* 12 May, 1975.
76. M. Skolnik, *Introduction to Radar Systems,* Chapter 13, McGraw-Hill, 1962.
77. J. Schlager, "Systems Engineering: Key to Modern Development", *IRE Trans.*, Vol. EM-3, pp. 64–66, July, 1956.
78. J. N. Warfield, *Systems Engineering,* Ordnance Res. Lab Report, Penn. State Univ., 10 August, 1955 (Available from Office of Technical Services, Wash. D.C.) PB 111801.
79. C. D. Flagle et al., eds., *Operations Research and Systems Engineering,* Johns Hopkins Press, Baltimore, 1960.

80. P. M. Morse and G. E. Kimball, *Methods of Operations Research,* Technology Press and John Wiley and Sons, 1950.
81. P. M. Morse, *Journal of Appl. Physics* 23, 165, 1952.
82. L. A. Brothers, *Journal of Ops. Rec. Soc. Am.* 2, 1, 1953.
83. P. J. Kahrilas, "Design of Electronic Scanning Radar Systems", *Proc. IEEE,* Vol. 56, No. 11, 1968.
84. L. J. Cantafio, "Prediction of the Minimum Investment Cost of Phased Array Radar", Supplement to *IEEE Trans on Aerospace and Electronic Systems,* Vol. AES-3, No. 6, November, 1967.
85. O. H. Oakley, "Trends in Marine Technology", *Astronautics and Aeronautics,* April, 1966. Particular attention is directed to the brief discussion of systems design which describes the merits of integrated ship design.
86. R. Shuster and D. Blattner, "The Worth Concept: An Aid to System Planning", Paper presented at the 1961 Winter Convention IRE, Los Angeles, California.
87. "Top 50 Defense Contractors", *Electronic Warfare,* June, 1978.
88. "Armament Decade", Science and the Citizen, *Scientific American,* November, 1973. The cost of the arms race for the world as a whole and for the vast majority of countries has continued its rise, while the problems of development and the urgency of social needs are as acute as ever.
89. J. Anderson, "Billions Squandered on Military Duds", 22 September, 1974.
90. J. Anderson, "Military Contractors make as much as 240% Hidden Profits", 1974.
91. "Aerospace Crisis", *Wall Street Journal,* Four article series, 1974.
92. R. Levine, "U.S. Military Strategy/Weaponry Begins to Draw Fire", *Wall Street Journal,* 3 June, 1975. Rarely challenged in the past, basic Pentagon planning is studied by lawmakers.
93. "Out of Control", Science and the Citizen, *Scientific American,* April, 1976. In a period of global economic recession and simultaneous price inflation one sector of the world economy—the arms race—appears to be exempt from the general squeeze on the public purse.
94. "50 and 100 years ago", *Scientific American,* March, 1977.
95. T. Jones, "The Flaws in Defense Contracting", *Wall Street Journal,* 25 November, 1977. Only when defense companies are required to assume the kind of management responsibility demanded elsewhere in the private sector will our national security benefit from the advantages of the market system.
96. S. Melman, "Who decides Technology?", *Astronautics and Aeronautics,* 1969.
97. "L. Nevin, "Strategic studies map influence of global armaments and militias" Associated Press news item, 27 August, 1976. Gives a profile of the International Institute for Strategic Studies. See References 13 and 14.

98. P. Morrison and P. Walker, "A New Strategy for Military Spending", *Scientific American*, October, 1978. An analysis of U.S. military forces finds that they so far exceed actual military needs as to be unsafe for the nation and the world. A program for prudently decreasing these forces is described.
99. "Off the Beam", Science and the Citizen, *Scientific American*, December, 1978. Efforts to revive the U.S. ABM program, which was severely limited under the terms of the 1972 SALT ABM treaty, have gained momentum in some circles over the past year or two, energized in part by recurrent public assertions that the Russians are engaged in a secret multi-billion ruble program of their own to develop a new kind of ABM system, based this time not on nuclear armed interceptor missiles but on intense beams of charged particles.
100. "USSR—Soviet demands for SALT II limitations on U.S. cruise missiles have been dropped", *ME/C* (Military Electronics/Countermeasures) December, 1978. Russian tests with MIG-25 Foxbat fighters shoot down drones simulating U.S. weapons.
101. "USSR—A new Generation of ICMBs is Cause for Concern", *World News ME/C,* October, 1978.
102. "Breakout System Tested for MX", *Technology News ME/C,* October, 1978.
103. F. Berry, Jr., "TACTAS", *ME/C,* November, 1978. The Tactical Towed Array Sonar (TACTAS) concept differs from hull-mounted passive sonars by deploying a towed array of hydrophones on a cable beneath and behind the ship.
104. B. Walsh, "The Multirole Cruise Missile", *ME/C,* September, 1978. As a major weapon system, the new generation of cruise missiles represents a powerful and versatile addition to the U.S. military's strategic and tactical force structure.
105. M. Stoiko, "*Soviet Rocketry*", Holt, Rinehart and Winston 1972.
106. L. Martin, "*Arms and Strategy*", David McKay Co., 1974.
107. "Carter endorses mobile missile plan" *Microwaves,* July, 1979.
108. "Can the Infants Survive" *Astronautics and Aeronautics* August, 1969.

2

ABM Defense

The functions of a defense system against ballistic missiles are to detect, discriminate or select, designate, track and intercept unfriendly missiles. A modern defense system must complete several hundreds of intercepts while at the same time tracking perhaps thousands of targets.

The general tactical situation of a missile attack may be analyzed by considering unfriendly intercontinental ballistic missiles (ICBMs) and sea-launched ballistic missiles (SLBMs), including various penetration aids and radar countermeasures and nuclear bursts, being committed against the defense system —which consists of a number of radar sites and employs surface-to-air guided missiles.

A modern defense system consists of six main subsystems: search radars, tracking radars, data processors, launching equipment, guided missiles, and communications, and control (C&C) networks. The information which is obtained from the search radars is analyzed in the data processor which in turn selects targets and designates tracking radars. Extremely accurate data, indicating target location and speed, is then provided to the data processor by the tracking radars. The data processor chooses an optimum missile trajectory and intercept point to engage each target, designates and assigns radar trackers and interceptors. Interceptors are launched by launch equipment and are guided to their targets by the radar trackers and the data processor. The C&C subsystem interconnects the various other subsystems in the defense system. This chapter will discuss the basic features of ABM defense systems.

Historically, defense radar designs have progressed from slow, mechanically-steered, single-function radars to fast, electronically-steered, multifunction radars. The functions of searching, recognizing, selecting, and tracking targets, and of designating and controlling trackers and interceptors have progressed from multiradar systems with individual operators, to single combined-function radars with automatic operation.

The evolution of defense radar systems begins in 1938 with

the installation of the first antiaircraft fire control system, the SCR-268 used in conjunction with searchlights. This remained the standard fire-control equipment until 1944 when it was replaced by the SCR-584 microwave reflector radar. This radar system could control an antiaircraft battery without the necessity for searchlights or optical angle tracking.

Search radars in operation during the 1950s were the CPS-6B, the FPS-3, and the FPS-6 heightfinder, used for long-range surveillance. In air defense at this time, most of the functions of recognizing and plotting aircraft tracks—previously performed by an operator—were carried out automatically by electronic digital computers such as those used in the SAGE (Semi-Automatic Ground Environment) system.

Radar defense against ballistic missiles began with the introduction of the Nike-Zeus system in 1956. It consisted of four radars: the acquisition radar, the discrimination radar, the target-tracking radar, and the interceptor radar used in conjunction with the Zeus interceptor missile.

The Nike-X system was introduced in 1962 to cope with the increased sophistication of penetration aids. This system employed but two radars: the multifunctional array radar and the missile site radar. The slow Zeus missile was used as the main interceptor with the Nike-X system, although the much faster Sprint missile was introduced in limited numbers. Also introduced for the first time was the phased array antenna.

The Sentinel system appeared in 1967 with two radars: the perimeter acquisition radar (PAR) used with the long-range Spartan interceptor missile, and the missile site radar (MSR) used with the fast short-range Sprint interceptor missile. The Sentinel system was intended to operate in a low-traffic environment. A direct outgrowth of the Sentinel system is the present Safeguard system designed for a high-traffic environment.

Two basic defense concepts have evolved in the present Safeguard system: area defense and terminal defense. Area defense involves the search, acquisition, and tracking of incoming ballistic missiles using the long-range perimeter acquisition radar (PAR), and the interception of these missiles with the long-range Spartan while they are still high above the earth's atmosphere. Terminal defense involves the precision target-tracking of incoming ballistic missiles using the high-capacity, short-range missile site radar (MSR), and the interception of these missiles with the short-range, high-acceleration

Sprint after they have reentered the atmosphere, and after they have been sorted by the atmosphere from decoys, chaff, and other offensive confusion devices. Generally, these two types of defense complement each other and provide an optimized defense that places maximum stress on the attacking missile force.

2.1. Safeguard System

References 1–12, 41–49

The technical objectives for the Safeguard ABM system are to intercept and kill a nuclear missile weapon with a nuclear missile weapon. To accomplish these technical objectives the Safeguard system is designed to:

1. Establish continuous surveillance of threat corridors with phased-array radars.
2. Sense high-traffic engagement conditions with phased-array radars.
3. Control radars and interceptors with high-capacity data-processing systems.
4. Initiate or veto system response to an attack via a command chain headed by the U.S. President.
5. Maintain alertness and availability for possible threats.
6. Be exercised at frequent intervals at full capacity.
7. Be operable while undergoing maintenance.
8. Be capable of being tested for alertness and availability.

The major components of the Safeguard system are as follows:

1. The Perimeter Acquisition Radar (PAR) is a UHF phased-array radar for long-range search, acquisition and tracking of enemy ballistic missiles.
2. The Missile Site Radar (MSR) is the battle-management radar, primarily a precision target-tracking and missile-guidance radar. It is a high-traffic capacity phased-array that operates in the S band region of the electromagnetic spectrum. It has less range capability than the PAR but higher resolution.
3. The Spartan is a long-range exoatmospheric interceptor capable of making intercepts at ranges of several hundred kilometers from the launch battery. It carries a high-yield nuclear warhead that can kill enemy reentry vehicles (RVs) at a range of several kilometers.

4. The Sprint is a short-range high-performance terminal interceptor. This interceptor has an operating range of several tens of kilometers against enemy RVs. Greater radar tracking precision is required for Sprint intercepts due to the much smaller Sprint nuclear warhead.
5. The Data-Processing System (DPS) is an integral part of each of the Safeguard phased-array radars and is of sufficient fundamental importance to system operation that it is considered a major component. Functions performed include radar control, incoming target selection and tracking, interceptor guidance, message formatting and data reduction for command-and-control displays and consoles. The DPS is configured as a multiprocessor and is denoted as the central-logic-and-control (CLC) system. Several processors with a common memory can operate in parallel on serveral different tracks or on different parts of the same track.

2.1.1. Perimeter Acquisition Radar (PAR)

The PAR is a UHF phased-array radar and is used in the Safeguard ABM system to search, acquire and track ICBMs and SLBMs at long range. The primary function which determines the PAR power and aperture requirements is the specification for wide-angle search of the horizon. These requirements are also determined to a lesser extent by the PAR's traffic-handling capability. The principle factors that have determined frequency selection and waveform requirements for the PAR are: (1) cost of power and aperture versus frequency, (2) radar cross-section (RCS) of RVs versus frequency and (3) propagation effects in both natural and nuclear environments versus frequency. Secondary factors considered in selecting frequency and waveforms included radio frequency interference (RFI) and available bands for frequency assignment.

The PAR design was originally conceived as a VHF phased-array with two apertures, one for transmit and one for receive. In this original design the receive aperture was about 30.5 meters in diameter but the transmit aperture was much smaller. In the present design PAR has only one aperture, common for both transmit and receive.

Figure 2.1 illustrates the sequence of events needed to intercept a missile at long range. The PAR searches for and tracks

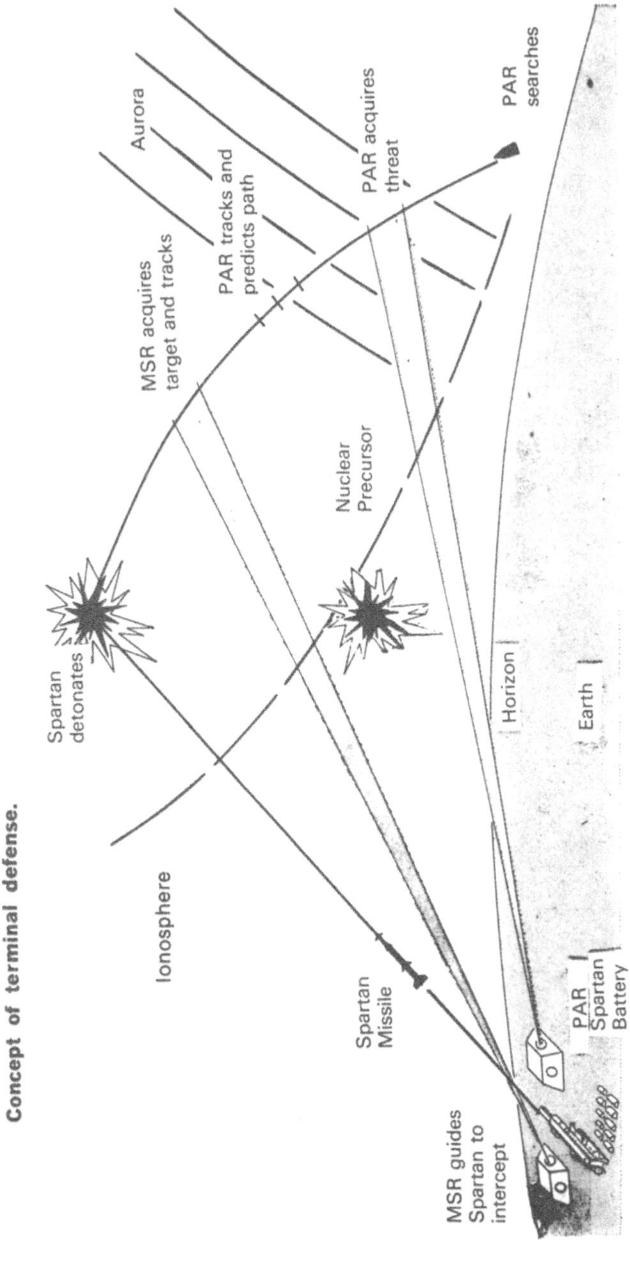

SOURCE: See reference 1.

Fig. 2.1. *Events for area defense intercepts*

the ICBM at long range at or near the radar horizon. To accomplish this the radar beam must penetrate the ionosphere and, consequently, it is subject to refractive and attenuative effects. Also, the radar must operate in the natural aurora and enemy nuclear environments. Operation in such environments becomes less pronounced at the higher frequencies. Thus, the shift from VHF to UHF significantly reduces the effects of natural and nuclear environments on PAR performance.

Figure 2.2 is a block diagram of the PAR. It consists of a conventional corporate-fed array with traveling wave tube output stages for the transmitter and parametric amplifiers for the receivers. The antenna array consists of a number of subarrays arranged to minimize couplings between the transmitter and receiver elements. The phase shifters are 4-bit diode types; orthogonal steering is used and a special beam steering computer controls the array of phase shifters. The PAR central processor (CLC) furnishes the beam pointing coordinates.

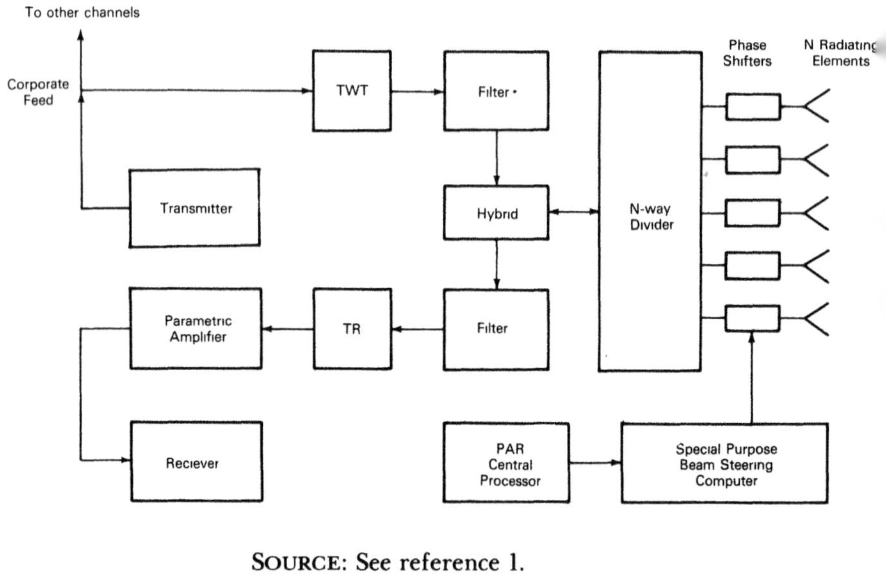

SOURCE: See reference 1.

Fig. 2.2. *PAR functional block diagram*

The PAR utilizes a number of receivers matched to the search and tracking functions of the radar. The search receiver consists of multiple channels that permit simultaneous generation of multiple receiver beams. The track receivers utilize monopulse tracking to achieve accuracies that are a small fraction of a beamwidth.

The PAR must operate continuously monitoring a wide field of view. PAR transmitters and receivers and their supporting equipment are sufficiently redundant to provide virtually 100 percent availability.

2.1.2. Missile Site Radar (MSR)

The MSR is an S-band radar and is used in the Safeguard ABM system to precision track RVs and guide both the Spartan and Sprint interceptors in a high-traffic environment. In a degraded mode of Safeguard system operation the MSR is capable of autonomous search and RV acquisition. The MSR is a phased-array radar with a 4.1 meter diameter aperture and has a range of several hundred kilometers against typical RVs. A single aperture serves for both transmitting and receiving. The MSR is generally configured as a four-faced radar to provide complete 360 degree azimuthal coverage. The MSR building has about the same floor space area as the PAR building but the MSR has only about 1 percent of the antenna aperture area of the PAR.

Figure 2.3 is the sequence of events needed to intercept a missile at short ranges using the concept of terminal defense. The MSR searches for and acquires the ICBM at short range above the atmosphere. Upon reentry into the lower atmosphere, the RVs become separated from the much lighter confusion and concealment devices. The MSR acquires and tracks RVs within the atmosphere, launches and guides Sprint missiles to intercept the RVs. The MSR is also used to launch and guide Spartan interceptors in long range intercepts.

Figure 2.4 is a block diagram of the MSR. The antenna is a space-fed phased-array with separate transmitting and receiving feed horns located at the rear of an anechoic chamber. The antenna thus acts as a focusing lens for the diverging primary pattern produced by the focal point feeds. The phase shifters located in the antenna face serve both to collimate the microwave energy from the feed and to steer the resulting

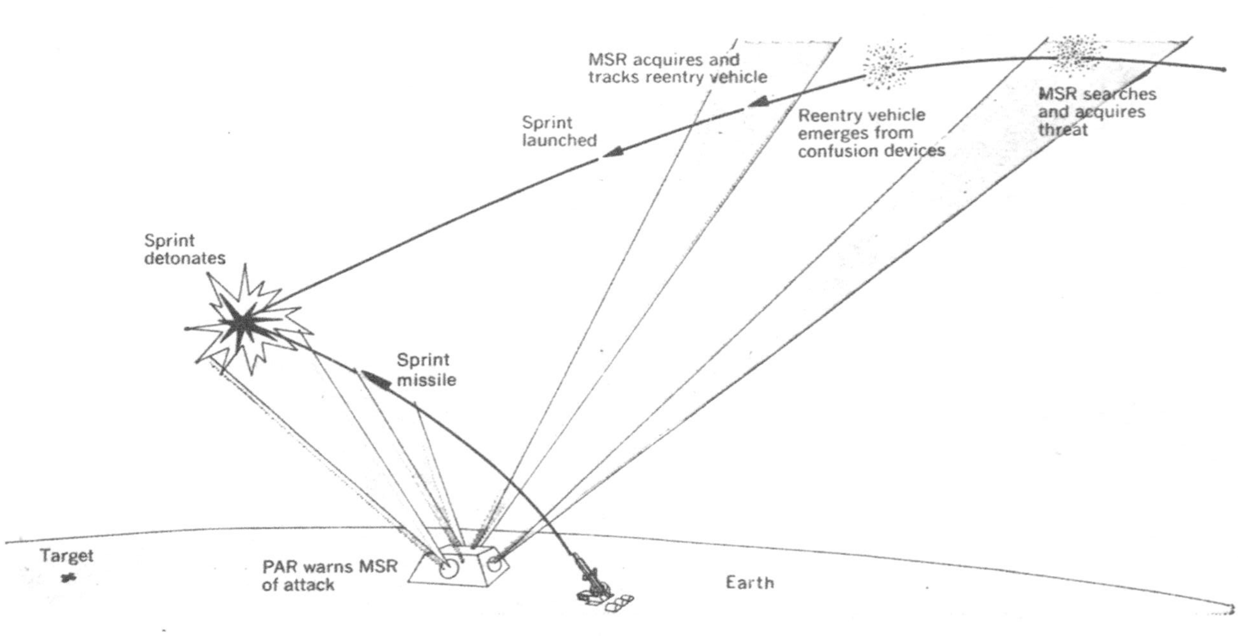

SOURCE: See reference 1.

Fig. 2.3. *Events for terminal defense intercepts*

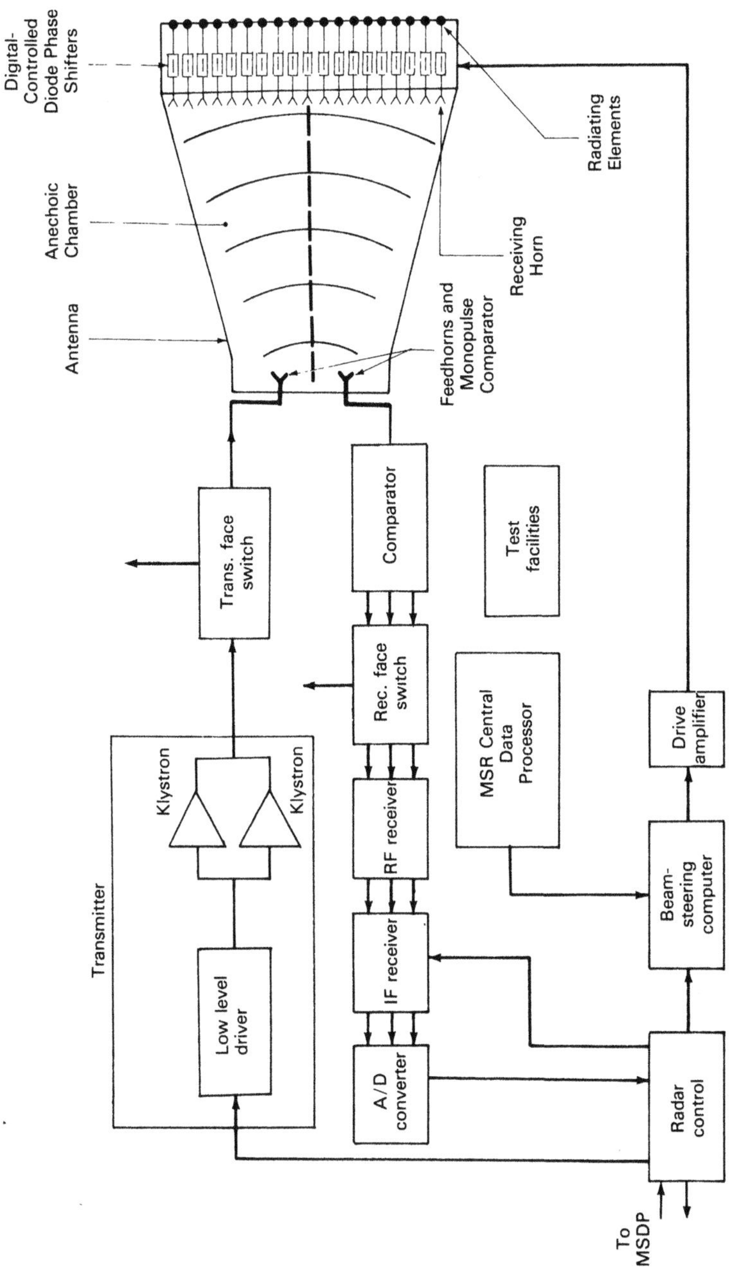

SOURCE: See reference 1.

Fig. 2.4. *MSR functional block diagram*

beam to the desired spatial location. A special-purpose computer with appropriate drive amplifiers controls the array phase shifters. The MSR central data-processor furnishes the beam pointing coordinates. The MSR transmitter consists of redundant, high-power klystrons capable of generating high peak power in a variety of pulse widths. The transmitter also produces exceptionally large amounts of average power, thus permitting the MSR to perform both search and tracking and guidance functions. Because of its completely redundant design, it is possible to perform maintenance on the transmitter while it is in operation. The MSR receiver uses redundant channels to achieve virtually 100 percent availability. There is a set of receivers matched to each set of the possible transmitted waveforms. These receivers have a common RF section featuring a cooled parametric amplifier. The MSR has both search and track receivers. The track receiver utilizes amplitude monopulse to determine accurately the position of targets within a fraction of the beamwidth. All receivers can be switched to handle inputs from the different array faces.

A prototype of the MSR is installed at the U.S. Army's White Sands Missile Range in New Mexico. The basic design of the MSR was accomplished in 1965 and a second prototype is now being operated at Meek Island, a part of the Kwajelein test site and the Pacific Missile Range.

2.1.3. Spartan Interceptor

The Spartan interceptor is a three-stage solid-propellant missile 16.7 meters in length and is used in the Safeguard ABM systems to intercept missiles at long range. It çarries a 1 megaton (MT) nuclear warhead and has a range of 600 km and speed of 3.6 km/sec. The PAR acquires the ICBM at long range and assigns the MSR which in turn acquires and tracks the ICBM. Following launch, the Spartan missile is guided by the MSR to intercept the ICBM. Figure 2.5 shows the Spartan missile. It should be noted that the first two stages of the missile have the same diameter.

Spartan missiles could in principle defend large areas. Approximately twelve PAR sites are needed to cover the entire U.S.A. This is the concept for area defense. However, a defense based solely on PARs and Spartans could be rendered ineffective if the Russians use fairly simple countermeasures, in

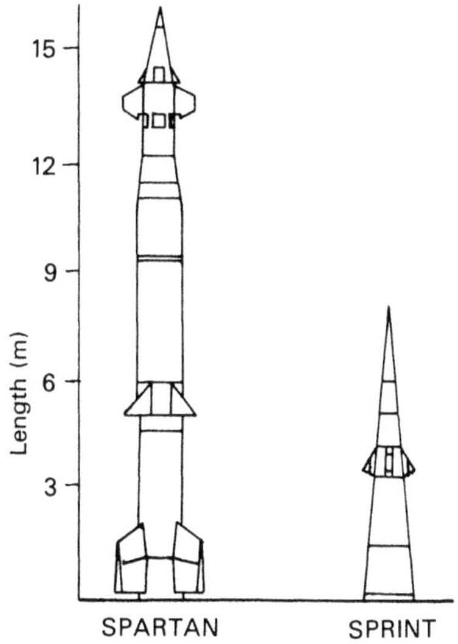

SOURCE: See reference 4.

Fig. 2.5. *U.S. defensive missiles deployed with Safeguard ABM system*

particular if large numbers of lightweight decoys—which would be indistinguishable to the PARs from actual RVs containing warheads—were used or other measures that would make the PAR ineffective, for example by the use of nuclear explosives, electronic jammers or light, widely-dispersed metal chaff.

2.1.4. Sprint Interceptor
Reference 4

The Sprint interceptor is a two-stage solid-propellant missile 8.2 meters in length and is used in the Safeguard ABM system to intercept missiles at short range. It carries a 1 KT warhead and has a range of 40 km. The speed of Sprint has not been released but it has been described as a very fast missile. The MSR acquires and tracks the ICBM above the earth's atmosphere and acquires and tracks the RV following its reentry and after it has been separated from the lighter confusion and concealment devices. Following its launch the Sprint missile is guided by the MSR to intercept the RV. The Sprint missile is also shown in Fig. 2.5.

Sprint missiles could in principle defend large areas. However, a large number of sites would be needed to cover the entire U.S.A. Sprint missiles essentially defend small areas about the target. This is the concept for terminal defense. The effectiveness of a Sprint or terminal defense system is less degraded by countermeasures which can be employed against a Spartan or area defense system. Light decoys can be distinguished from actual RVs because they would be disproportionately slowed down by the atmosphere and possibly because their detectable wakes in the atmosphere would be considerably different. Radar blackout would also be much less of a problem. Because of their short range, however, Sprints could defend only those targets in their immediate vicinity. Thus an enemy could choose to attack some cities with enough weapons to overwhelm the defense while leaving other cities untargeted. Heavy radioactive fallout could also be produced over large parts of the country by the enemy's delivering large-yield weapons outside the areas covered by the Sprint defenses. A nationwide Sprint or terminal defense system would therefore require a nationwide fallout-shelter program.

2.1.5. Data-Processing System (DPS)

Following search and acquisition, the PAR computes the predicted ICBM trajectory. Each track measurement is used to update and refine the predicted trajectory with the entire operation taking about 1–2 milliseconds. The trajectory can be obtained to the required accuracy with about 100 pulse hits and these can be spread over several minutes of the ICBM's trajectory. Therefore, the average time required of a processor is less than a millisecond per missile or object per second. On this basis, a single processor may perform trajectory predictions for over 1,000 objects simultaneously. Additional processors may be added if needed. Since a massive attack would last for minutes and not all objects would appear simultaneously to the PAR, the processor would gradually build up to this rate. As it completed a particular track the processor would drop that object and start a new object track. MSR calculations follow a similar procedure but are made at an even faster rate. Spartan and Sprint guidance calculations take a longer time per calculation and therefore are made at a higher rate. However, the number of defensive missiles that must be in guidance simultaneously from a single MSR is small compared with the total number of objects a PAR must handle. Therefore much less computer capacity is needed for the guidance function.

The actual number of processor units required for each defense site is selected based on the expected traffic demand. Additional processor units are added to handle all of the other functions required to support these basic functions and to supply the necessary number of spare processors to meet availability requirements.

The DPS hardware for the PAR, MSR and BMDC sites uses common hardware modules or racks interconnected as a multiprocessing system. The central processor (CLC) which is the heart of the system uses at least two of the following units: program storage units each containing two 8K 64-bit word modules, operand storage units each containing 16K 64-bit words, input/output controllers and timing and status units. Additional processor units and storage units may be added as required. Multiprocessing, in addition to providing assurance of large throughputs by adding more processors, has the further advantage of modularity, thus permitting the most economical size of DPS installations at each site as well as providing through redundant units a high degree of fail-safe operation.

The software weapons process structure is composed of tactical routines plus radar and computer tests as well as maintenance, diagnostic and operating-system routines. These run on a real-time basis with fixed operating deadlines tied to the major radar cycles. Software facilities are based upon third-generation hardware architecture and include base-register addressing, double-level indexing, indirect addressing, data editing and registers for internal central-processor-unit (CPU) operations. The software structure is broken down into tasks which are in turn controlled through a task-list table. Task precedence is controlled through conditional-enablement tables and run priority is controlled through position in an absolute-enablement table.

The Safeguard ABM system DPS hardware and software are both representative of large-scale computer systems of the present. Multiprocessors have been used in several data-processing systems as the logical way to provide large data-processing throughputs when uniprocessors with adequate capacity are not available. The software for the Safeguard ABM system is a complex real-time process but, again, these functions have been carried out before. Except for peak tracking requirements, the Air Force's FPS-85 phased-array radar at Eglin AFB in Florida, which is controlled by a duplex IBM MOD-65 installation, performs tasks quite similar to those required of the PAR DPS.

2.1.6. Costs

The most powerful argument against building an ABM system is the cost of phased-array radars. By costs are meant complete costs of buildings, power plants, installation and testing as well as the costs of the electronic portions of the system. In the late 1960's, PARs and MSRs are estimated to cost about $200 million each when all costs, from site activation to final checkout as an operational system, are included. This is the so-called investment cost. Table 2.1 presents the cost breakdown for Safeguard ABM system radars.

In addition to the costs that have been designated as investment costs, there are operating and maintenance costs for these radars that amount to about 5 percent of the initial investment per year. Also, the costs of interceptors are variously estimated at between $1–$2 million each and these have not been included in Table 2.1.

Since the costs of defense are between 5–10 times the costs of

Table 2.1. *Cost breakdown of PAR or MSR radar*

RADAR COMPONENT	% OF INVESTMENT COST
RADAR ELECTRONIC COMPONENTS	25
DATA PROCESSOR	12
INSTALLATION AND TEST	11
SPARES	5
OTHER SITE EQUIPMENT	5
RADAR BUILDING	20
POWER PLANT AND POWER EQUIPMENT	15
SUPPORT FACILITIES, REAL ESTATE, AND DESIGN	7
TOTAL	100%

SOURCE: See reference 1.

offense it appears that on the basis of cost alone the present costs of defense would have to be reduced by these same factors to produce a parity in costs between the two. Thus the costs of radar electronic components, data processors, installation and testing, spares, miscellaneous equipment, radar buildings, power plants and power equipment, support facilities, real estate and design should be reduced by a factor 5–10 from the estimate cost of about $200 million. Although there is no doubt that costs will be reduced in the future, particularly in the higher-cost-contributing components of the ABM system, it is rather hard to believe that these costs can be reduced across the board by the indicated amounts to achieve a cost parity between the defense and the offense.

A mitigating factor to the assumption of costs in building an ABM system are the benefits to be derived therefrom. Thus, it appears more appropriate to compare the defense and offense systems not merely on the basis of costs alone but also with regard to system value, which includes the system availability, anti-penetration capability and effectiveness as well as the costs. The theory of system value presented in the previous chapter applies equally well to both defensive and offensive systems and should be used to compare such systems on the basis of total value and not just costs alone.

2.2. Radar Operations

References 1–12

Radar operations involve the detection, tracking, resolution and accurate measurement of missiles and other objects of concern. In a determined missile attack, a variety of objects are likely to appear in swarms along the threat corridors leading to the continental U.S.A. Along with RVs such objects as decoys, chaff, ECM, booster sections and fragments can be expected to be seen by the defensive radars. Nuclear bursts may also be used by the enemy to blind defensive radars. Two useful tools are available for the assessment of a radar's performance: first, the general radar equation and, second, waveform analysis.

The significance of the generalized radar equation is that it permits the comparison of all types of radars, from the simplest to the most advanced. Special forms of this equation may be used to evaluate target detection and tracking, the measurement precision, the effects of jamming, chaff, beacon and telemetry operations, nuclear bursts, constant false alarm, and angle tracking systems. Pulse, pulse-doppler, and continuous-wave systems may be analyzed in terms of dropoff from optimum performance, propagation, attenuation, and digital integration effects.

The significance of waveform analysis is that it permits the comparison of all types of radar receiver responses from the simplest to the most sophisticated. Special waveforms provide different responses to the measurement of objects, effects of jamming, chaff, nuclear bursts and other effects. The precision of radar measurements may be analyzed in terms of the postulated or known waveform used by a radar.

2.2.1. General Radar Equation

References 9, 13–24

The general radar equation describes radar performance when the signal is competing with thermal noise, in terms of the performance parameter common to all types of radar: the ratio of received energy-to-noise power per hertz. This parameter determines the detection range or, more specifically, the probability of detection and the probability of false alarm as a function of range. Along with signal and antenna characteristics, it determines the precision of measured target position and velocity. This particular aspect of the energy-to-noise ratio will be discussed later.

The radar cross-section (RCS) of a target is best defined by considering how a radar operates. Power is generated in a transmitter, radiated from an antenna, and propagated through a medium to a target. It is then reradiated by the target, returned to a receiving antenna, and delivered to a receiver where it is detected and processed. Assuming free space propagation:

$$\text{Power density at target} = \frac{P_T \cdot G_T}{4\pi R_T^2} \text{ watts/meter}^2 \qquad (2.1)$$

where

P_T = average transmitter power in watts,
G_T = gain of transmitter antenna,
R_T = range between target and transmitter in meters.

The reradiated power is given by:

$$\text{Power reradiated by target} = \frac{P_T \cdot G_T}{4\pi R_T^2} \cdot \sigma \text{ watts} \tag{2.2}$$

The RCS σ is defined from this last equation. It is the equivalent reflecting area, in square meters, which when multiplied by the power density at the target, gives the power which is reradiated by the target in the direction of the receiving antenna.

The power density which arrives at the receiver is given by:

$$\frac{P_T \cdot G_T \cdot \sigma}{(4\pi)^2 \cdot R_T^2 \cdot R_R^2} \text{ watts/meter}^2 \tag{2.3}$$

where

P_T = average transmitter power in watts,
G_T = gain of transmitter antenna,
σ = RCS of target in square meters,
R_T = range between target and transmitter in meters,
R_R = range between target and receiver in meters.

If the capture area of the receiving antenna is A_R, in square meters, the power received by the radar P_R is:

$$P_R = \frac{P_T \cdot G_T \cdot \sigma \cdot A_R}{(4\pi)^2 \cdot R_T^2 \cdot R_R^2} \text{ watts} \tag{2.4}$$

The capture area of the receiving antenna A_R is an effective area and is given in terms of the product of the gain G_R of the receiving antenna and the RCS of a dipole $\lambda^2/4\pi$:

$$A_R = G_R \cdot \frac{\lambda^2}{4\pi} \text{ meters}^2 \tag{2.5}$$

Making this substitution into the equation for received power P_R gives the following form:

$$P_R = \frac{P_T \cdot G_T \cdot G_R \cdot \lambda^2 \cdot \sigma}{(4\pi)^3 \cdot R_T^2 \cdot R_R^2} \text{ watts} \tag{2.6}$$

If the transmitting and receiving antennas are at the same location, $R_T = R_R = R$.

Each of the R^2 terms may be modified by a propagation factor F to take into account the effects of propagation through the atmosphere with its possibilities of interfering reflections off the surface of the earth. Graphs are provided in the appendices which give values for the attenuation and refraction of RF waves in the earth's atmosphere and ionosphere. Ohmic and similar losses in the antenna, transmission line, etc. may be accounted for by a loss factor L. Thus, the received power is given as:

$$P_R = \frac{P_T G_T G_R \lambda^2 \sigma}{(4\pi)^3 R_T^2 F_T R_R^2 F_R L} \text{ watts} \tag{2.7}$$

The received energy is the product of the received power and the time of observation τ. If the noise power per hertz N_o, is defined as the product of Boltzmann's constant K and the effective temperature T of the receiving channel, we have:

$$\frac{E}{N_o} = \frac{P_T G_T G_R \lambda^2 \sigma \tau}{(4\pi)^3 K T R_T^2 F_T R_R^2 F_R L}. \tag{2.8}$$

The value of K, Boltzmann's constant, is 1.38×10^{-23} joules per degree. When $T = 290°$ Kelvin, $KT = 4 \times 10^{-21}$ watts per hertz bandwidth. This value of T is the commonly accepted reference temperature, although slightly different computed values are sometimes used. The radar frequency f is related to the wavelength through the equation $f\lambda = c$, where c is the velocity of light, a constant of nature numerically equal to 3×10^8 meters per second or 186,000 miles per second.

In an optimum matched-filter detection system, the ratio of received energy-to-noise power per hertz determines system performance. Specifically, if a threshold is set in the decision mechanism of the radar receiver for a specified probability of false alarm, then the ratio of received energy-to-noise power per hertz establishes the maximum probability of detection.

If we designate by S/N the detected ratio of signal-to-noise power, by B the receiver bandwidth, by F the receiver noise figure and by $D \geq 1$ the system degradation, representing the ratio of the signal power required in the actual system to the signal power that would be required in the matched-filter system, the equation becomes:

Table 2.2. *Definitions*

E/N_o—Ratio of received energy to noise power per hertz or the noise out of a filter one hertz wide. N_o equals KT, where K is Boltzmann's constant and T is the effective noise temperature of the system.

S—Mean signal power.

N—Mean noise power.

$S/N \cdot (B\tau)$—Ratio of received energy to noise power per hertz required for detection in an optimum system corresponding to specified probabilities of detection (P_D) and false alarm (P_F) or required for a specified precision in the measurement of target position or velocity. Figures 2.6 and 2.7 plot this ratio for the two most frequent "standard" problems: a signal of constant amplitude in noise and a signal that for any single observation is constant but whose amplitude has a random (Rayleigh) probability distribution.

D—Departure from the matched-filter system requirement defined by $S/N \cdot (B\tau)$.

P_D—Probability of detection. For the Gaussian signal P_D equals $P_F^{1/(1+S/N)}$.

P_F—Probability of false alarm, or that the noise will exceed the detectivity threshold and be considered the signal.

P_T—Average transmitter power.

G_T—Gain of the transmitting antenna in the direction of the target.

λ—Wavelength.

σ—RCS or equivalent reflecting area of an isotropic collector.

Only for a very simple point target is this a constant. The apparent cross-sectional area of a more complex target fluctuates as a result of destructive interference by reflections from various portions of the target. Radar performance is affected by this fluctuation.

τ—Time of observation. For a tracking antenna, it is determined by the time constant of the tracking system or the allowable data rate for a scanning antenna (a search radar or a tracking radar in the acquisition phase) it equals AM, where A is the time between the three-decibel points on the one-way antenna pattern and M is a factor used to make the area under the plot of G_TG_R as a function of time equal to G_TG_RAM. For the usual antenna patterns, M is about 1.6 decibels.

K—Boltzmann's constant: 1.37×10^{-23} joules per degree K.

T—Effective noise temperature in degrees K:

$$T = T_A/L_R + T_R + T_E$$

where T_A is the antenna noise temperature (a composite temperature of the radiating sources within the field of view of the antenna); L_R, the loss between antenna and receiver; T_R, the effective noise temperature of the transmission line between antenna and receiver and equal to $T_T(1-1/L)$, where T_T is the thermal temperature of the transmission line; and T_E, the effective receiver noise temperature and equal to $(F-1)T_o$, where T_o is 290 degrees K, the reference temperature at which the noise figure F is defined.

R_T—Range from transmitting antenna to target.

R_R—Range from target to receiving antenna.

F_T, F_R—Propagation factors used to take into account absorption in the medium and the effects of reflections from boundary surfaces, such as the surface of the earth.

L—Ohmic losses, such as caused by waveguide, antenna, and duplexing equipment. They include losses in the receiving system between antenna and receiver. If P_T is the transmitted power at the transmitter, L must also include losses between transmitter and transmitting antenna.

Any consistent sets of units, such as watts, meters, seconds, and joules, can be used.

$$\frac{E}{N_o}=\frac{S}{N}\cdot(B\tau)\cdot F\cdot D=\frac{P_T G_T G_R \lambda^2 \sigma\tau}{(4\pi)^3 KTR_T^2 F_T R_R^2 F_R L} \tag{2.9}$$

This is the *general radar equation* and is applicable to all types of radar including pulse, pulse-doppler, continuous wave, and monostatic. Table 2.2 gives a summary of definitions for the symbols used in equation (2.9).

2.2.2. Probability of Detection and False Alarm

References 9, 13–26

In the absence of a signal, the output of a radar receiver fluctuates as a result of noise present at its input. Normally, the noise is due to the combination of: (1) the antenna noise temperature (a composite temperature of the radiating sources within the field of view of the antenna), (2) the effective noise temperature of the transmission line between antenna and receiver, and (3) the effective receiver noise temperature. This noise is called thermal noise in contrast to unwanted radar echoes within the antenna field of view, which is called clutter.

If a threshold is set at the output of the receiver, occasional noise will exceed the setting and result in a false indication, or false alarm. Increasing the threshold to a higher value diminishes the probability of this occurrence.

If a signal is now introduced into the system, it too must exceed the threshold setting to produce a true indication or detection. Therefore, although increasing the threshold decreases the probability for a false alarm, it also decreases the probability that a small signal can be detected. This is true since the signal, when added to the noise at the receiver input, assumes a fluctuating character whether or not the signal itself fluctuates. In general, the detection is assured as the signal becomes much greater than the noise. Only when the signal is small enough to compete with the noise does the probability of its detection become an important consideration.

The two most frequent detection problems, a signal of constant amplitude in noise and a signal whose amplitude has a random probability distribution from pulse to pulse, are illustrated in Figs. 2.6 and 2.7. As shown in both figures, the

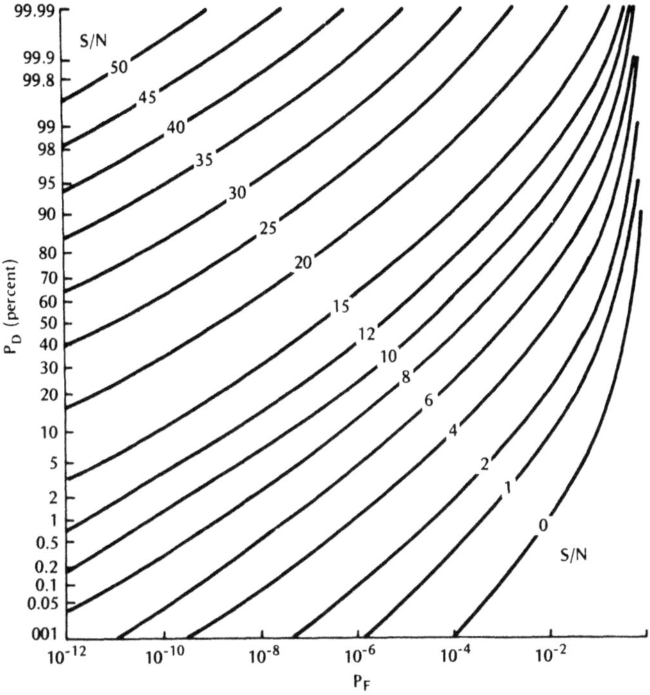

SOURCE: See reference 13.

Fig. 2.6. *Probability of detection P_D versus probability of false alarm P_F for a single look at a sinusoidal signal in noise*

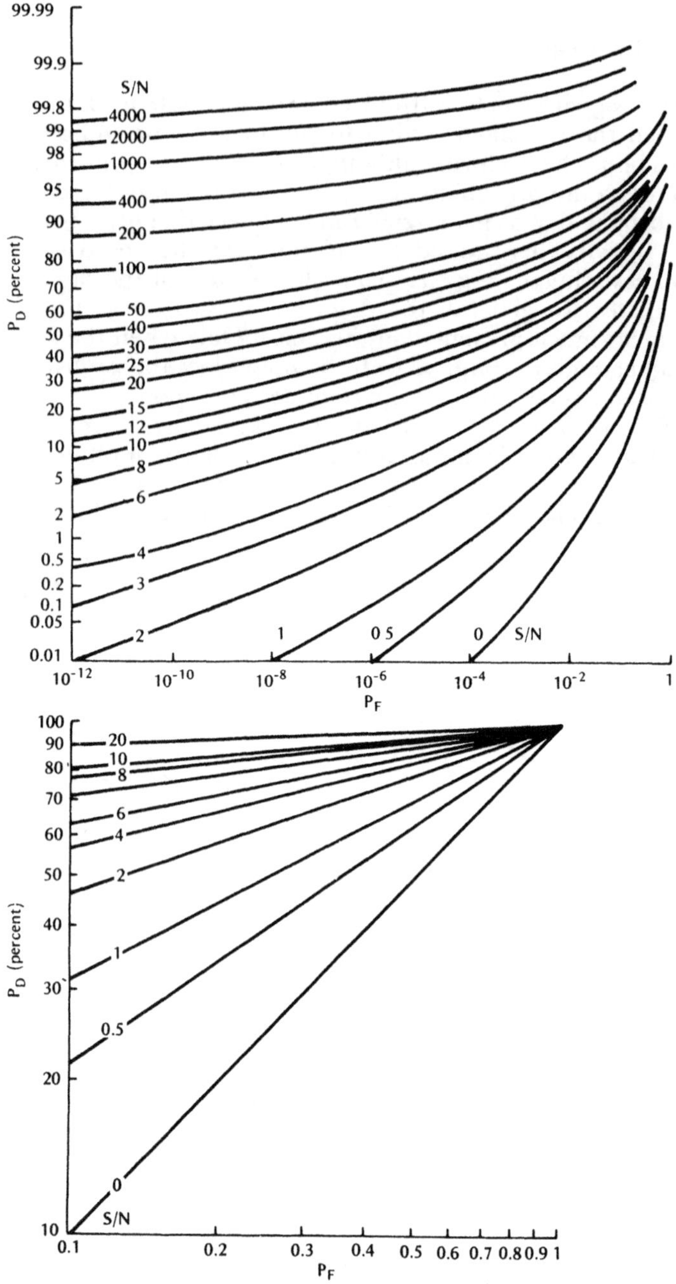

SOURCE: See reference 13.

Fig. 2.7. *Probability of detection P_D versus probability of false alarm P_F for a single look at a gaussian signal in noise* probability of detection P_D is plotted versus the probability of false alarm P_F with the detected signal-to-noise power ratio S/N plotted as a parameter. It is seen that a given probability of detection can be achieved by either decreasing the probability of false alarm or increasing the system S/N ratio. In both figures $S/N = E/N_o$, i.e., an optimum system is considered with $B\tau = 1$.

2.2.3. Target Detection

References 9, 13–23, 27

As can be seen from equation (2.9), an upper bound to the ability of a radar to detect a target is the ratio of the energy received from the target to the noise power density in which the target must be detected, i.e., the energy-to-noise per hertz ratio. As discussed previously, this quantity is designated E/N_o and bounds the signal-to-noise power ratio in the receiver, i.e., $S/N \leq E/N_o$. The minimum signal-to-noise ratio* required is in turn fixed by the desired probability of detection P_D, and probability of false alarm P_F. A typical design minimum for the signal-to-noise ratio S/N is of the order of 10 decibels. The signal-to-noise ratio S/N is related to the radar parameters and the target range R by the general radar equation, equation (2.9).

If detection is the only requirement and the radar problem is stated in terms of a requirement that a solid angle Ω must be searched every T_S seconds, the "surveillance radar equation" may be obtained from the general radar equation. For this case, equation (2.9) can be rearranged into a form which is more suitable for a scanning radar. This is done by noting the distinction between the information period T_S and the time of observation of a single target τ. These two times are related to the solid angle of the radar beam ω and the solid angle of the space which is to be scanned Ω by:

$$\frac{\tau}{T_S} = \frac{\omega}{\Omega}$$

The solid angle of the beam that defines the solid angle within

*In the following, signal-to-noise ratio refers to either ratio S/N or E/N_o implying an optimized receiver in which case $S/N = E/N_o$.

which a target can be detected with the antenna looking in any one direction is related to the gain of the antenna, approximately by the equation:

$$\omega = \frac{4\pi}{G_R} \tag{2.10}$$

Substitution of these expressions into equation (2.9) gives the so-called surveillance radar equation, for monostatic radar:

$$\frac{E}{N_o} = P_T \cdot A_R \cdot \frac{T_S}{\Omega} \cdot \frac{\sigma}{4\pi R^4 KTL} \tag{2.11}$$

where A_R = effective area of the receiver antenna.

This equation points up the interesting fact that, for a scanning antenna and a point target, the signal-to-noise ratio E/N_o is determined by the antenna aperture, the information or scanning period and the solid angle covered in the search. The wavelength has dropped out of the equation, except as it appears indirectly through the target RCS σ.

The RCS σ is independent of the wavelength for large spherical bodies, but is proportional to λ^2 for conical bodies viewed from the front. From these considerations, it can be seen that it is generally advantageous to operate a surveillance or search radar at as low a frequency (long wavelength) as possible, so long as the radar operating wavelength is similar to but not less than the expected dimension of the target.

2.2.4. Tracking Operation

References 9, 13–19

If the radar is to perform tracking, the signal-to-noise ratio given by the general radar equation, equation (2.9), must be maximized for best detection. The gain G_R of the receiver antenna must also be maximized for the highest angular measurement accuracy. Thus, the quantity of interest which must be maximized in a tracking radar is the product EG_R/N_o. Substitution of appropriate values for these two quantities in the general radar equation gives the "tracking radar equation" for monostatic radar:

$$\frac{E}{N_o} \cdot G_R = P_T \cdot A_R^2 \cdot G_T \frac{\sigma\tau}{4\pi R^4 KTL\lambda^2} \tag{2.12}$$

This equation indicates that, for a tracking radar and a point target, the signal-to-noise receiver-gain product EG_R/N_o is determined by the receiver antenna aperture and the transmitter gain. The wavelength dependence strongly favors short wavelength (high frequency) operation if the target RCS is independent of frequency. However, for conical shapes where the RCS σ is proportional to λ^2, the wavelength dependence drops out of equation (2.12).

Additional requirements for the attainment of radar accuracy will be discussed later.

2.2.5. Beacon and Telemetry Operations

References 9, 13–22

In beacon operation, only one way-transmission is involved, first from the ground transmitter to the beacon transponder and then from the transponder to the radar receiver. In either case, the signal-to-noise ratio can be written as follows:

$$\frac{E}{N_o} = \frac{P_T G_T G_R \lambda^2 \tau}{(4\pi)^2 KTR^2 L}. \tag{2.13}$$

Since the product $G_R\lambda^2$ in equation (2.13) is independent of the wavelength, the signal-to-noise ratio E/N_o is determined primarily by the transmitter gain and the receiver antenna effective aperture area.

In the usual beacon pulse system, enough energy must be received at the beacon transponder to trigger the transponder on each individual pulse, otherwise no pulse-to-pulse integration is possible. The received signal at the radar, on the other hand, may under certain circumstances be integrated.

In telemetry operation, as is the case for beacon operation, the transmission is one-way so that equation (2.13) again applies.

Telemetry may be used both as a part of the operational defense system and as an important subsystem in the R&D testing of system components. Therefore, the importance of telemetry system operation cannot be underemphasized in any case.

2.2.6. Jamming Environment

References 3, 4, 9, 14

In a jamming environment, the noise power per hertz, $N_o = KT$ which appears in equation (2.9) is no longer the background against which the signal competes, and accordingly is replaced by the jamming noise power per hertz which is received from the jammer. The noise power per hertz received from a jammer is:

$$J_o = \frac{J_T G_J G_a \lambda^2}{(4\pi)^2 R_J^2} \text{ watts/hertz} \tag{2.14}$$

where

J_o = noise power/hertz received from jammer,
J_T = jammer transmitter power/hertz,
G_J = gain of jammer in direction of radar,
G_a = gain of receiving antenna in direction of jammer,
λ = wavelength,
R_J = range to jammer.

The equation which most appropriately describes radar performance in the presence of noise jamming (i.e., the energy-to-jamming noise per hertz ratio) can therefore be written as:

$$\frac{E}{J_o} = \frac{P_T G_T G_R \sigma \tau R_J^2}{4\pi J_T G_J G_a L R^4} \tag{2.15}$$

where all quantities have been previously defined in the discussion of equations (2.9) and (2.14).

The self-screening range of an onboard jammer R_{ss} is the maximum range at which the target is visible in the main beam of the radar. For this condition, $G_a = G_R$ and $R_J = R_{ss}$. Substitution of these values in equation (2.15) and solving this equation for R_{ss} gives the self-screening range:

$$R_{ss}^2 = \frac{P_T G_T \sigma \tau}{4\pi J_T G_J L(E/J_o)} \tag{2.16}$$

The screening range R_s of the radar against targets not carrying jammers but in the presence of jammers in other directions and at other ranges can also be expressed in terms of the self-screening range R_{ss}:

$$R_s^4 = R_{ss}^2 \cdot R_J^2 \cdot \frac{G_R \sigma}{G_a \sigma_J} \tag{2.17}$$

Equations (2.16) and (2.17) indicate that the self-screening and screening ranges are determined by the transmitter gain. The wavelength has dropped out of these equations, except as it appears indirectly through the target and jammer RCS. Therefore, in both main beam and sidelobe jamming applications, it is generally advantageous to operate the radar at as low a frequency (long wavelength) as possible, so long as the radar operating wavelength is similar to but not less than the expected target dimension. This is the same requirement for maximizing the signal-to-noise ratio E/N_o in a search radar, as discussed previously. Thus, the low frequency operation of a radar is required for both maximizing the available signal-to-noise ratio and minimizing vulnerability to noise jamming.

2.2.7. Chaff Environment

References 3, 4, 9, 14

In a chaff or distributed target environment, the noise power per hertz KT which appears in equation (2.9) is no longer the predominant background and accordingly is replaced by the chaff energy C which is received from the chaff. The RCS of the chaff is:

$$\sigma_c = \left(\frac{\sigma_d N}{V_c}\right) \cdot V = \bar{\sigma} V$$

where

σ_d = average scattering cross section of a single dipole (Rayleigh statistics) in square meters,
N = total number of dipoles in chaff cloud,
V_c, V = volumes of chaff cloud, and radar resolving cell, respectively,
$\bar{\sigma}$ = average scattering cross section density of the chaff volume V_c in square meters/cubic meter.

If the chaff, or any distributed target, completely fills the radar beam the chaff cross section can be expressed by:

$$\sigma_c = \bar{\sigma}\tau(R\phi)^2 \tag{2.18}$$

where the volume of the radar resolving cell has been given as $V = \tau(R\phi)^2$, and:

R = range to target in meters,
ϕ = the 3 decibel beamwidth of the radar beam in radians,
τ = pulsewidth in meters.

The product $\bar{\sigma}\tau$ is commonly referred to as the reflection coefficient of the chaff, a dimensionless number. As shown, the reflection coefficient is proportional to average RCS density of the chaff and the radar pulse length. In the case of airborne or elevated radars looking downward, the reflection coefficient turns out to be a function of the angle of incidence, nature of the ground surface, etc.

Since the solid angle $\phi^2 = \omega$ is related to the gain again by equation (2.10), equation (2.18) can be rewritten as follows:

$$\sigma_c = \bar{\sigma}\tau \frac{4\pi}{G_R} R^2 \tag{2.19}$$

For a surveillance radar and a distributed target, the chaff signal-to-noise ratio is obtained by using equations (2.11) and (2.19),

$$\frac{C}{N_o} = P_T \cdot \frac{T_s}{\Omega} \cdot \frac{\bar{\sigma}\tau\lambda^2}{4\pi R^2 KTL} \tag{2.20}$$

Interestingly, antenna aperture does not appear in equation (2.20).

The visibility of a target in a chaff, or distributed target, background is given by the signal-to-chaff energy ratios,

$$\frac{E}{C} = \frac{\sigma}{\sigma_c} = \left(\frac{\sigma}{\sigma_d N}\right) \cdot \left(\frac{V_c}{V}\right) \tag{2.21}$$

The ability therefore, to detect a target in a distributed target volume that exceeds the volume of the radar resolution cell is inversely proportional to the size of the radar resolution cell V.

If the distributed target is completely contained in a single radar resolution cell, it behaves more like a point target. Assuming that no mutual couplings exist between the individual scatterers, the RCS for this case can be stated as:

$$\sigma_c = \bar{\sigma} \cdot V_c = \sigma_d \cdot N$$

For this case, the visibility of a target is simply:

$$\frac{E}{C} = \frac{\sigma}{\sigma_c} = \frac{\sigma}{\sigma_d N} \tag{2.22}$$

The ability to detect a target in a distributed volume of chaff contained in a single radar resolution cell is therefore independent of the radar resolution cell size.

Equations (2.21) and (2.22) indicate that the signal-to-chaff energy ratio is independent of the wavelength for targets whose RCS varies as λ^2 and depends only upon the relative RCS of the target and the chaff. This is true because the dipole RCS varies as λ^2. In this case, advantage can perhaps be taken of the fact that large objects resonate at low radar frequencies, thus requiring long and numerous dipoles for effective masking of target visibility. In many applications, the effective implementation and deployment of dipoles at the required lengths and numbers becomes technically difficult to achieve. In such applications it may be advisable to operate the radar at as low a frequency (long wavelength) as possible, so long as the radar operating wavelength is similar to but not less than the expected dimension of the target. On the other hand, if the target RCS does not vary with frequency, then the signal-to-chaff energy ratio becomes wavelength dependent. In this case, it is desirable to operate the radar at the highest possible frequency (smallest wavelength). It can be seen from this brief discussion that a very careful evaluation of the expected target shape is needed prior to selection of the radar frequency in a chaff environment.

Other examples of distributed targets in three dimensions, are rain, snow, foliage, and clutter. The last two are of course confined to the earth's surface.

2.2.7.1. *Chaff Glint*

A chaff cloud is an aggregate of wire dipoles scattered at random inside the chaff cloud and with random orientations with respect to the viewing radar. The signal seen by the radar fluctuates, as a result of the random orientations of the individual wires in the cloud. At any instant, the probability of seeing an occasional chaff glint above or below the average chaff value given by equation (2.18) is a well established phenomenon and is given in Fig. 2.8. This figure shows the

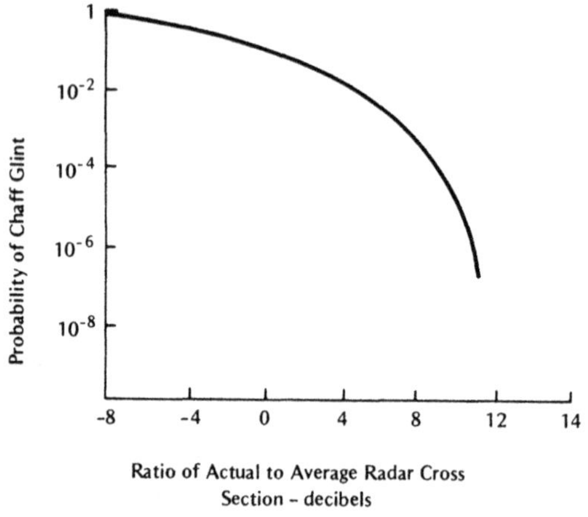

Fig. 2.8. *Probability of chaff glint*

chaff distribution function and is based on the Rayleigh distribution for amplitude (exponential distribution for RCS). Thus, the probability of seeing a chaff glint 10 decibels above the average chaff level is about 5 in 100,000. If the chaff cloud contains 1,000 radar resolution cells, then the probability of seeing a chaff glint 10 decibels above the average chaff level is about 5 percent, if all cells are being observed at a given instant. Prolonged observation of the chaff cloud will therefore certainly increase the probability of seeing chaff glint. This occasional glint from the chaff may be confused by the radar with the occasional specular returns from a target.

2.2.8. Nuclear Environment

References 3, 9

Radar blackout is caused by the large number of free electrons released by a nuclear explosion. All but a few of these electrons are removed from atoms or molecules of air, which thereby become ions. There are two main causes for the formation of ions; the fireball of the explosion, which produces ions because of its high temperature, and the radioactive debris of the explosion, which releases beta rays (high energy electrons) that ionize the air they traverse. The second mechanism is important only at high altitude.

The electrons in an ionized plasma cloud of gas have the property of bending and absorbing electromagnetic waves, particularly those of low frequency. These phenomena have been observed in long-distance communications using the ionosphere. Attenuation can reach such high values that the defensive radar is prevented from seeing any object behind the ionized cloud. This is not the case for chaff, which blocks the radar only at the chaff range and not beyond.

Blackout is a severe problem for an area defense designed to intercept missiles above the upper atmosphere. The problem is aggravated because area defense radar is likely to employ low frequency radar waves, which are the most suitable for detecting missiles at long range. Such radars are, however, more vulnerable to blackout.

Blackout can be caused in two ways: (1) by the defensive nuclear explosions themselves, and (2) by deliberate explosions set off at high altitude by the attacker. Although the first are not avoidable, the defense does have a choice of setting them off at altitudes and in locations that will cause minimal blackout of its radar. The offense can sacrifice a few early missiles to effect blackout at strategic locations.

Total reflection from the plasma cloud occurs when the radar frequency equals or is less than the so-called plasma frequency which is given as:

$$f_p = 8970 \quad N_e^{1/2} \text{ hertz} \tag{2.23}$$

where N_e is the number of free electrons in a cubic centimeter. Much smaller electron densities will, however, suffice for considerable attenuation. The equation for attenuation in decibels per kilometer is given as:

$$F_a = 1.4 \times 10^{-5} \frac{(2\pi f_p)^2}{(2\pi f)^2 + f_c^2} \cdot f_c \text{ decibels/kilometer} \qquad (2.24)$$

where

f_p = plasma frequency given by equation (2.23),
f = radar frequency in hertz,
f_c = collision frequency of an electron with atoms of air in hertz.

Collision frequency is a function of altitude and is given as:

$$f_c = 2 \times 10^{11} \frac{\rho}{\rho_0} \text{ hertz} \qquad (2.25)$$

where

ρ = air density at altitude,
ρ_0 = air density at sea level.

The atmospheric density ρ at h feet altitude decreases exponentially with altitude:

$$\frac{\rho}{\rho_0} = e^{-\frac{h}{22000}} \qquad (2.26)$$

The size of the fireball is easily estimated. Its diameter D is given by the simple scaling law:

$$D = \left(\frac{Y\rho_0}{\rho}\right)^{1/3} \text{ kilometers} \qquad (2.27)$$

where Y is the yield in megatons.

The second of the two mechanisms producing an ionized cloud, the beta rays issuing from the radioactive debris of a nuclear explosion, can be even more effective than the fireball mechanism. If the debris is at high altitude, the beta rays will follow the lines of force in the earth's magnetic field, with about half of the beta rays going immediately down into the atmosphere and the other half traveling out into space before returning earthward. These beta rays strike the atmosphere and ionize the air molecules. Beta rays of average energy penetrate to an altitude of about 60 kilometers; some of the more energetic rays go down to about 50 kilometers. At these altitudes then, a high altitude explosion will give rise to sustained ionization as long as the debris of the explosion remains in the vicinity.

The fission yield y deposited per unit horizontal area in the ionized disk (created by reentering beta rays), which is needed for total blackout is given as:

$$yt^{-1.2} < 10^{-2}\ \text{tons/unit area/second}^{1.2} \tag{2.28}$$

where t is the time after the high altitude explosion, in seconds.

The fission yield y is related to Y, probably to within a single order of magnitude, by the relation:

$$y = \frac{Y}{2\pi d^2}\ \text{tons/unit area} \tag{2.29}$$

where d is the diameter of the ionized disk which is produced by the beta blackout mechanism.

Fireball blackout effects may be determined by using equations (2.23) through (2.27) in conjunction with the general radar equation, equation (2.1). Beta blackout effects may be estimated using equations (2.28) and (2.29).

2.2.9. Radar Waveforms

References 9, 13–22, 28

The radar waveform employed defines the performance of the radar receiver in terms of its response to the presence of an object in the radar's field of view. The receiver response determines the radar's ability to resolve between objects and, in conjunction with the energy-to-noise ratio, to precisely measure the position and velocity of objects. The receiver response is characterized as being a function of the time delay and frequency shift relative to the radar's time of transmission and frequency. It is a maximum at the shifted time and frequency coordinates, relative to the radar, and rapidly falls to zero away from this position. The time of arrival of the radar pulse, its frequency, rms duration and rms bandwidth all characterize the receiver response. The maximum receiver response denotes the objects's echo and doppler position and equals the signal energy at the receiver's input. This description of the receiver response is quite general in that the actual response varies with the particular waveform which is being used by the radar.

The general radar waveform has a period T between pulses, a pulse repetition frequency (PRF) of $1/T$, and individual pulses which are characterized as being either:

a. single-tone carrier,
b. pulse-compression (FM carrier),
c. coded in amplitude, frequency, or phase.

This classification of waveforms is made on the basis of the response produced by each class at the output of an optimum receiver. Thus, the single-tone carrier and pulse-compression waveforms produce, for an extended observation period $T_0 \gg T$ of many pulses, a bed of nails a type of highly ambiguous response, somewhat as shown by Fig. 2.9. The actual response is maximum at position (0, 0) and drops off rapidly, so that the spikes decrease in amplitude with distance away from position (0, 0). This type of response has essentially clear areas between the individual spikes which have a finite thickness $\delta\tau\delta\nu$. When only one pulse is observed $T_0 = T$ and the response seen in Fig. 2.9 reduces to a single spike at (0, 0). On the other hand, coded waveforms tend to produce, for extended observation periods, the inverted thumbtack-type of response shown by Fig. 2.10, and which is conspicuously nonambiguous. The responses shown in Figs. 2.9 and 2.10 are for a single target and are called ambiguity functions.

The area occupied by each type of response, to a first approximation, corresponds to the duration-bandwidth product of the particular waveform. Each response peak has an area $\delta\tau\delta\nu$ which is inversely proportional to this product. The thumbtack response has a pedestal of height $\delta\tau\delta\nu$. The significance of the pedestal for the thumbtack response is that for many slightly separated targets, the pedestal contributions increase the "self-noise" of the system. The actual number of ambiguities in a bed of nails response is approximately $1/\delta\nu$ for range, $1/\delta\tau$ for doppler, and $1/\delta\tau\delta\nu$ for total.

In a bed of nails response, each spike is centered on an area equal to unity. Consequently, the unambiguous capacity per spike is $1/\delta\tau\delta\nu = T_0F_0$, which is the duration-bandwidth product for the waveform. In a thumbtack response, each spike is centered on the entire area T_0F_0. The capacity of a thumbtack type response is apparently therefore equal to $(T_0F_0)^2$, the square of the duration-bandwidth product of the waveform. Coded waveforms typically have higher bandwidth-duration products than do the simpler single-tone and pulse-compression waveforms. Coded waveforms would therefore appear to accommodate a greater number of targets. This is true except as limited by the self-noise which is inherent with coded waveforms. Their utility is determined by target strength and the number of targets inside the area T_0F_0.

2.2.10. Resolution

References 9, 13–22, 28

The ability of a radar to distinguish closely separated targets from one another depends on (1) the signal-to-noise ratio returned from each target, and (2) the distance between the targets.

The response of a radar receiver to a target possessing a particular set of parameters (e.g., range, doppler shift, angular position) is described by the ambiguity function of the receiver response. (Examples of ambiguity functions, were seen in Figs. 2.9 and 2.10). The ambiguity function is under the control of the radar designer. The ability to resolve targets in

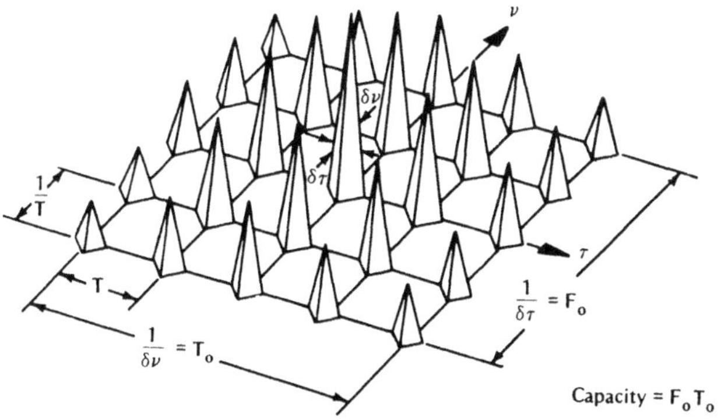

Fig. 2.9. *Approximate response (ambiguity function) of a repetitive waveform*

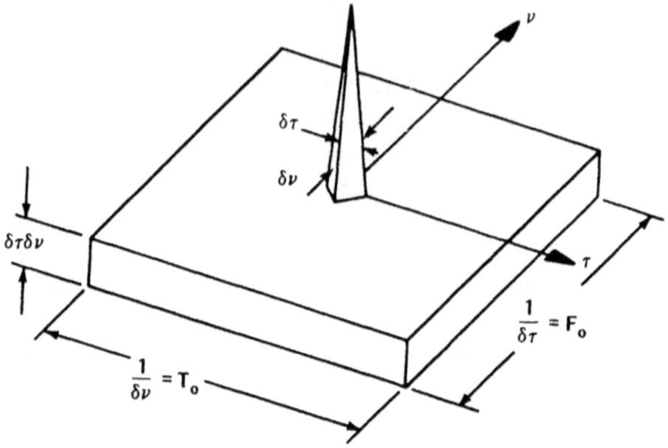

Fig. 2.10. *Approximate response (ambiguity function) of a coded waveform*

only one dimension depends primarily on that dimension, assuming a nondistributed target environment, as follows:

1. in range, as the auto- or crosscorrelation function of the transmitted signal (i.e., the range pattern),
2. in doppler, as the Fourier transform of the envelope of the transmitted time waveform (i.e., the frequency pattern),
3. in angle, as the Fourier transform of the antenna illumination function (i.e., the antenna pattern).

In terms of the rms duration α, rms bandwidth β of the required signal, and rms antenna length γ of the aperture, the range, doppler, and angle resolving capability of the radar system may be given as:

$$\delta\tau = \frac{1}{\beta}$$

$$\delta\nu = \frac{1}{\alpha} \qquad (2.30)$$

$$\delta\mu = \frac{1}{\gamma}$$

These equations give the capability of the radar waveform and aperture to resolve targets which are separated in time delay, frequency, and angle. A preliminary inspection of equations (2.30) suggests that improvements in resolution may be obtained through main lobe shortening of range, frequency, and antenna patterns. This holds only if the resulting sidelobes are found to be acceptable.

The quantities α, β, and γ are measures of the spread of the range, frequency, and antenna patterns, respectively.

If a radar must operate in a dense target environment it should have an ambiguity function approximating the spike in Fig. 2.10. There is, however, a constraint in the use of such an ambiguity function, "self-noise", i.e., achieving low sidelobe levels. Any function which is the Fourier transform of another function limited in extent (e.g., the physical dimension of an antenna or the length of envelope of a time waveform), has "sidelobes" which respond to the dense target environment and become additive, thus creating "self-noise". In other words, proximity targets interfere with detection and resolution by virtue of the presence of their sidelobes. In any case, useful degrees of resolution may be obtained by the more common waveforms using pulse and pulse-compression waveforms, by careful shaping of the signal and antenna illumination to produce low sidelobes in the ambiguity function in regions where the interfering targets may be present.

Because of the self-noise effect, the resolution of targets of grossly (>40 db) different radar cross section is possible only with pulsed radar. Therefore, the ultimate resolution of grossly different targets is peak power limited. In the case of moderately (<40 db) different radar cross sections, a variety of common waveforms may be used.

In practice $\delta\tau$, $\delta\nu$, $\delta\mu$ become identified with the waveform pulsewidth τ, bandwidth B, and antenna half-power beamwidth θ_B. For rectangular pulses of width at least equal to $1/B$, after envelope detection has removed the carrier frequency, the rms bandwidth can be found from:

$$\beta = \left[\frac{2B}{\tau}\right]^{1/2} \quad \text{where} \quad \tau \geq \frac{1}{B} \tag{2.31}$$

For a single rectangular pulse of duration τ, the rms duration is given as:

$$\alpha = \frac{\pi\tau}{\sqrt{3}} \tag{2.32}$$

For a uniformly illuminated aperture of half-power beamwidth θ_B, the rms antenna length is given as:

$$\gamma = \frac{\pi}{2\theta_B} \tag{2.33}$$

Equations (2.30) may be rewritten as follows for the special cases given by equations (2.31), (2.32), and (2.33):

$$\delta\tau = \left[\frac{\tau}{2B}\right]^{1/2} \text{ seconds} \quad \text{for rectangular pulses where } \tau \geq \frac{1}{B}$$

$$\delta\nu = \frac{\sqrt{3}}{\pi\tau} \text{ hertz} \quad \text{for a single rectangular pulse of duration } \tau \tag{2.34}$$

$$\delta\mu = \frac{2\theta_B}{\pi} \text{ radians} \quad \text{for a half-power beamwidth } \theta_B$$

Equations (2.34) now express the resolving capability of the radar in range, (coherent) velocity, and angle in terms of the radar parameters of pulsewidth, bandwidth, and beamwidth.

2.2.11. Accuracy

References 13–22, 28

The optimum ambiguity function of a measurement radar is nearly the same as the ambiguity function of a high resolution radar. While we cannot quite realize optimum resolution and optimum accuracy* with the same signals and antenna illumination, the difference in measurement accuracy is likely to be less than a small fraction of one percent. Although this difference in accuracy may be small, its existence, plus the inherent difficulty of designing signals to give high simultaneous resolution in range and doppler, suggests that accuracy can be enhanced by the use of a sequence of signals. Such a

*The terms *accuracy* and *precision* are used interchangeably in this section. Technically, of course, *accuracy* incorporates both systematic and random errors, whereas *precision* is due only to errors of a random nature.

technique could be useful where only well isolated targets are expected.

One possibility would then be to construct a radar capable of a sequence of waveforms as follows:

1. A target is detected with a waveform providing only modest resolution, under the assumption that only a single target is present.
2. Knowing the target's approximate range, a burst of very short pulses is then transmitted and received, yielding a very accurate but ambiguous range estimate (the ambiguity being removed by the previous course measurement). It is usually necessary to transmit a burst rather than a single pulse in order to get sufficient energy to make an accurate measurement.
3. A long, constant frequency, pulse is then transmitted for accurate measurement of doppler, and perhaps used for angular measurement as well.

This technique may be economically desirable when the radar has time, and consequently the energy, to spare, such as when the number of targets involved is relatively small.

This technique achieves high accuracy with a single observation of each measurable quantity. However, if the variation in the parameters of the target is small, or if their form of behavior as a function of time is known in advance, the accuracy of estimating the values of interest can be enhanced by the number of repeated observations and in many important instances, with the time and distance spanned between observations. For example, if the target is a satellite, the form of the orbit, neglecting perturbations, is known to be a geocentric ellipse. Rather than attempt to predict its orbit using only a few, closely spaced, highly accurate measurements, it might be more desirable to use only a few measurements, widely separated in time and space, with decreased accuracy on each observation.

The question then of accuracy becomes not only a matter of high single-hit measurement capability, but also of the number and kind of targets which the radar must measure. If the number of simultaneous targets is large, and if the targets themselves are capable of rapid variation, it will be necessary to observe each one at high rate. On the other hand, if the form of the target motion is known a priori, fewer observations may be necessary.

To the extent that improved resolution capability implies an increased capability for measurement accuracy, the remarks of the previous section apply here also.

The error in the measurement of range, doppler, and direction is given as:

$$\sigma_\tau = \frac{1}{\beta(2E/N_o)^{1/2}}$$

$$\sigma_\nu = \frac{1}{\alpha(2E/N_o)^{1/2}} \qquad (2.35)$$

$$\sigma_\mu = \frac{1}{\gamma(2E/N_o)^{1/2}}$$

where $\dot{E}/N_o$ is the energy-to-noise ratio.

These equations give the capability of the radar waveform and aperture to accurately measure target position, velocity, and angular direction.

Inspection of equations (2.35) indicates that the accuracy of measurement depends on the signal-to-noise ratio as well as on the waveform descriptors α and β and the aperture descriptor γ.

Finally, combining equations (2.30) and equations (2.35) results in the following relationships between the accuracy or precision of the measurement and the quantity being measured:

$$\sigma_\tau = \frac{\delta\tau}{(2E/N_o)^{1/2}}$$

$$\sigma_\nu = \frac{\delta\nu}{(2E/N_o)^{1/2}} \qquad (2.36)$$

$$\sigma_\mu = \frac{\delta\mu}{(2E/N_o)^{1/2}}$$

As previously stated, equations (2.35) indicate that the limiting precision of radar measurements depends not only on the waveform of the received signal, as was the case for radar resolution given by equations (2.30) but also on the signal-to-noise ratio.

In practice, equations (2.36) can, given equations (2.34), be written as

$$\sigma_\tau = \left[\frac{\tau}{2B}\right]^{1/2} \frac{1}{(2E/N_o)^{1/2}} \text{ seconds,}$$

$$= \frac{c}{2}\left[\frac{\tau}{2B}\right]^{1/2} \frac{1}{(2E/N_o)^{1/2}} \text{ meters } (c \text{ in meters/second}),$$

$$\sigma_\nu = \frac{\sqrt{3}}{\pi\tau} \frac{1}{(2E/N_o)^{1/2}} \text{ hertz,} \tag{2.37}$$

$$= \frac{\lambda}{2}\frac{\sqrt{3}}{\pi\tau} \frac{1}{(2E/N_o)^{1/2}} \text{ meters/second } (\lambda \text{ in meters}),$$

$$\sigma_\mu = \frac{2\theta_B}{\pi} \frac{1}{(2E/N_o)^{1/2}} \text{ radians } (\theta_B \text{ in radians}),$$

$$= \frac{\theta_B}{90} \frac{1}{(2E/N_o)^{1/2}} \text{ degrees } (\theta_B \text{ in degrees}).$$

In equations (2.37), the conversion multipliers $c/2$ and $\lambda/2$ have been used to convert time to length, and frequency to velocity, respectively.

Equations (2.37) express the radar precision of measurement when we know the energy-to-noise density ratio E/N_o, and the waveform parameters of (1) pulsewidth (or time of observation in a coherent system) τ, (2) bandwidth B, and (3) half-power beamwidth, θ_B. It should be noted that in general for a coherent system $\tau > 1/B$, whereas for a noncoherent system $\tau = 1/B$. In the latter case τ is identified as being the actual pulsewidth; in the former case τ is identified as being the period of observation which may include a number of pulses.

Since the energy-to-noise ratio can be expressed in terms of the observation time τ and bandwidth B as follows:

$$\frac{E}{N_o} = B\tau \frac{S}{N} \tag{2.38}$$

the precision of the measurement can also be obtained by combining equations (2.37) with equation (2.38). The result is:

$$
\begin{aligned}
\sigma_\tau &= \frac{0.5}{B(S/N)^{1/2}} \text{ seconds,} \\
&= \frac{0.25c}{B(S/N)^{1/2}} \text{ meters,} \\
\sigma_\nu &= \frac{0.39}{\tau^{3/2}B^{1/2}(S/N)^{1/2}} \text{ hertz,} \\
&= \frac{0.19\lambda}{\tau^{3/2}B^{1/2}(S/N)^{1/2}} \text{ meters/second,} \\
\sigma_\mu &= \frac{\sqrt{2}\,\theta_B}{\pi(B\tau)^{1/2}(S/N)^{1/2}} \text{ radians,} \\
&= \frac{\sqrt{2}\,\theta_B}{180(B\tau)^{1/2}(S/N)^{1/2}} \text{ degrees.}
\end{aligned}
\tag{2.39}
$$

Equations (2.39) express the radar precision of measurement when the signal-to-noise power ratio S/N, and the waveform parameters of (1) pulsewidth (or time of observation in a coherent system) τ, (2) bandwidth B, and (3) half-power beamwidth θ_B are known. Again, for a coherent system $\tau > 1/B$, whereas for a noncoherent system $\tau = 1/B$.

It should be noted that equations (2.37) and (2.39) are valid only for those classes of waveforms given by equations (2.31), (2.32), and (2.33), respectively.

2.3. Computer Operations

References 1, 5–7, 9, 11, 12, 29–31, 50

Computer operations involve the purely tactical functions of target trajectory prediction, discrimination analyses, interceptor guidance and the operating functions of controlling the radar, message formatting, data reduction and the testing functions of both radars and computers as well as the service functions of maintenance and diagnostic testing.

The radar operation centers around the computer program because of the many functions that are carried out concurrently. Although the operator makes requests which influence operation and observes the system activity, the immediate radar control is completely without human intervention. The computer program responds in an adaptive manner to the dynamic environment. The tactical situation is assessed at typical rates of about twenty times per second and radar power is allocated in accordance with a complex priority scheme. Using this priority scheme the program controls radar operations as a function of (1) the operator request, (2) the functional phase of operation, (3) the target environment, (4) the status of the radar equipment and (5) the time the target is available for viewing.

A simplified flow diagram of an ABM radar system is shown in Fig. 2.11. The main subsystems are (1) the antenna, (2) the beam-steering unit (special purpose computer), (3) the transmitter, (4) the receiver, (5) the general purpose computer, (6) the display and control console and (7) the input-output interface.

In operation, designation data is supplied to and/or taken from the radar installation through the input/output interface equipment. The interface equipment uses these data to generate an input to the general purpose computer, representing the command. The computer references the designated search or location coordinates to the local radar site coordinates and then transforms them to directional cosine array coordinates. This information is then supplied to the special purpose computer which sets the phase in each element of the array, passive or active, to form a beam in the designated direction. The video processor then sends a trigger to the transmitter causing it to fire a burst of energy through the array in the designated direction.

A search or location function is performed in the designated sector. Target amplitude, range, and angular error signals are obtained in analogue form at the output of the receiver. The video data processor transforms this information to a digital format and sends it to the general purpose computer. The acquisition mode starts when the signal exceeds some previously established threshold. The general purpose computer does the necessary data smoothing for generation of best estimates of the target's present and future position and velocity. The tracking and AGC operations are equivalent to conventional Type II or Type III servos. The targets are tracked in three coordinates, in range, and in two of the array beam's direction cosines. The search and track functions are continued on a simultaneous (track-while-scan) or time-shared basis, depending on the system design.

The beam steering unit includes computer controlled self-checking equipment to permit the detection and reporting of equipment failures to the system computer.

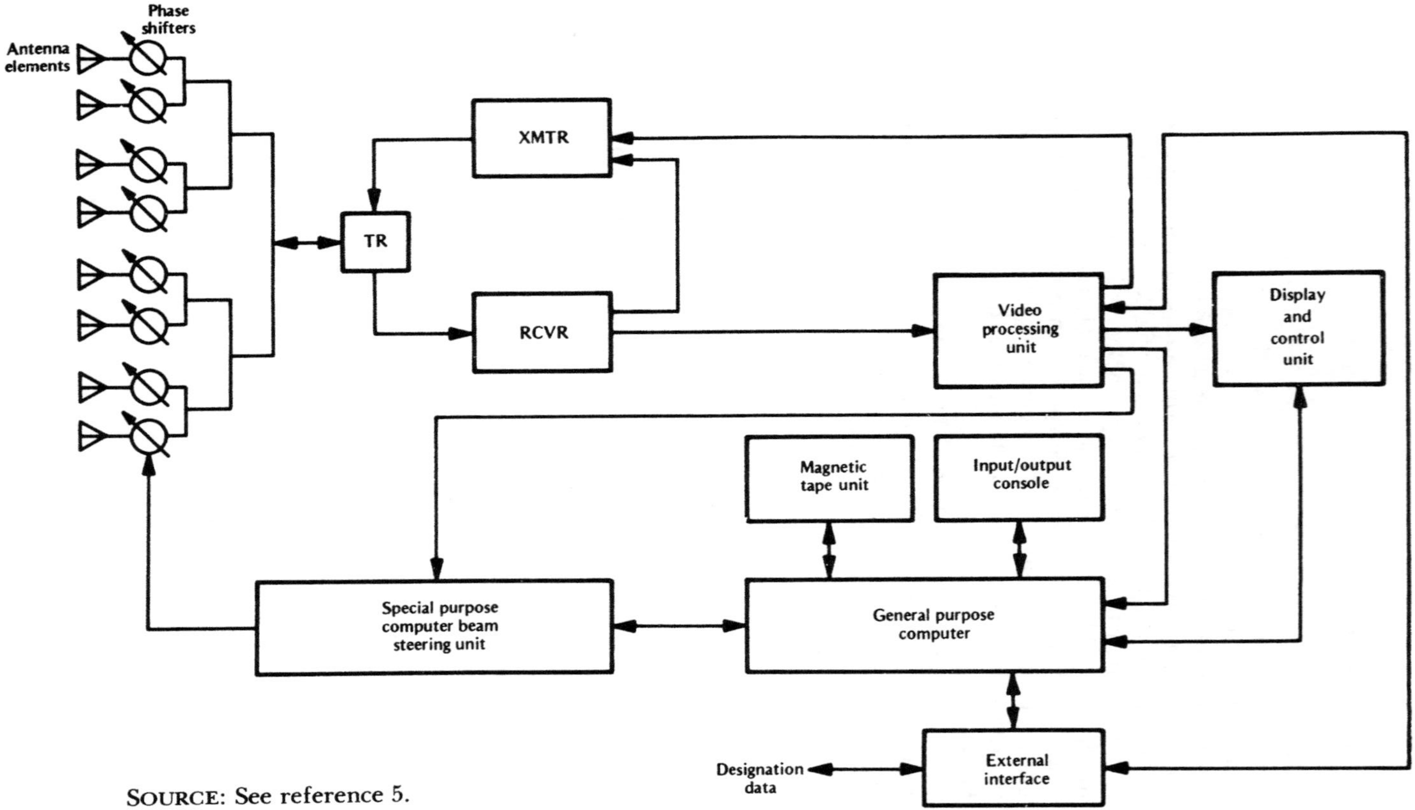

SOURCE: See reference 5.

Fig. 2.11. *Functional block diagram of an array radar using parallel feed phase shifters*

The control console contains displays to observe the search sector and to monitor the angular error signals and the range-gated amplitude of the tracked targets. Operational, test, and calibration modes are selected at the console.

The mission data output of the radar is sent back to a control center through the interface equipment and/or is placed on magnetic tape for further analysis.

When operating the radar is tied to the Defense Control Center (DCC) and operates as a member of the ABM system which consists of many such radars similarly connected to the DCC. The simplified flow diagram of Fig. 2.12 illustrates the interconnection between functions of the computer program. Thus, the functions controlled by the computer are grouped in eight areas: (1) communications, (2) search, target selection and tracking, (3) radar control, (4) interceptor guidance, (5) signal processing, (6) testing, (7) local display and input–output and (8) man–machine.

The most pressing demands on the computer program are related to the traffic handling capacity of computers when both targets and defensive missiles are being tracked simultaneously. In peak traffic situations the data loads build up quite rapidly at both the PAR and MSR sites. The actual number of processor units required for each defense site is selected based on the postulated threat. Additional processor units are added to handle all the other functions required to support these basic tactical functions and to supply the necessary number of spare processors to meet availability requirements. The computer functions will now be discussed in more detail.

2.3.1. Communications

Prime communications between the ABM radar and the Air Defense Control Center (ADCS) at Cheyenne Mountain in Colorado are maintained over a 2,400 bit per second data link. The general purpose (GP) computers at the two facilities communicate directly. However, as shown in Fig. 2.12, the operator at the ABM site is informed of message transfers and of radar sensor activity through the situation display and hardcopy printouts. Administrative and deferable tactical information is relayed over teletype and immediate voice contact is always available via telephone.

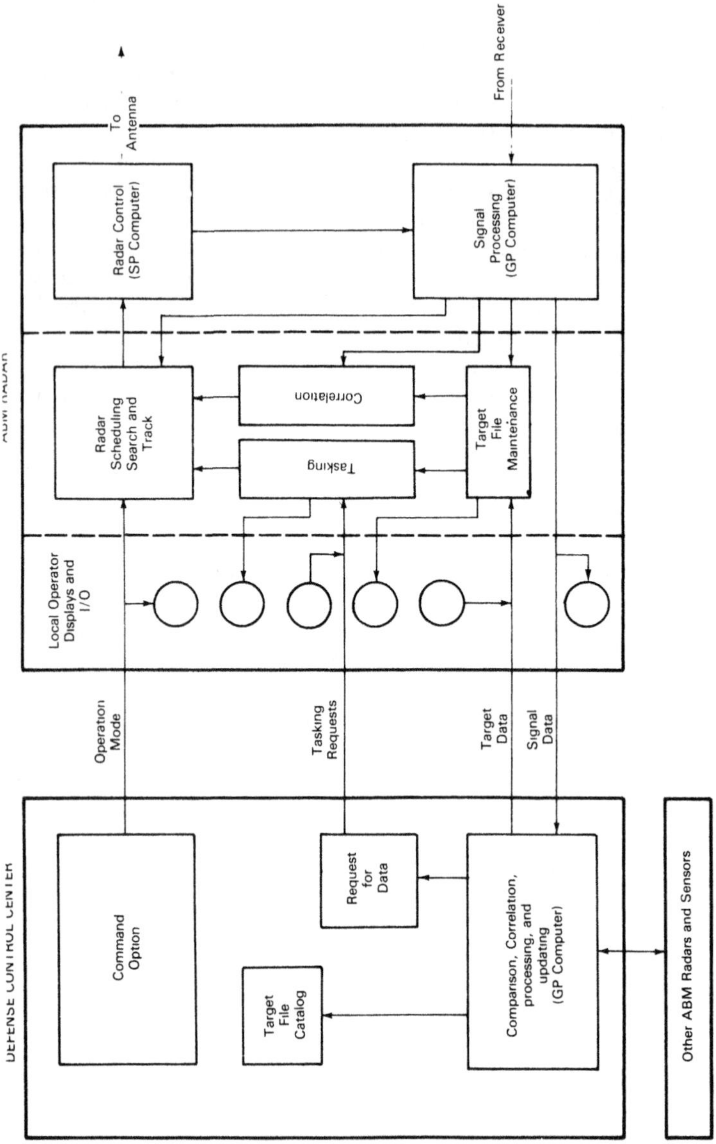

SOURCE: See reference 6.

Fig. 2.12. *Functions of ABM computer program*

2.3.2. Search, Target Selection and Tracking

The ABM radar is in continuous surveillance of the horizon. Search mode changes are dependent upon the defense situation. Allocation of radar power to the search function as well as changes in search surveillance patterns can be made through the operational mode change. Three basic system functions can be selected with no interruption to the system operation, namely search, target selection and target tracking.

2.3.2.1. Search

In the search operation the computer controls the priority scheduling of targets in the radar's field of view. Data is obtained by the radar on all targets and this data is used to compute target impact points. When targets appear threatening, requests are made by the DCC for more data on these targets via the tasking requests. As indicated in Fig. 2.12, the local operator can also request data. Over 150 request combinations are available for each target. Because requests may be received over the data link at any time, they are stored in the computer memory and then honored on a priority basis. Requests may be in the form of continuing requests or in the form of one-time requests specifying the amount of data required.

Upon entry of the tasking request in the local computer the target file is checked to insure that the target is on file. The time-ordered file is generated listing the entry time into the local radar coverage. The file is repetitively reviewed at high rate to see which tasking requests can be honored. Time-ordered tasking files are generated automatically in the computer upon receipt of target data and are displayed to the operator on the situation display.

Where there is no tracking activity all radar power is allocated to search. This increase in search power expenditure increases the probability of detection. The priority of each task requiring radar power is reviewed at typical rates of about twenty times per second and as many tasks as possible are scheduled by the GP computer within the limitations of the radar. If radar pulse burst periods are set to provide coverage to 7,500 km (50 milliseconds) and data from the previous period is used to generate the next period, the real-time response to the changing environment is 100 milliseconds.

2.3.2.2. Target Selection

The scheduling, search and tracking functions are the most complex problems encountered in the real-time computer program. All activity in the computer is based on priority. The design philosophy is to allocate the radar power and time to the highest priority tasks on a real-time basis. Thus all available time and energy is expended with peak efficiency.

The target selection is made on the basis of confirmation of the threat through impact-point computation and discrimination analysis. Priority may dictate partial computations be made and even temporary deferral of target selection. For example, a low-priority target impacting in the countryside may be deferred until a later time because a peak of high-priority targets impacting at vital locations requires the use of all available power. Thus overload or saturation, both with regard to internal computations and target selection, are handled by deferring the action to a later time.

2.3.2.3. Target Tracking

When the GP computer makes a track measurement on a target it uses this measurement to update and refine the predicted trajectory and signature of the target. Each such calculation requires very many operations in the data-processor (GP computer). However, the data-processor is such that a single unit can perform the calculation in about 1–2 milliseconds. Because the radars were designed to give high accuracy measurements on each track pulse, only about 100 pulse hits per target are needed and these can be spread over several minutes of the target trajectory. Therefore, the average time required of a processor is less than one millisecond per target per second. Thus, a single processor unit may perform trajectory predictions for over 1,000 targets simultaneously. Since a massive attack would conceivably last for minutes and not all targets would appear simultaneously, a single processor unit would build up to this rate. As it finished with a particular track it would drop the target and start on a new target. As example, if the attack lasted thirty minutes with several hundred targets entering within the radar's coverage each minute, a single processor unit could handle several thousand such targets over the duration of the attack. It is known, however, that each legitimate target may well be accompanied by other objects such as penetration aids. If each legitimate target is accompanied by ten other objects such as decoys, then it can be seen in the case of our example that ten processing units will be needed to handle the expected load.

Tracking schemes divide generally into two basic approaches: low-data-rate tracking and high-data-rate tracking. Low-data-rate tracking takes advantage of the fact that a target on a ballistic trajectory requires only periodic updating whereas high-data-rate traffic requires continuous updating. In low-data-rate tracking the known trajectory is used to maintain track by correcting the next predicted target position by the last smoothed data point obtained. In high-data-rate tracking the tracking is maintained continuously at high rate with each data point used to update the data. In tracking operations the use of a second-order filter is usually satisfactory for midcourse tracking but is not satisfactory for reentry tracking where high rates of change of acceleration of the target exist. Thus, for reentry tracking the target trajectory is defined more closely by a higher order filter. Filters use up available capacity in the computer in almost direct proportion to the order of the filter. The approximate filtering of data can also be made using algorithms in the computer in the effort to conserve computer capacity.

2.3.3. Radar Control

After the tasks are scheduled by the GP computer, either the search or track function is entered into the GP computer for generation of the radar commands for the next pulse burst period (PBP). Interlacing of functions can be obtained by programming holes in the search period or by the addition of a second-beam steering system. It may not be necessary to interlace the search and track functions within a single PBP to meet system requirements.

The search patterns are generated in the SP computer from a table of preprogrammed patterns. Depending on the frequency selected for the particular PBP length and the internal calibrations activity, the next search PBP is generated. A major part of the search program is used to synchronize information to the return processing subroutine. The predicted target position, target size, target type, expected inaccuracy of the return, pulse-type, frequency and fixed-type commands are made up by the GP computer and are used to generate the radar controller list that consists of a time-ordered list of commands for each PBP. After the next list is generated by the GP computer it is sent to the radar interface and control equipment core memory for storage until the end of the present PBP. Upon completion of the present PBP, the radar controller accesses the new list of commands for sequential execution of each command upon time coincidence of its counter with the time computed for that of the command.

2.3.4. Interceptor Guidance

Interceptor guidance calculations take longer time per calculation and are made at a higher rate than are similar calculations in target measurements. Balancing this effect is the relatively low number of defensive missiles that must be in guidance simultaneously from one MSR compared with the total number of targets and other objects that a PAR must handle. The functions of the computer in the guidance of interceptors is quite similar to that for tracking of targets, with the added function of computing the so-called hit equation which may be applicable to the particular guidance scheme in use. Scheduling, readiness and launch commands are also made by the GP computer for interceptors as well as performing the normal guidance function while the interceptor is en route to its target.

2.3.5. Signal Processing

The return processing in the GP computer is highly sophisticated. Although the program can limit the amount of return data through the signal processor via threshold control, transient or failure conditions tend to saturate the computer. The computer programs must respond to all such abnormal conditions, inform the operator and continue with radar operation. In the return processing, data is associated with the proper target; corrections for range, monopulse angle and doppler are made; filtering of range sidelobes from nearby resolved targets is made; and conversion to desirable units and the queueing of corrected terms to the appropriate processors is made. At this point in the functional data flow, several actions may occur depending on the target type and status. Data may be either routed to the search or track routines, the correlation section, the target file and/or the DCC. Since all data to be passed from one program to the next resides in queue storage and only the location of this data is transmitted to the programs which use it, each queue block of data may be used by several programs and progressive processing may change portions of the data, set condition flags and transmit the next sequence. The scheduling, search and track routines use the data placed primarily in track file. It is the duty of the data-processing section to update the quality and status indicators of a target data group upon smoothing it into an observation. When an observation consisting of many smoothed returns is completed it is transmitted to the display, to the DCC and to the target file maintenance sections of the program. The correlation function is started upon the detection and confirmation of a target from the search function. Two types of correlation occur. First, the position of the target is correlated with the predicted target location for each target in the surveillance coverage. The target is identified as a known track if it falls within the error thresholds set for each coordinate of the local coordinate set. If the target does not correlate it is identified as an unknown track and the program proceeds to the second type of correlation. Unknowns are then correlated with all targets in the track file to prevent the accumulation of multiple unknown assignments.

Since a complete up-to-date file must be maintained for correlation and tasking purposes, DCC inputs are forwarded to the local GP computer each time they are updated using the inputs from other radars and sensors. In addition, the local GP computer incorporates target data of its own to maintain the file current.

2.3.6. Testing

Other functions carried on by the radar control area of the SP computer program include the status checking of all radar equipment. In addition to the hardware checks which indicate ready status, a series of go-no-go tests are run on the radar and radar interface and control equipment at the rate of about one per second. Failures of particular tests result in the automatic switchover of redundant equipment. In addition to the automatic testing for availability of the equipment, the desired configuration is also manually inputted through the system status console and is checked by the program at regular intervals or whenever the status read button on the system status console is depressed. The desired system configuration is maintained unless catastrophic failures exist which cause the program to revert to a test condition and cause an error message to be presented to the operator. Data regarding radar control is forwarded to the return processing section of the program. Queues and pointers locating all transmit and expected-return information are passed on to the return signal processing section.

2.3.7. Local Display and Input/Output

In addition to an overall system status wall display and two situation displays, a TV video console displays the pointing of the radar beam in azimuth and elevation and the video outputs of the signal processors to the tactical operations personnel. Other displays, such as the computer operations room display, the maintenance off-line display, the receiver calibration display, the system status console, the transmit calibration display, the radar status console, the radar monitor console, the radar interface and control equipment maintenance console are manned by radar personnel as required for preventative maintenance or repair.

2.3.8. Human Operator

As stated earlier, the local operator is informed of all incoming and outgoing communications. He also has available to him information regarding the tactical situation and operational status of the system. The role of the tactical operations operator is to monitor the situation. He can assess system performance because he is informed of both the tactical environment and the immediate status of the radar equipment. The operator makes requests for action by tasking the system on the same basis as does the DCC. The GP computer program decides when to honor the request. The situation is displayed primarily on two IBM 2250 computer-derived displays. Observations are selectively displayed on a geographical background with the target number and altitude. The same display presents statistics and operator call-up information around the edge of the track display.

2.4. Interceptor Operations

References 1, 5, 7, 9, 32–38

There are three possible guidance techniques and trajectories useful for steering an interceptor to a target missile: preset, command and terminal guidance. Each technique has its own peculiar advantages, especially with respect to its jamming susceptibility and accuracy.

Preset guidance may be utilized when the target missile is on a deterministic course and its future position can be computed accurately. The interceptor is then launched along a course to intercept the target missile. A principal advantage of preset guidance is that the interceptor can operate in an adverse jamming or nuclear environment with relative immunity.

In the case of a command guidance system the target missile and interceptor are tracked separately. The tracking information is then fed into a computer which is usually part of the direction equipment and the interceptor is programmed to intercept the target missile at some optimum computed point. The programming may be automatic or semi-automatic. This technique can be made extremely accurate; however, it is rather susceptible to various types of active countermeasures. Foremost among the many advantages of command guidance is that most of the weight is on the ground and therefore the defensive payload capability can be increased. Command guidance is presently the most active operational guidance available. Beam-rider guidance is a form of command guidance in which the interceptor rides up the guidance beam. The beam may be pointed directly at the target missile or may be programmed to provide any suitable trajectory. Very accurate range information is not required for this system and therefore it is quite useful in jamming environments.

In terminal or homing guidance the interceptor senses the

target missile directly by means of radar illumination, heat, visible light or other electromagnetic sensors. Some types of terminal guidance systems are very susceptible to jamming. However this type of system can be very accurate because of the nature of the tracking. Terminal guidance can be very useful in the terminal phase of an interceptor's flight after the interceptor has been directed to the general vicinity of the target missile.

The two basic defense concepts in missile interception are area defense and terminal defense. Area defense involves the interception of missiles at very long range, while they are still high above the earth's atmosphere. Terminal defense involves the interception of missiles at very short range, after they have reentered the atmosphere and after they have been sorted out by the atmosphere from decoys, chaff, and other confusion devices. Nuclear warheads are used to destroy target missiles in both the area and terminal defense strategies. As a consequence it appears that the preset guidance technique is particularly suitable for the guidance of ABM interceptors. However, some benefits are possible using non-nuclear warheads and, as a consequence, the more accurate guidance techniques cannot be completely discarded, especially for use in terminal defense where the friendly bursts may be too near the sites being defended.

2.4.1. Parabolic Trajectory

References 9, 32, 33

The trajectory of a short-range missile becomes a parabola when the eccentricity is equal to unity. This represents an ideal case because, for a missile traveling in the atmosphere, the variations in aerodynamic drag make it impossible to achieve exactly $e = 1$. Nevertheless, the assumption of $e = 1$ is often convenient to establish the approximate velocity and flight time for an interceptor to reach a particular altitude and range under no-drag conditions. This establishes an estimate of the limits of performance that cannot be exceeded by an actual missile.

The geometry of the parabolic path is depicted in Fig. 2.13. Neglecting the effects of air drag and the motion of the earth, the path of the interceptor over a flat earth may be deduced from the following conditions:

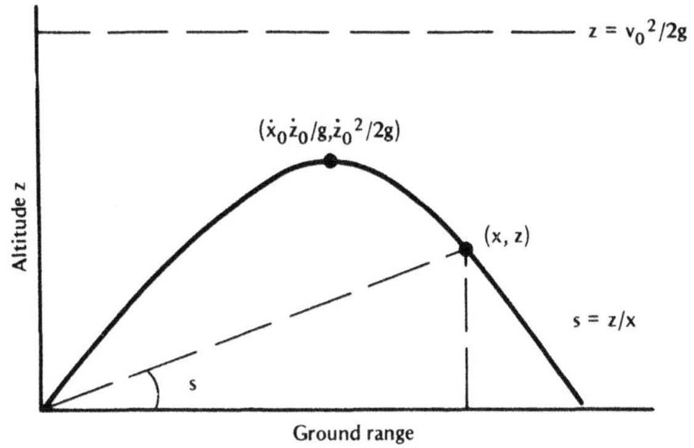

Fig. 2.13. *Ideal parabolic trajectory for no drag*

$$\ddot{x} = 0 \quad , \quad \ddot{z} = -g$$

Upon integrating with respect to time:

$$\dot{x} = \dot{x}_0 \quad , \quad \dot{z} = -gt + \dot{z}_0,$$

and

$$x = \dot{x}_0 t + x_0 \quad , \quad z = -\frac{1}{2}gt^2 + \dot{z}_0 t + z_0$$

The constants $\dot{x}_0$ and $\dot{z}_0$ are the respective initial velocities while x_0 and $\dot{z}_0$ are the initial positions.

The equation of the path follows from these equations when the time t is eliminated and $x_0 = z_0 = 0$.

$$z = \left(\frac{\dot{z}_0}{\dot{x}_0}\right) x - \left(\frac{g}{2\dot{x}_0^2}\right) x^2.$$

This is the equation of a parabola in Cartesian coordinates. If the equation is rearranged to be put in the form:

$$\left[x - \left(\frac{\dot{x}_0\dot{z}_0}{g}\right)\right]^2 = -\frac{2\dot{x}_0^2}{g} \cdot \left(z - \frac{\dot{z}_0^2}{2g}\right),$$

it may be seen that the vertex of the parabola is at the point $\dot{x}_0\dot{z}_0/g$ and $\dot{z}_0^2/2g$, and the distance from the focus to the vertex is $\dot{x}_0^2/2g$.

The velocity and angle between the flight path and the earth are given in terms of the x and z components of the velocity:

$$v^2 = \dot{x}^2 + \dot{z}^2,$$

$$\cos\gamma = \frac{\dot{x}}{v}, \text{ or } \sin\gamma = \frac{\dot{z}}{v}$$

The rates of change of the velocity v and angle γ are obtained by differentiating the above equations:

$$\dot{v} = -g\sin\gamma$$

$$\dot{\gamma} = -\frac{g}{\dot{x}_0}\cos^2\gamma$$

Thus, the interceptor decelerates at its highest rate during the initial portion of its trajectory, when γ is at its largest value. Its highest turning rate occurs as it approaches the vertex of the parabola.

Finally, the time of flight to reach a point on the parabolic trajectory with coordinates x, z such that the ratio z/x equals the slope between x, z and the center of coordinates in Fig. 2.13 is given by:

$$t_f = \frac{2(\dot{z}_0 - s\dot{x}_0)}{g},$$

where s is the slope.

2.4.2. Navigational Triangle

References 9, 32, 33, 37

The closing of the navigational triangle is the geometrical requirement for the interception of a moving target by the interceptor. The coplanar case is illustrated in Fig. 2.14. In the figure, the missile is fired from the center of coordinates C and assumes its flight path along the parabolic trajectory until the intercept is made at some point I at the future time t_f, which is both the interceptor time of flight, and the target time of flight between its positions T and I along its trajectory. If the line of sight (LOS) between the launch point C and the target position at

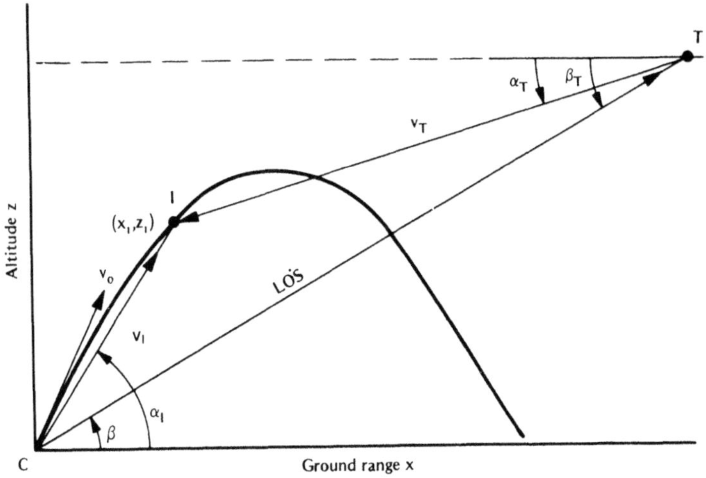

Fig. 2.14. *Navigational triangle at launch*

launch T is assumed to be a reference at the instant of launch, then the triangle CIT is called the navigational triangle at that instant. Once the interceptor is launched, the requirement for interception demands that the navigational triangle close in time. At any instant the requirement is met by applying the law of sines to the navigational triangle:

$$\frac{v_T}{\sin(\alpha_I - \beta)} = \frac{v_I}{\sin(\alpha_T - \beta)}$$

or

$$\sin(\alpha_I - \beta) = q \sin(\alpha_T - \beta) \tag{2.40}$$

where q is the scalar speed ratio v_T/v_I and the volocity v_I is expressed in terms of the intercept coordinates x_I, z_I and the time of flight t_f discussed in Section 2.4.1:

$$v_I^2 = \frac{x^2 + z^2}{t_f^2} = \dot{x}_I^2 + \dot{z}_I^2$$

The angle $\alpha_I - \beta$ determines the angle of aim at any instant during the intercept, and is the solution to the fire control problem.

In actual practice $\alpha_I, \alpha_T, \beta$ and q are functions of time, so that equation (2.40) gives the instantaneous requirement for interception. All quantities in this equation except α_I (the interceptor aim angle) are measured so that the equation simply states the required interceptor aim direction at that instant.

If the aim angle $\alpha_I - \beta$ is plotted as a function of the target angle $\alpha_T - \beta$ with the ratio of velocities q as a parameter, the curves of Fig. 2.15 result. These curves are useful in illustrating

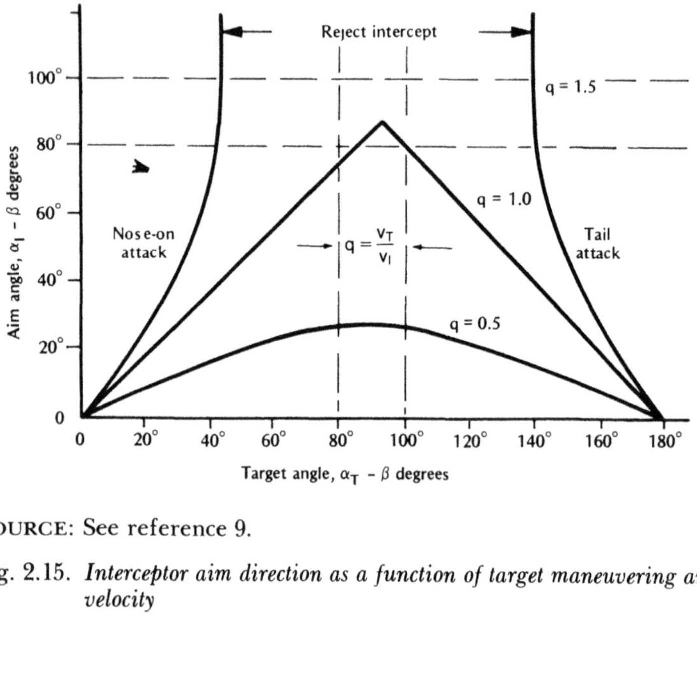

SOURCE: See reference 9.

Fig. 2.15. *Interceptor aim direction as a function of target maneuvering and velocity*

the effect on the required aim angle of small changes in the target velocity or angle.

For the cases illustrated (namely, when $q = 0.5$, 1.0, 1.5) it is seen that the less speed advantage the target has over the interceptor, the less violent will be the requirement for continually aiming the interceptor. In fact, when the target has a speed advantage, the interceptor is constrained to nose-on attacks and launch points near the defended areas.

2.4.2.1. *Pursuit Course*

A pursuit course is described when the interceptor is always aimed along the LOS to the target. The condition for this course is $\alpha_I - \beta = 0$ and the turning rate is given by:

$$\dot{\alpha}_I = \dot{\beta}$$

Thus, the lateral acceleration required of the interceptor is given by

$$a_I = v_I \dot{\alpha}_I = v_I \dot{\beta}$$

Since the LOS between the interceptor and the target is turning due only to the target velocity component $v_T \sin(\alpha_T - \beta)$, the lateral acceleration is given by:

$$a_I = -\frac{v_I v_T \sin(\alpha_T - \beta)}{r} \tag{2.41}$$

where r is the length of the LOS.

It can be seen from inspection of equation (2.41) that the lateral acceleration in the interceptor is minimized as the target angle $\alpha_T - \beta$ tends to zero, which is the case for a nose-on or tail chase of the target. However, the inverse dependence of lateral acceleration on the closing range r imposes a severe demand for lateral acceleration at very short ranges. This demand may be too great to be practicable in a homing interceptor, whose response time may not be sufficient to make the required correction.

2.4.2.2. *Lead Pursuit Course*

A lead pursuit course is obtained when the interceptor velocity vector maintains a constant angle with the line of sight between the interceptor and the target. The condition for interception is that the time rate of change $\dot{\beta}$ of the line of sight must be zero.

The lateral acceleration required of the interceptor is obtained by differentiating equation (2.40) and is given by:

$$a_I = \frac{v_T \dot{\alpha}_T \cos(\alpha_T - \beta)}{[1 - q^2 \sin^2(\alpha_T - \beta)]^{1/2}} \tag{2.42}$$

Examination of this equation reveals that the interception is possible only if the following condition holds:

$$|q \sin(\alpha_T - \beta)| \leq 1 \tag{2.43}$$

This condition dictates a nose-on attack as the target's speed advantage increases. This forces the highest possible demand on the lateral acceleration, since the cosine term in the numerator, and the denominator of equation (2.42) both tend to unity as the value of q increases.

Assuming that the condition given by inequality (2.43) holds, it is seen that equations (2.41) and (2.42) indicate a less severe requirement for lateral acceleration for the lead pursuit course, as compared to the pursuit course. This implies that less guidance equipment is needed for the lead pursuit interceptor.

2.4.2.3. Lead Collision Course

The pursuit courses described previously may be considered as particular cases of a general navigational equation that can be expressed by:

$$\dot{\alpha}_I = c\dot{\beta} \tag{2.44}$$

where c is defined as the navigational constant.

The value of c may extend from unity to infinity. When $c = 1$ equation (2.44) becomes the statement for the pursuit course. When $c = \infty$ the rate of change of the LOS between interceptor and target $\dot{\beta}$ as given by equation (2.44) must be zero for a finite angular rate of change $\dot{\alpha}_I$. This condition is the requirement for a lead pursuit course. The lead pursuit course is the fastest course between interceptor and target.

2.4.3. Guidance

References 9, 32, 33, 37, 39, 40

At the completion of the launch phase and when the interceptor is under the control of the guidance system, the heading of the interceptor may be required to change continually at rates determined by equation (2.44). The determination of a new aim direction may be performed by a remote site, following which change commands are transmitted to the interceptor, or the interceptor may itself perform this function. Thus, a number of navigational triangles may be computed during the course of the flight. At any particular instant during the flight, the momentary rate of change of the line of sight is given by the following equation which may be deduced from Fig. 2.14:

$$\dot{\beta} = \frac{v_I \sin(\alpha_T - \beta) - v_T \sin(\alpha_T - \beta)}{r} \tag{2.45}$$

The actual kinematic path of the interceptor will depend upon the choice of the navigational constant c, i.e., the type of course chosen.

Guidance of the interceptor may be accomplished by the following scheme: (1) precompute the course to be followed, prior to interceptor launch; (2) use a preset on-board guidance system to keep the interceptor on its programmed course, and (3) keep the interceptor on course by continually upgrading the course at a remote site and commanding changes to the interceptor, and by having the interceptor perform this latter function, i.e. homing.

2.4.3.1. *Preset Guidance*

The function of a preset guidance system is to maintain the stability of a missile in flight along a given course. The use of preset guidance may be considered if the target can be precomputed and the navigational triangle selected prior to launch of the interceptor. In other words, where the target is nonmaneuvering and is flying a deterministic trajectory.

A typical missile autopilot consists of displacement and/or rate gyros, associated amplifier and correcting circuits, and perhaps a programmer or computer. The autopilot may be regarded as being independent of the servomechanism functions which actually aim the interceptor. The use of displacement and rate gyros for components of autopilots varies with the application. Accurate control necessitates damping of the gyro displacements about one or more axes. A decision is therefore required in a given application whether to employ displacement gyros with suitable differentiating circuits, or to utilize rate gyros to perform the function directly. The characteristics of the gyros, e.g., sensitivity, accuracy, and reliability, are significant factors which must be considered in the choice along with weight and volume. A gyro with two degrees of freedom occupies less volume than two displacement gyros each with one degree of freedom.

If the interceptor is to be launched vertically and then subsequently pitched to its final trajectory, the autopilot must stabilize the interceptor about its roll, pitch, and yaw axes. For this purpose displacement gyros with two degrees of freedom will be utilized in order to keep volume at a minimum, in which case two gyros and two differentiating circuits will be needed. The gyro platform is fixed to the internal structure of the interceptor and the gyro axes are oriented in such a way as to place the spin axis of each gyro initially parallel to the interceptor pitch and roll axes, respectively. Angular displacements of gyro gimbals represent interceptor roll, pitch, and yaw and are measured by appropriate electrical pick-off units which are used in a conventional manner to command the servomechanisms of the control surfaces.

Prior to interceptor launch, the electronic circuits are energized, and the gyros brought up to their operating speeds. The preset positioning of the gyros is then made. The steering program is commenced immediately after launching, and ends when the interceptor achieves its preset course.

Because of the random forces and torques arising from wind gusts, thrust variations, aerodynamic misalignments or interactions, and structural vibrations, it is not possible to accurately predict the performance of an inertial autopilot. These effects are best assessed through extensive flight tests designed to provide sufficient and statistically significant data. Another major error source in an autopilot is gyro drift. The cumulative effect of gyro drift can be computed knowing the average natural drift rate.

Inertial autopilot guidance alone is practicable only for relatively long-range interceptors which are to be employed against precomputed targets. The time required to preset an autopilot and the inability to change the interceptor's course in flight are the principal reasons why this type of guidance may not be suitable for high-speed short-range small interceptors. A principal advantage of the inertial autopilot is its relative immunity when operating in an adverse jamming or nuclear environment.

2.4.3.2. Command Guidance

A typical command guidance system employs separate radars to track the interceptor and the target. The intercept point is computed and commands are then sent to the interceptor, which closes the appropriate navigational triangle.

A principal advantage of this type of system is the fact that the interceptor can be guided along a lead collision course, unless the interceptor is constrained to a vertical launch. Thus, command guidance using separate radars affords the shortest time of flight to interception. If the interceptor is launched vertically, a time delay is introduced which may be of such magnitude that the interceptor cannot intercept a fast target.

Apart from its principal advantage for optimum navigation, command guidance has several general limitations. First of all, the accuracy of the guidance decreases with increase in range because of longer time delays in processing data, the greater magnification of measurement errors, and the poorer resolution of single targets in the presence of multiple targets. Other limitations are higher susceptibility to adverse jamming and nuclear environments, and the greater number and cost of ground radars and computers.

A version of command guidance is beam riding, in which the interceptor follows a radar beam that tracks the target. The interceptor describes a path about the axis of the radar beam, which is already tracking or can be directed at the target. This version of command guidance requires a capture procedure that places the interceptor in the target tracking beam. The most satisfactory procedure is to first boost the interceptor to a speed close to its flight speed in an auxiliary beam of very short range but of much greater width than the guidance beam. Another procedure is to launch the interceptor directly into the guidance beam at the most favorable position and then to move the beam to acquire the target. This technique requires a longer time before the guidance becomes effective; consequently, with a fast target, the interceptor may be required to maneuver at an unacceptably high value.

Beam riding is generally employed as midcourse guidance, although it may be continued to the point of interception. Since the interceptor is a captive of the beam that tracks the target, the interceptor essentially flies a pursuit course. The beam rider is particularly capable of utilizing homing for terminal guidance, since a seeker in the interceptor could take over and navigate just prior to the intercept. In any case, the beam rider should have the capability for rapidly correcting its course, otherwise a narrow fast-moving beam might lose the interceptor.

2.4.3.3. *Terminal Guidance*

Homing is employed for terminal guidance when it is desired to obtain a faster response to the navigational requirements set forth by equation (2.44). To obtain an elementary solution to this equation, it is necessary to assume that the rate of change of the LOS given by equation (2.45) can be simply expressed. For instance, in a nose-on attack, with $v_T \gg v_I$, the equation may be written:

$$\beta = \frac{\alpha_T}{1 - t_c p} \tag{2.46}$$

since $t_c = r/v_T$ is the closing time to interception. The number p represents the differential operator d/dt.

If this value of β is the input to the guidance system whose time constant is τ, then the ratio of the output to the input of the guidance system is expressed as:

$$\frac{\alpha_I}{\beta} = \frac{1}{1 + \tau p}$$

and upon substitution of equation (2.46):

$$\frac{\alpha_I}{\alpha_T} = \frac{1}{(1 + \tau p)(1 - t_c p)}$$

This last equation resembles the transfer characteristic of a simple servo system with gain factor:

$$K = \frac{1}{(t_c/\tau)(1 - \tau/t_c)^2} \tag{2.47}$$

The guidance system is critically damped when $K = 1/4$ which when inserted in equation (2.47) results in the condition:

$$t_c = \tau$$

This says that the closing, or flight time, t_c for a critically damped system equals the guidance system time constant τ. Since the time constant of a terminal guidance system is essentially the interceptor time constant, normally of the order of 0.1 second, it is seen that such a guidance technique has considerable advantage over a command system whose time constants are much longer, especially at long range.

2.4.4. Kill Mechanisms

The physical mechanisms by which an ABM missile can destroy or damage an incoming warhead depend on the type of warhead which is used by the defending missile, i.e., conventional or nuclear.

2.4.4.1. Conventional Warheads
Reference 27

Conventional warheads are usually explosive types in order to achieve the maximum damage probability. The damage resulting from the explosion may be inflicted in a number of ways. The oldest and best understood is the disintegration of a warhead into a number of pieces of metal which may have been pre-formed into balls or other shapes that would be ejected with high energy by the explosion, or the shell simply might be broken up into a number of variously shaped fragments. The damage inflicted to the incoming warhead is due to the kinetic energy and momentum of the individual pieces of metal which enable them to penetrate and tear the RV. The volume of potential damage in space is defined by the so-called kill radius and can be made large when the proper explosive is selected and when the individual pieces of metal are given the proper mass and shape. This type of warhead is effective particularly against aircraft and missiles, since their structures are especially vulnerable to fragments, and less effective against RVs. In addition to the mechanism of a direct hit by a fragment, a secondary kill mechanism is due to blast—providing the target is located within the range of appreciable pressures and temperatures generated by the explosion. The relative contributions to the damage in this type of warhead depend on the weight ratio of metal-to-explosive charge. The main disadvantage of the conventional warhead is its small kill radius against RVs and practical implementation therefore would require an extremely high degree of guidance accuracy. The main advantage of a conventional warhead is that the defense operates in a clean non-nuclear environment, i.e., the elimination of blackout and other nuclear effects due to friendly nuclear explosions.

2.4.4.2. Nuclear Warheads
References 3, 9

The defensive nuclear warhead can be considered as a conventional warhead of enormously greater capability than a fragmenting- or blast-type warhead of comparable weight. In contrast with the conventional warhead, the nuclear warhead kill mechanisms involve the emission of neutrons, the emission of X-rays and, when the weapon is exploded in the atmosphere, blast. In this latter mechanism both the conventional and nuclear warheads are similar.

Neutrons have the ability to penetrate matter of any kind. Those released by defensive weapons could penetrate the heat shield and outer jacket of an offensive warhead and enter the fissile material itself, causing the atoms to fission and generating large amounts of heat. If sufficient heat is generated the fissile material will melt and lose its carefully-designed shape, after which it can no longer be detonated.

The kill radius for neutrons depends on the design or hardness of the offensive weapon and the yield or energy release of the defensive weapon. The miss distance, or distance of closest approach between the defensive and offensive missiles, can be made small enough to achieve a kill by the neutron mechanism. This is particularly true if the defensive missile and radar have high performance and the interception is made no more than a few tens of kilometers from the ABM launch site. The neutron kill mechanism is therefore practical for the short-range defense of a city or other important target. It is highly desirable that the blast yield of the defensive warhead be kept low to minimize the effects of blast and heat on the defended area. Figure 2.16 illustrates the neutron kill mechanism in which neutrons are shown penetrating an enemy warhead, causing the uranium 235 or plutonium to melt and

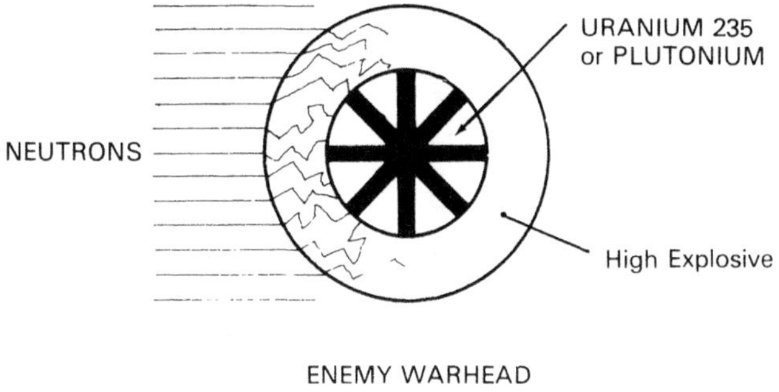

SOURCE: See reference 3.

Fig. 2.16. *The neutron kill mechanism*

lose its shape. As a consequence it can no longer be assembled for firing.

X-rays carry off most of the energy emitted by nuclear weapons, especially those in the megaton range. If sufficient X-ray energy falls on an RV, it will cause the surface layer of the vehicle's heat shield to evaporate. This in itself may not be too damaging but the vapor leaves the surface at high velocity in a very brief time and the recoil sets up a powerful shock wave in the heat shield. The shock may destroy the heat shield material or the underlying structure. X-rays are particularly effective above the upper atmosphere where they can travel to their target without being absorbed by air molecules. The defense can therefore use megaton weapons without endangering the population below since the latter is protected by the intervening atmosphere. The kill radius for X-rays is very large and this reduces the accuracy required of the defensive missile and allows successful interceptions at ranges of hundreds of kilometers from the ABM launch sites. The X-rays make possible an area defense and provide the key to exoatmospheric nuclear defense. Figure 2.17 illustrates the X-ray kill mechanism. In Fig. 2.17a, the X-rays travel unimpeded to the target. On striking the RV they are absorbed by, and intensely heat up, a thin layer of the RV's heat jacket. This creates a shock front that travels through the jacket, as shown in Fig. 2.17b. Figure 2.17c shows the jacket breaking up and detaching from the RV.

The third nuclear kill mechanism can operate only in the atmosphere, much like a conventional blast except of a much

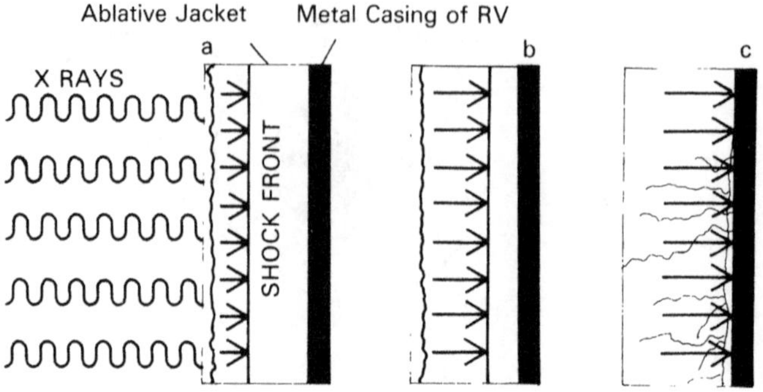

SOURCE: See reference 3.

Fig. 2.17. *X-Ray kill mechanism*

higher intensity. The increased atmospheric density reached within a shock wave from a nuclear explosion in air can produce several hundred g deceleration of a RV. This compares with a maximum force of about 100 g's normally encountered by the RV during its reentry. Figure 2.18 illus-

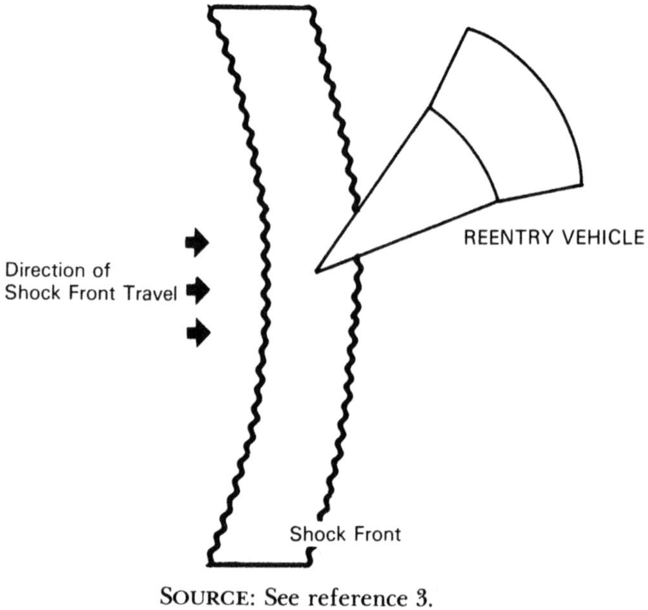

SOURCE: See reference 3.

Fig. 2.18. *Blast kill mechanism*

trates the blast kill mechanism. If the defensive warhead is exploded inside the atmosphere, the resulting shock front of air can cause the incoming RV to decelerate with a force equivalent to several hundred times the force of gravity, thereby leading to its destruction or malfunction.

References

1. C. M. Johnson, "Ballistic Missile Defense Radars", *IEEE Spectrum,* March, 1970. Describes the Safeguard ABM system and its major components, the PAR and MSR radars, the Spartan and Sprint Interceptors and the Data-Processing system.
2. M. Eleccion, "Safeguard—systems experiment or politicians delight?", *IEEE Spectrum,* 1975.
3. R. L. Garwin and H. A. Bethe, "Anti-ballistic Missile Systems" *Scientific American,* March, 1968. Discusses nuclear effects in particular and the general effects of adverse countermeasures environments on a defense system.
4. J. Rathjens, "The Dynamics of the Arms Race", *Scientific American,* April, 1969.
5. P. J. Kahrilas, "Design of Electronic Scanning Radar Systems", *Proc IEEE,* Vol. 56, No. 11, 1968. Presents a format for the design of electronic scanning radar systems. Particular attention is directed to the discussion of automatic multitarget acquisition logic, computer-controlled iterative tracking equations, interpolative null tracking and computer functions.
6. J. Emory Reed, "The AN/FPS-85 Radar System", *Proc. IEEE,* Vol. 57, No. 3, March, 1969. The AN/FPS-85 is a large operational UHF radar. Describes the radar's operation, some insights into the system design, pictures of actual hardware, a summary of system testing and a project summary.
7. P. J. Kahrilas, "HAPDAR—An Operational Phased-Array Radar", *Proc. IEEE,* Vol. 56, No. 11, 1968. The HAPDAR (Hard Point Demonstration Array Radar) described is a multi-function phased-array radar in operation at the White Sands Missile Range. The principle design feature is the TACOL (Thinned Aperture Computed Lens) array. The array and system design concepts are discussed and a detailed physical, functional and analytical description of the radar is presented. Recently obtained operational test results on array patterns and multitarget tracking performance are discussed and compared to design specifications.
8. H. York, "The Great Test Ban Debate", *Scientific American,* November 1972.
9. J. Constant, *Introduction to Defense Radar Systems Engineering,* Spartan/Hayden Books, 1972. Presents topics on ABM radar systems.

10. W. Armstrong, "History of Electronics in U.S. Army", *Countermeasures,* August 1975.
11. F. Moncrief, "Radar 77", *Microwave Systems News,* October 1977. London's Radar 77 follows the two and one-half year sequence established at Radar 75 in Washington. Papers at this year's conference emphasize current radar trends, with phased arrays, bistatic or multi-radar systems, and computer applications sharing the limelight.
12. R. Kyle, "Phased-Array Antenna Fundamentals", *Countermeasures,* March 1978.
13. W. M. Hall, "General Radar Equation", *Space/Aeronautics R&D Handbook,* Space/Aeronautics Magazine, 1962–1963. Derivation of the general radar equation is given, and special forms of this equation are used to illustrate various radar problems and systems.
14. M. I. Skolnik, *Introduction to Radar Systems,* McGraw-Hill, 1962. A detailed reference. Particular attention is directed to Chapter 2 for a discussion of the radar range equation and the detection of targets, and to Chapters 9 and 10 for the theoretical treatment of detection and the extraction of information from radar signals. A discussion of ionized media is given in Chapter 14.
15. D. K. Barton, *Radar System Analysis,* Prentice-Hall, 1965. A detailed reference on radar systems. Particular attention is directed to Chapters 1, 4, 5, 6 for the theory of radar detection and to Chapter 2 for the theory of radar measurements. Chapter 13 provides a discussion for the processing of radar data.
16. C. E. Cook and M. Bernfield, *Radar Signals,* Academic Press, 1967. A detailed study of the theory of radar signals.
17. M. H. Carpentier, *Radars: New Concepts,* Gordon and Breach Science Publishers, 1968. Provides a comprehensive theoretical basis for modern radar. Particular attention is directed to the discussion of ideal receivers, radar performance, ambiguity functions, and clutter elimination.
18. M. I. Skolnik, *Radar Handbook,* McGraw-Hill, 1969.
19. F. E. Nathanson, *Radar Design Principles,* McGraw-Hill, 1969. Gives extensive discussion of radar target characteristics, atmospheric effects, and the backscattering properties of sea and land.
20. W. M. Brown and C. J. Palermo, *Random Processes, Communications and Radar,* McGraw-Hill, 1969. Introductory text. Particular attention is directed to the brief discussion of detection in Chapter 6.
21. J. V. Di Franco and W. L. Rubin, *Radar Detection,* Prentice-Hall, 1970. Gives a powerful and highly mathematical treatise on detection theory.
22. W. S. Burdic, *Radar Signal Analysis,* Prentice-Hall, 1968.
23. "Detection Theory", *Proc. IEEE,* Vol. 58, No. 5, May 1970. A detailed reference and summary of detection theory. Gives basic

outlines, recent theoretical developments and areas of application.
24. F. E. Terman, *Radio Engineers Handbook*, McGraw-Hill, 1943. Reference summarizing basics of radio and electronics. The discussion of the propagation of radio waves is given in Chapter 10.
25. J. I. Marcum, "A Statistical Theory of Target Detection by Pulsed Radar," and "Mathematical Appendix", *IRE Transactions on Information Theory*, April 1960.
26. P. Swerling, "Probability of Detection for Fluctuating Targets", *IRE Transactions on Information Theory*, April 1960.
27. D. Howard, "Predicting Target Caused Errors in Radar Measurements", *Electronic Warfare*, February 1978.
28. John L. Allen, "Array Radars, A Survey of Their Potential and Their Limitations", *Microwave Journal*, May 1962. Reviews fundamental array types. Particular attention is directed to the discussion on detection, resolution, and accuracy of targets.
29. T. C. Bartlee, *Digital Computer Fundamentals*, McGraw-Hill, 1966. A general reference for modern digital computers.
30. R. Leach, "Evaluating the Quality of Prediction for a Position Predicting or Tracking System", *ARS Journal*, November 1962. Discusses how radar data should be processed to give a best prediction of position and assessment of the prediction.
31. B. Widrow, *et al.*, "Adaptive Antenna Systems", *Proceedings of the IEEE*, Dec. 1967. A summary of computer techniques which use algorithms to solve simultaneous equations.
32. R. B. Dow, *Fundamentals of Advanced Missiles*, Wiley, 1958. A useful engineering reference covering many subjects on a first principle basis. Recommended in its entirety for follow-on to the material in this chapter.
33. A. S. Locke, *et al.*, *Guidance*, ed. G. Merril, Van Nostrand, 1957. An excellent in-depth technical treatise on guided missiles and related topics.
34. O. R. S. Elnan and H. Ho, "Interception of High Speed Target by Beam Rider Missile", *AIAA Journal*, July 1963. This is a specialist reference. The trajectories of a line of sight beam rider missile are determined for the case where the target is moving at suborbital speeds and the missile speed is smaller by a factor of 2 or 3. Of primary interest are the conditions under which interception may take place, i.e. the minimum target engagement range and interception range as functions of missile-to-target speed ratio and missile normal acceleration capability. The results of this analysis show that interception will take place for an arbitrary speed ratio with a reasonable value of lateral acceleration, if the range at which the missile engages the target can be made sufficiently large.
35. D. B. Duncan, "Analysis of an Inertial Guidance System", *Jet Propulsion*, February 1958. Describes the characteristics of an inertial navigation system for a simplified guidance problem. Error

equations relating component performance to guidance accuracy are derived. These equations exhibit the 84-minute period oscillations which typify inertial systems.

36. W. T. Russell, "Inertial Guidance for Rocket Propelled Missiles", *Jet Propulsion*, January 1958. Explains the basic elements of inertial guidance systems primarily intended for rocket propelled missiles. Precision gyroscopes, accelerometers, and stabilized platforms are described. Guidance schemes appropriate to ballistic missiles are discussed and a trajectory for the V-2 rocket is presented as an example. The problem of gravity computation is considered and the effect of instrument errors is explored. A missile steering computer and a rocket motor shutoff computer are given for the V-2 rocket. Numerical values of coefficients are calculated for the sample V-2 trajectory.
37. A. Ivanov, "Improved Radar Designs Outwit Complex Threats", *Microwaves*, April 1976. Three guidance techniques for air-defense guided missiles are described and the widely-used CW semi-active homing system is detailed.
38. "Army will conduct demonstrations of the homing overlay intercept portion of its layered defense ballistic missile defense concept in the early 1980s", *Electronic Warefare*, April 1978.
39. H. Rainwater, "Weather Affects MM-Wave Missile Guidance Systems", *Microwaves*, September 1977. A survey of propagation studies reveals that millimeter-wave radar and radiometer missile guidance systems are both strongly affected by adverse weather conditions, but each in a different way.
40. "Missile Warhead Tracking Concepts Probed", *Electronic Warfare*, June 1978.
41. "Latest Soviet Threat Intelligence", *International Report EW/DE* (Electronic Warfare/Defense Electronics), May 1978. The addition of the SA-5 and SA-8 SAMs further complicates the dense geographic, frequency and pulse density problems that would be faced by NATO aircraft attempting to penetrate an Eastern European environment.
42. S. Tilson, "The Antiballistic Missile", *IEEE Spectrum*, August, 1969. See Reference 1.4.
43. "Army Missile Intercepted by ABM System", UPI news item, 31 August, 1970.
44. P. Friel, "The Merits of a Limited Ballistic Missile Defense", *Astronautics and Aeronautics*", October, 1970. See Reference 1.8.
45. Ed., "Strategic Warfare", *Space/Aeronautics*, January, 1969. See Reference 1.16.
46. T. Tiede, "$5 billion defense site abandoned, left to decay", NEA news item, June 1, 1977.
47. "Extraterrestrial Eavesdropping Seen Possible via Radar Leakage", *Microwave Systems News*, June, 1978. See Reference 1.39.

48. "Off the Beam", Science and the Citizen, *Scientific American*, December, 1978. See Reference 1.99.
49. "Can the Infants Survive", *Astronautics and Aeronautics*, August, 1969.
50. "Radar Computer Sorts Waveforms for ICBM Defense", *MSN* (Microwave Systems News), September 1978.

3

ABM System Design and Performance

The intercept or tracking capability of a radar system is affected by the errors introduced by the tracking, data-processing, and interceptor subsystems. In this chapter we will develop the basic considerations most involved in the design and performance of an ABM system radar within the context of an analytical procedure for determining the total error contributions of subsystems. All the subsystem errors are combined in vector root sum square (rss) form at the target's nominal position representing the total system tracking error. The magnitude of the total error vector can be represented as the radius of a sphere around the nominal target position, in which the target position is indeterminate. We will analyze each subsystem error contribution. Finally, we will build an error-model block diagram based on the complete system configuration. Parametric curves of vector error versus range may be generated, allowing system error to be read directly from the curves. The design and performance analysis of a simple radar mission in this chapter is by way of example of a Perimeter Acquisition Radar (PAR) which can reliably detect a 0.1 m^2 ballistic target at the radar range of 2,000 km.

3.1. System Requirements

References 1–10

To evaluate the capability of a defense radar system to intercept missiles, the target model and system parameters must be defined over their entire spectrum of possible variation. This discussion of target model and system parameters is included to emphasize those factors that cause errors in the tracking and computation process.

To evaluate system tracking capability, we must first know the system requirements. In addition, the spectrum of target characteristics and dynamics, tracking geometry, and nominal trajectories must be specified. Typical system requirements may be imposed that require tracking capability during the three different phases of a target's trajectory: (1) launch phase (powered flight), (2) midcourse phase (orbital flight), and (3) terminal phase (reentry).

3.1.1. Data Accuracy
Reference 1

The data accuracy specifications which affect the basic defense system configuration include (1) position and velocity tracking accuracy requirements, (2) resolution and operation in jamming, chaff, and nuclear environments, and (3) number of targets to be tracked. These system parameters are established by design studies (operational analyses and tradeoffs) of the total environment in which the system must operate.

Each phase of the missile trajectory imposes different burdens on the tracking system in terms of accuracy and rate of data-processing capability. During launch phase, the accuracy of the predicted point of impact is less important than the rate of data availability. At this time, immediate readout of data is required for alerting and for threat evaluation. The impact prediction need only be sufficiently accurate to confirm the attack. During midcourse flight, the accuracy of the predicted point of impact assumes major importance. For the terminal phase, both high accuracy and high computation rates are required for a successful interception at a time of high deceleration, as well as dynamic effects of the atmosphere associated with reentry.

Accuracy of the computed trajectory is of particular importance during the launch and terminal phases, when the trajectory may have higher order derivatives which affect position and velocity filter accuracy. For acquisition or reacquisition of the target, the radar must be directed at a point which is computed from the predicted trajectory. During the midcourse phase the missile follows the classic Keplerian motion of orbiting bodies under the influence of gravity. This validates the assumption of a second-order equation of motion for predicting the vehicle path during the midcourse part of the trajectory.

3.1.2. Target Characteristics
References 1

Tracking radar performance is directly related to return signal strength. The signal-to-noise ratio is inversely related to radar accuracy. For skin (bare body) tracking the signal-to-noise ratio is a function of RCS area, tracking radar radiated power, receiver sensitivity, and range, all as given by the general radar equation. For beacon tracking, which may be used with friendly interceptors, the signal-to-noise ratio is a function of beacon power, tracking radar receiver sensitivity, and range.

Figures 3.1 and 3.2 show the effective signal-to-noise ratio of a typical long-range reflector radar for skin tracking and beacon tracking, as a function of slant range. The target RCS and the beacon power appear as parameters in these figures. Accurate tracking is achieved when the signal-to-noise ratio is greater than about 10 decibels. This condition is shown in the figures. Inspection of the figures shows the limited range performance of this particular radar for low target RCS or low beacon power.

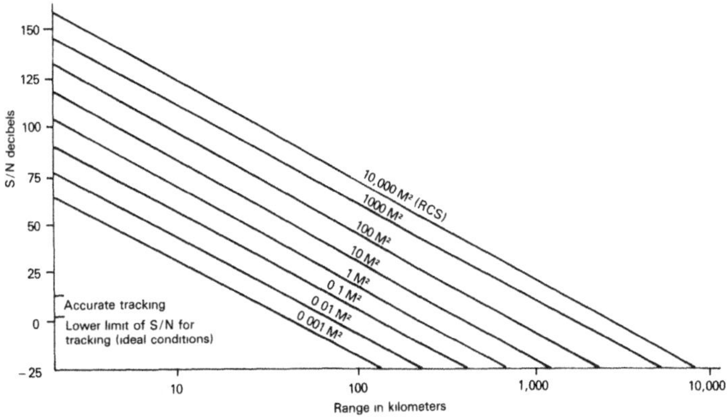

SOURCE: See reference 1.

Fig. 3.1. *Signal-to-noise ratio for skin tracking, long-range radar*

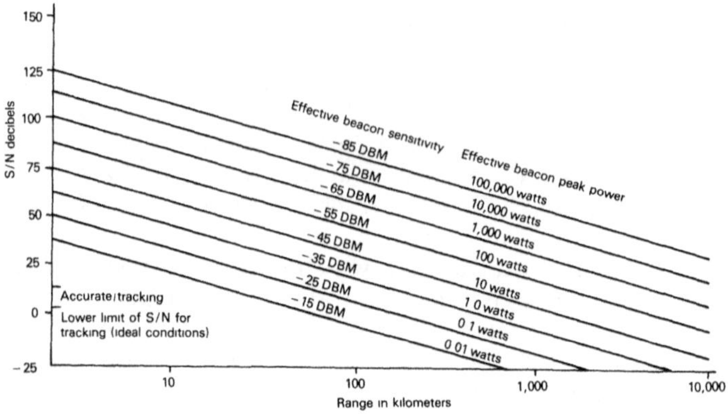

SOURCE: See reference 1.

Fig. 3.2. *Signal-to-noise ratio for beacon tracking, long-range radar*

3.1.3. Target Trajectory and Dynamics

References 1

Nominal target trajectory, maximum and minimum range, and elevation to the target and pointing angles for acquisition must all be determined from a trajectory analysis in relation to defense radar positions. Together with target trajectory, the target dynamics determine the ability of the radar antennas and servos to track the target.

The target ballistic coefficient W/C_DA is required to determine (1) the deceleration rates of the vehicle during reentry, and (2) the effects of the atmosphere on the terminal trajectory. This coefficient determines the effects of the atmosphere on the dynamics of a reentering body and closely predicts vehicle velocity as a function of reentry angle, reentry velocity, and altitude for various ballistic coefficients. Empirical equations may thus be developed, based on such data, to assist the radar in tracking and reacquisition without necessitating programming of complex equations of motion.

3.1.4. Tracking Geometry

References 1

The expected target trajectory, trajectory phase (launch, midcourse, or terminal), and location of the defense radars relative to the trajectory define the range, azimuth angle, and elevation angle for search and tracking. With the target's ballistic coefficient, the various derivatives can also be specified. We can use these results to determine the optimum radar parameters to minimize target-dependent errors. The errors are the target dependent range and angle noise and the dynamic lag errors computed by the use of the position, velocity, and acceleration constants of the positioning servo loops. The tracking azimuth and elevation for most radar installations, especially those using arrays, is limited by the tracking error, off broadside for arrays, and gimballing for conventional antennas. The minimum tracking angle for early downrange acquisition is limited, depending on antenna height, the radar beamwidth, the receiver noise, clutter conditions, and line of sight.

3.1.5. Subsystem Location

References 1

Proper location of subsystems in the defense operating area is essential to increase the target position and velocity accuracy, and to minimize effects of jamming, chaff, and nuclear explosions. In a point defense system, subsystems are located at or near the defended point. In an area defense system, they are located throughout the area being defended.

3.1.6. Array Characteristics

References 6–8, 10

The first operational radar, the British CH, was an elementary array. Many vehicular antennas have taken the form of arrays. They are particularly suited for missile defense radar systems.

The primary difference between mechanical and electronic scanning is that in electronic scanning the electrical shape is independent of the mechanical shape. Shape flexibility for the array allows the choice of mechanical shape either to minimize blast damage or to conform to a vehicle's shape.

A second difference is that the array radar can scan its beam aperiodically, or rapidly change from one scan direction to another quite widely removed. In most mechanically-scanned radars there is a periodic movement for search and acquisition.

Also, tracking of more than one target on a time-sharing basis is severely limited by the mechanical inertia of the antenna reflector. In contrast, the array radar is limited in these abilities only by the small, but nonzero, inertia of the phasing scheme and by the amount of RF power available. Coupled with other attributes, this relative independence from inertia allows the phased-array radar to be at one instant a search, and at the next a track, radar. Scan flexibility allows the array radar to cover wide angular segments at high speed.

A third difference between the array and the reflector is that the array radiation pattern is much easier to modify. This is because the array aperture illumination is completely controllable, e.g., in phase and, in principle at least, in amplitude, although control is obtained at the expense of increased complexity of the controls.

An elementary example occurs in the consideration of search-radar data rates for fixed requirements on resolution. The rate at which a given solid angle can be searched under these circumstances is directly proportional to the amount of average power which can be efficiently transmitted and subsequently received. High-resolution, high-gain radars such as might reasonably be expected to be used in modern defense systems cannot efficiently use available amounts of power unless some means is found to allow inspection of angular resolution cells in parallel.

Still a fourth characteristic of arrays which differentiates them from mechanical reflectors is their ability to radiate vast amounts of power. Singly-fed mechanical antennas are limited to the amount of power transferable through a single feed. Even with extreme pressurization, this severely restricts mechanical radars. Attempts to use multiple feeds for mechanical reflectors result in unsatisfactory aperture illumination (coma or aberration), aperture blocking (sidelobe deterioration), or both. These effects arise whenever an attempt is made to scan a mechanical radar without moving the reflector surface, such as might be necessary for sufficiently large, or sufficiently hard, apertures where resolution in two angular dimensions is required.

The principal limitation of the array radar is its relative high cost and great complexity. Unless the competing reflector is unusually expensive (as when the effective aperture is to exceed 1,000 wavelengths and if extremely low sidelobe performance is required), an array is probably not competitive if one or two mechanical radars will meet the requirement. Intuitively, it seems reasonable that a mechanical antenna would be less expensive than an array of separate elements. Similarly, the feed structure of mechanical radars is much less complex than that dictated by the requirement of feeding an array of phased elements. In addition, the problem of positioning a mechanical antenna is a relatively simple electromechanical problem, while controlling the phase of the elements of an array is a complex control problem.

An inherent limitation of the array is the broadening of the beam as it is scanned away from broadside. This effect generally requires that the radar be designed to have the desired resolution at or near the maximum angle of scan off-broadside. For wide-angle systems, this results in the requirement of a much larger aperture than would be the case if the beam did not broaden. At angles of scan not too far removed from broadside, the variation of beamwidth with scan angle is inversely proportional to the cosine of the angle off-broadside. Thus, if an array radar must be designed to provide coverage $\pm 60°$ off-broadside with no worse than $1°$ angular resolution within the covered volume, the aperture linear dimension must be at least twice that to provide a $1°$ beam at broadside. Since the complexity of the required array increases approximately as the square of the increased linear dimension, this means a fourfold increase of complexity. For many applications, such an increase in the number of receiving beams (hence receiving and beam forming apparatus) is also required.

A further requirement arises when the array is used for tracking. Tracking error is proportional to (among other factors) the angular beamwidth of the tracking beam, and signal-to-noise ratio. Both of these quantities are relatively degraded as the array radar is scanned off-broadside. Therefore, the tracking accuracy is a function of the angular position of the tracked target. Such a condition is difficult to cope with, in many systems, by any other means than "worst case" design, and this will require an even greater aperture in many cases.

A further limitation of virtually all arrays proposed to date is in bandwidth. In most arrays, the phasing scheme operates modulo 2π. That is, if the path delay required for a given element is $7\pi/2$ wavelengths, the actual phase shift effected for that element is $3\pi/2$. This does not cause any difficulty unless the signal bandwidth is so broad and the array so large that the array linear dimension is about a wavelength at the signal bandwidth frequency. In this case it becomes necessary to provide true envelope delays at each element in order to preserve the sense of the phase front. In modern radars, which often must provide both extreme range resolution (requiring a broadband signal) and a high degree of angular resolution (implying a large aperture), this limitation may become bothersome.

From a system designer's point of view, array antennas are contenders for selection:

1. In high-traffic environments.
2. In practically all shaping applications.
3. When it is desired to use a search radar for a reserve tracking system in the event of battle damage.
4. For volume search, where high resolution is required in both angular dimensions, or where a high data rate is specified.
5. For tracking, where extreme accuracy is required on violently moving targets, or where it is desired to combine search and tracking apertures.
6. If an octant or more must be searched periodically to more than about 2,000 kilometers with complete coverage in one second or less.
7. In virtually all applications where the beamwidth is varied in a single aperture.
8. Where the resolution or gain requirement is so stringent that tolerance cannot be maintained in reflecting structures.
9. For applications requiring peak powers in excess of about 100 megawatts at S-band, with $1/f^2$ scaling for other frequencies. Singly-fed mechanical reflectors are limited above such power levels.

The freedom from periodicity enjoyed by the electronically scanned array is a distinct advantage in many applications. However, in some applications it does not add a great deal of capability to the system even when phased-array radar is used.

The first possibility which occurs to the system designer is that of combining search and track functions in a single phased-array radar. If this were not done, a great many tracking reflector radars, plus one or more search radars, would be required. A combination of functions is generally desirable, particularly in applications where a system's total weight and complexity are of paramount importance.

The question of whether the two functions can be combined for a given radar application depends on a number of factors relating to the nature of the targets one wishes to acquire and track. We first assume that tracking can be effected by switching the position of the beam, requiring position changing time τ_s, after which the beam is allowed to dwell for time τ_d. Then the total time T used for tracking n targets is $T = n(\tau_d + \tau_s)$. If some time T_S is reserved for search, then the total time T_o used for both search and tracking is:

$$T_o = n(\tau_d + \tau_s) + T_S \tag{3.1}$$

It is seen that the tracking time T is dependent on the number of targets expected to be handled by the radar. Times τ_d and T_S are determined by the nature of the target, specifically by its relative violence of motion and its RCS. Only time τ_s is strictly a radar design factor, and it may not be entirely under the control of the designer. Time T_S, however, can be changed to some degree by varying the average radiated power for the search function.

A first approximation of the combined function capability can be had by assuming that the target simultaneity statistics are such that, with the maximum number of targets being tracked, no search function will be performed. Alternatively, other portions of the defense system (number of interceptors available) may be such that only n targets can be engaged simultaneously in any event. Such a first estimate defines the number of targets which can be simultaneously tracked, as given in equation (3.1). Figure 3.3 is a plot of this equation for a few reasonable values of the parameters involved. It can be seen from these curves that the multiple tracking capability of the array radar is generally quite sensitive to the type and motion of the target and target arrival characteristics for which the radar is designed. For dwell times which are more than 0.2 to 0.4 of the total tracking time T, it is not particularly helpful to use an array, even if the speed of changing beam position is zero. On the other hand, if ratio of dwell time to total tracking time is low (below about 0.1), the number of targets which can be tracked is greatly dependent on the inertia of the phasing scheme.

The range of values for τ_d, τ_s, T, and T_S that might be encountered in modern systems, particularly those pertaining to ballistic missile targets, may be obtained by considering a

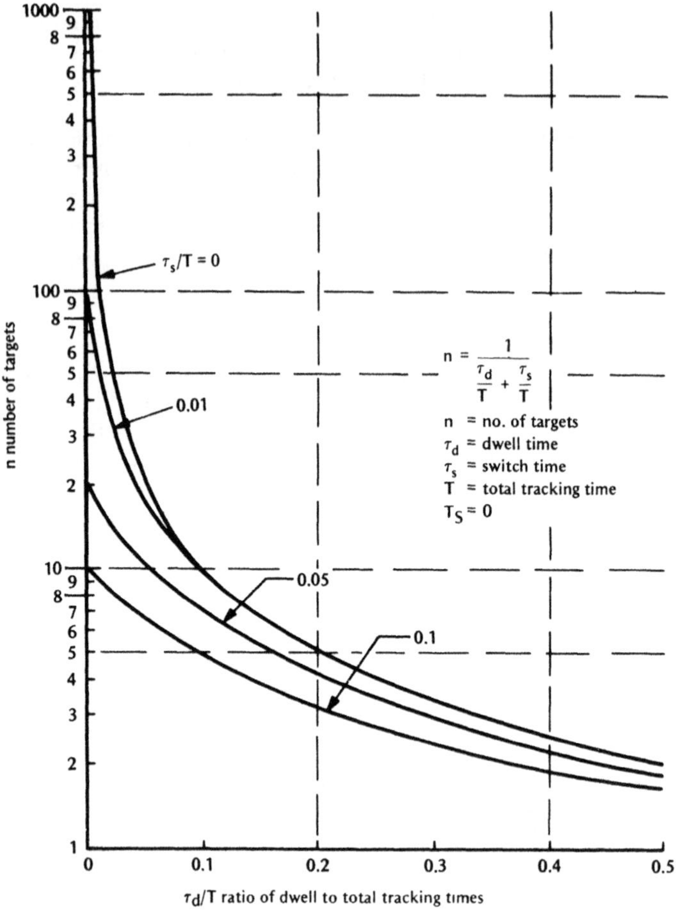

SOURCE: See reference 7.

Fig. 3.3. *Function sharing for phased-array radar*

reentering missile. For this case, the radar problem is mor severe since targets undergo violent movements as the decelerate on reentry. Although we do not know the exa speed of any operational ballistic missile, we can safely assum that it is less than the earth's escape velocity. A conservati guess would be about ten kilometers per second.

In the case of several ballistic missiles intended to hit th same target simultaneously, we assume that all n missiles arrive at the outer limits of our detection system at the same time. Suppose that this outer bound is 2,000 kilometers range and the system requires that all targets are to be tracked as their ranges become less than 1,500 kilometers, down to 500 kilometers. This allows approximately fifty seconds for detection and acquisition by the radar, and 100 seconds for tracking, i.e., $T_S = 50$ second and $T = 100$ seconds.

The time T in this example is the time corresponding to radial passage within the tracking perimeter. This is probably also a representative figure for nonradial trajectories as well. In a given application, we may decide from tracking, signature, and energy considerations that a two-second dwell time is needed and that the beam position switching time is 200 microseconds, a value compatible with published ferrite phaseshifter data, i.e., $\tau_d = 2$ seconds and $\tau_s = 2 \times 10^{-4}$ seconds. Referring to Fig. 3.3, we see that the curve of interest is very close to that for $\tau_s/T = 0$. For the example assumed, $\tau_d/T = 0.04$ from which Fig. 3.3 indicates a possibility of tracking about twenty-five targets.

This example indicates the type of approach used to determine the shared-function capability of an array radar. The example given is obviously oversimplified but it does illustrate the scan flexibility of an array radar. The task is of course much more difficult in actual practice.

Another possible advantage of the array radar implied by the above exposition of its possibilities in a multifunction role is that of "graceful degradation". Because it is independent of a periodically-varying scan function, it can be quickly converted from search radar to a specialized function such as tracking. Furthermore, in the tracking mode, it can be arranged to gradually vary its performance from entirely satisfactory to merely acceptable.

Its scan flexibility allows the array radar a means of gradually absorbing an otherwise saturating attack. For instance, consider the dwell time τ_d previously discussed. This parameter is originally determined by a desire to achieve a virtual 100 percent kill probability for the maximum number of targets which can be expected. However, should an enemy confront the system with a larger number of targets than anticipated, it would seem reasonable to relax the amount of dwell time so that all targets can be engaged, even though the probability of kill would be correspondingly lowered for some targets. Within the bounds set by other system parameters (e.g., T_S, T, τ_s, number of defensive missiles available, etc.), this reduction in dwell time can be accomplished in a gradual manner by an array, as implied in the foregoing example of scan flexibility. In comparison, a system which utilizes reflector trackers exclusively must at best degrade in quanta of τ_d/m, where m is the number of trackers. The difference is obviously academic if m is sufficiently large. For systems using a few trackers, this advantage may justify the use of the array radar.

3.2. The Radar Design Problem

References 1–37

The gain of an antenna is given by:

$$G = \frac{4 \times 10^4}{\theta_\beta \phi_\beta} \tag{3.2}$$

where

G = gain, and
θ_β, ϕ_β = half-power beamwidths in the antenna's azimuth and elevation planes, respectively; each beamwidth is defined as the ratio of wavelength-to-antenna diameter; in degrees.

The complexity of an array is proportional to the number of individual elements of the array. The number of elements in a flat planar-array, at half wavelength spacings, is given by:

$$N = \frac{4 \times 10^4}{\pi \theta_\beta \phi_\beta} \tag{3.3}$$

indicating that, for example, a pencil beam $1° \times 1°$ requires approximately 13,000 antenna elements. Furthermore, the usable angle of scan of a single planar array is limited by the decrease in the projected size of the antenna at large off-broadside angles. Thus, for typical arrays, the number of elements in an array is likely to be of the order of 10^4 and at least several array faces may be required to scan extended volume coverage, as for example a hemisphere.

It becomes apparent that arrays, particularly those of the

active element type, are complex and costly. We are then faced with the question of what advantages, if any, the array configuration possesses to justify the increased cost. To answer this question, let us review the factors that determine the performance capabilities of any radar.

3.2.1. Radar Performance Factors
References 6–8

The ten most common factors determining radar performances are:

1. Flexibility of shape,
2. Flexibility of scan,
3. Aperture illumination,
4. Power handling capability,
5. Complexity,
6. Detection of targets,
7. Resolution of targets,
8. Measurement accuracy of target parameters (e.g., position, velocity, RCS),
9. Reliability,
10. Operation in adverse environments.

The flexibility, illumination, power handling, and complexity aspects of array-radar performance have been discussed previously. In this section we will review array capability to reliably perform detection, resolution, and accurate measurement of targets in adverse environments.

3.2.1.1. Detection

Equation (2.11) may be employed to determine the detection advantage, if any, of an array radar over a conventional mechanical system. The available design parameters are the transmitted power P_T, the effective aperture area A_e, and the rate of scan Ω/T_S.

Although desirous of high search and data rates, the designer must consider the tradeoff between search rate and available power, since the cost of a radar is often dominated by the power requirements, particularly at the megawatt levels. The effect of the search rate parameter on detection is best determined by considering the product $\Omega R^4/T_S$ which appears in the surveillance radar equation. Targets at long range require a lower rate of search to insure that the minimum signal-to-noise ratio can be maintained at the receiver. The flexibility of scan inherent with an array antenna is particularly suitable for varying the scan time T_S with range R. Thus, at maximum range, an array may increase T_S, thus decreasing the required average power and therefore providing wide-angle, long-range surveillance at increased detection rates per watt, in comparison with a mechanical radar.

Due to lack of mechanical motion with the array, the effective antenna area A_e can be made quite large, when compared with mechanical antennas. This may be an economical method, trading power costs against array complexity costs, particularly at the lower frequencies where fewer elements are needed. This approach is clearly not practical in a distributed target environment.

The scanning flexibility of an array presents another useful method for obtaining the required signal-to-noise ratio. The procedure is to program the radar (1) to operate at a lower than required probability of false alarm, and (2) to stop the scan should a false alarm occur, in order to reexamine the space. This procedure results in a gain in signal-to-noise ratio E/N_0, which far exceeds the slight increase in scan time T_S which occurs from occasionally stopping the scan to increase the probability of target detection.

3.2.1.2. Resolution

The ability of a radar to resolve targets differing only in angle is a function of the antenna gain. This can be seen by taking the last of equations (2.34) and equation (3.2). Thus, for $\theta = \phi$:

$$\delta\mu = \frac{4 \times 10^2}{\pi G^{1/2}} \tag{3.4}$$

The gain achievable with conventional mechanical antennas is presently on the order of 70 decibels. For an array of elements with half-wavelength spacing, the maximum gain of a broadside beam is:

$$G = \pi N \tag{3.5}$$

Thus, 70 decibels for an array requires about $10^7/\pi$ elements—an exorbitant number in terms of number, elements, and cost. Since an economical array may have as many as 10^4 elements, the array can achieve a gain of $\pi \times 10^4 = 45$ decibels, which is much less than the achievable reflector gain. However, the array can match the reflector in beamwidth. This can be accomplished through the density, or space, taper technique mentioned earlier. In this technique the density of the active elements rather than their amplitude is varied, thus achieving a large area with corresponding small beamwidth, while retaining but a fraction of the elements. All this of course is accomplished at the expense of higher sidelobes. The principal difference between array and reflector antennas will lie in the far-out sidelobes. The conventional reflector sidelobe level will average only about -70 decibels, while the array far-out sidelobes will average only about -45 decibels. For many applications, this deficiency may be acceptable.

For resolving closely-spaced targets of about equal cross-section, the close-in sidelobe level is most significant. The improved control of the antenna illumination which is possible with many array types may permit realization of close-in sidelobes. Thus, for resolving only close-in targets, the array can be superior to the conventional reflector antenna of the same area (beamwidth), due to the better control over close-in sidelobes. For resolution at angles far removed from the main beam, the reflector excels often by a wide margin. However, density tapering techniques can improve the array disadvantage in far sidelobe performance.

It was pointed out in section 2.2.10 that the only waveform which can attain resolution of arbitrarily large radar cross-section differences between targets is the pulsed waveform. This resolution is obtained by the straightforward method of using extremely short pulses. Thus, the ability of a radar to resolve in range is limited only by the peak power obtainable, the component bandwidth, and the transmitting efficiency (ratio of on to off time). For many applications, the peak power limitation is the main restriction of the mechanical reflector; for these cases an active-element transmitting array offers promise. Passive-element receiving arrays may still be used, but it may be necessary to resort to delaying techniques to handle the wide bandwidth of the signals.

3.2.1.3. Measurement Accuracy

To the extent that the array cannot match the conventional reflector in gain, and therefore in resolving capability, the remarks of the previous section apply here. However, as previously discussed, the array can provide a higher signal-to-noise ratio per watt of transmitted power than a reflector radar. This may be used to overcome the higher gain capability of the reflector.

The additional features that arrays offer for enhanced measurement accuracy are connected with their rapid scanning capability and other flexibilities. When resolution is not a problem, sequential measurements may be profitably employed. The utility of the array configuration in this application lies in its ability to vary the amount of dwell time, scanning rapidly in areas where there are no targets requiring measurement, but dwelling longer where such targets occur. The ability of such a scheme to justify the use of an array clearly depends on the premium attached to efficient use of power as balanced against the increased control equipment necessary.

The enhanced accuracy that can be purchased by repeated measurement may or not be an argument for an array configuration, depending on the target characteristics involved. For example, if the targets are satellites with predictable dynamics, a few measurements well separated in time and space may suffice for attaining the necessary accuracy. Such measurements can be made with a simple constant scan rate surveillance radar in the normal course of its surveillance activities, i.e., track-while-scan operation. Array flexibility presents a concrete advantage for measurement accuracy only when a large number of highly maneuverable objects exists, which requires many measurements on many targets, each often in time to preclude confusion due to target motion.

It appears that achievement of highly accurate measurement is not significantly improved by the use of an array configuration, except in the situation where a large number of frequent measurements must be made, due to the simultaneous occurrence of many maneuverable targets. In such a situation, the scanning flexibility of the array is the factor of importance, and a passive element array may suffice.

3.2.1.4. Reliability Considerations

Since multibeam arrays are fabricated entirely of nonvariable passive circuitry, they can be more reliable, cheaper, and more reproducible than steerable arrays. On the other hand, as a result of the fixed nature of the beams, angular information is inherently less accurate than from a steerable array. This decreased accuracy arises from the fact that in a multibeam array the beam is fixed and therefore cannot be pointed directly at the target, thus decreasing the signal-to-noise ratio which would otherwise be available, and by equations (2.39) resulting in the lower accuracy for the measurement.

The tremendous redundancy associated with active-element arrays is frequently cited as an advantage from a reliability standpoint, in that the parallel operation of such an array allows a larger number of failures without serious system degradation.

The validity of this argument is seriously compromised, unless the original system is designed for adequate protection against the effects of failure of items which are not naturally redundant, such as control devices, prime power, etc. It is also necessary to exercise care in the design of redundant units for long life within the necessary phase and amplitude tolerances, and to provide means for rapid detection and easy replacement of those units failed or out of tolerance.

Further, if the radar is to be cycled on and off periodically, the transient failure rate of an array may become unmanageable, or one may find that the extreme differential stability demanded from unit to unit requires long warm-up times for its realization.

Successfully maintaining such a complex system depends largely on the extent to which the system can be automated for fault detection, fault location, and replacing components. The situation is quite analogous to that of the digital computer maintenance problem, but significantly complicated by the added complexities of phase and amplitude stability requirements.

3.2.1.5. Operation in Adverse Environment

To the extent that an array radar has a lower achievable gain and higher far-out sidelobes than a conventional reflector radar, it appears more vulnerable to operation in jamming, distributed target (chaff, clutter, rain, etc.), and nuclear environments. However, the potential exists for using arrays in two situations of interest for defense: operation in a jamming environment, and blast damage survival.

Examination of the radar equation in the form appropriate for operation under heavy jamming generally leads to one conclusion: the most universally useful defenses against jamming are effective radiated power (the product of transmitter power and transmitting antenna gain) and the radar's usable bandwidth. It is apparent that the tremendous power capability associated with an active-element transmitting array represents a real advantage in such an environment. Note, however, that the indication is thus that one should emphasize a large, powerful transmitting array to achieve sensitivity in such an environment rather than or in addition to, the economically more attractive large receiving array.

There seems to be little question but that the stationary nature of the array antenna permits hardening against the blast and thermal effects of bomb bursts greatly in excess of that obtainable with conventional mechanical reflectors. The question seems to be how hard can they be made and at how great a cost.

3.2.2. Array Advantages and Disadvantages

References 6–8, 10, 32–35

It has been pointed out that arrays offer potential for increased performance and flexibility over conventional radars in some areas.

Passive-element arrays can provide more detection capability in demanding situations, due to their scanning flexibility, but at the cost of control complexity.

Active transmitting element arrays offer the possibility of enhanced range resolution and accuracy and useful operation in a jamming environment, since the tremendous peak powers available allow the use of short pulses. It is not apparent, however, that arrays offer any advantage in doppler resolution; and for angular resolution, it is quite clear that conventional reflector antennas can provide more angular resolution in particularly demanding situations for reasonable expenditures and complexities. However, the slow angle rates necessary with such high-gain reflector antennas may again favor the array configuration for some tasks, particularly where extreme angular resolution is not required, and where any requirement for hardening exists.

The principal advantage offered by the array configuration lies in its ability to cope with a relatively large number of maneuverable targets, which is by virtue of its scanning capability. On small numbers of targets, or those with predictable dynamics, the array appears to offer no significant accuracy advantages over more conventional radars. If proper care is taken in array radar design, increased reliability due to the inherent redundancy may be obtained under continuous operating conditions, if high reliability components are used and the maintenance procedure is sufficiently automated and streamlined. If the radar must survive the effects of nearby bomb bursts, regardless of cost, the array configuration appears mandatory, but the cost undetermined.

Finally, an important point is that a single array radar, designed with flexibility in mind, can perform in a near optimum manner a variety of functions on a time-shared basis, serving as both a surveillance radar and a tracker, and do so reliably at a price.

3.2.3. Conclusions

It is apparent that the array configuration has great potential; it is also apparent that the array radar can become quite complex and costly. Since the installation of an array will usually be considered in the context of increasing system effectiveness, the subject of cost-effectiveness will no doubt receive considerable attention.

First, the configuration offering the greatest long-run economy for the task at hand should be selected. For example, if the task requires only versatile scanning at a modest power level, passive-element arrays or combined systems, such as active receiving arrays, should be investigated. If the task demands an active-element array, the cost is likely to be nearly a linear function of the number of elements, and the question is how to get the most from the least number of elements.

The effective area A_e of the array is proportional to the number of elements N and to the wavelength squared. Therefore, the least number of elements needed to achieve a given area is obtained at the longest wavelength or lowest frequency. Furthermore, the number of elements varies with the square of the frequency, so that a choice of a higher frequency will rapidly increase the number of elements required. Tailoring of the array geometry to the required scan can produce savings in elements. Where limited far-out angular resolution will suffice, density tapering reduces the number of active elements in an active-element array.

Second, array designers should strive to reduce the per-element long-range costs. One obvious technique is to reduce the amount of hardware associated with each element. For example, rather than use a driver stage and a final amplifier following the phase shifter in each element transmitter, it might pay to use phase shifters capable of handling higher input power levels and use only the final amplifier in each element package, thus using fewer higher-powered drivers, each driving several elements.

Another technique for reducing costs is associated with mass production of elements. Since sustained production and sustained consumption of element units may not be cost-saving, the building block concept suggests itself, in which general-purpose standardized-drive elements can be designed and produced in each frequency band of interest to meet specifications generated by consideration of system requirements at that frequency. Using purchased building blocks, arrays could then be constructed in a manner similar to that widely used to build special-purpose digital equipment.

3.3. Error Analysis

References 1–10, 28, 32–36

Errors of an independent statistical nature can be summed on a rss basis. These include both systematic and random errors of the tracking subsystems. Therefore, the total system tracking error for a typical intercept is determined by taking the rss of the errors due to each tracking subsystem. The error is read from a set of parametric curves providing a convenient method for presenting data which are functions of several parameters (e.g., S/N and off-axis angle). These curves are generated under the constraints of the target model and tracking conditions, according to the error model adopted for the evaluation. The parametric curves, which represent "single-look" accuracies, not only facilitate the evaluation of tracking capabilities for the assumed models and conditions, but may be utilized for other models and conditions to obtain reasonable evaluations if proper conversion factors are applied. Smoothed position and velocity accuracies are derived from single-look data by the application of the numerical filtering technique using least square fit. This technique, the error model, and the ground rules applied are discussed in the following sections.

3.3.1. Error Components

Errors can be categorized as either random or systematic. Random errors are characterized by random fluctuations about zero mean. They are associated with antenna noise, receiver noise, tracking jitter, antenna flexure due to wind load, target fluctuation noise, and noise on data pickoffs. Systematic errors are usually due to misalignments in the system. Depending on the nature of the misalignment, the result may be either a bias error only or a fluctuation superimposed on a bias. Some systematic errors are due to other than misalignment, for example, the refraction of the radar beam by the ionosphere. Another example is dynamic error due to radar lags caused by high velocity and acceleration of the target. These are characterized by both variable components and a constant component. They can be corrected by recording error signals.

The following partial list illustrates the variety of error sources which can affect the accuracy of the total system:

1. Tracking and data readout,
2. Antenna and receiver noise,
3. Uncertainties due to flexural and torsional misalignments of the antenna,
4. Stabilization data (pitch, roll, yaw) errors of inertial references which may be used on defense platforms and interceptors,
5. Uncertainties of interceptor position due primarily to wind currents,
6. Uncertainties associated with data refinement (e.g., correction for radar beam refraction or radar angle errors),
7. Uncertainties in subsystem location.

One example of how uncertainty in subsystem location can

affect the measurement of a target is shown by Fig. 3.4. In this case, the location error is amplified by the curvature of the earth. If P_R and P_T (in Fig. 3.4) are the true positions of radar and target at the time of measurement, and if the error in radar location is *d*, then we see that the error in the measured target location is *D*. The amplification factor is therefore

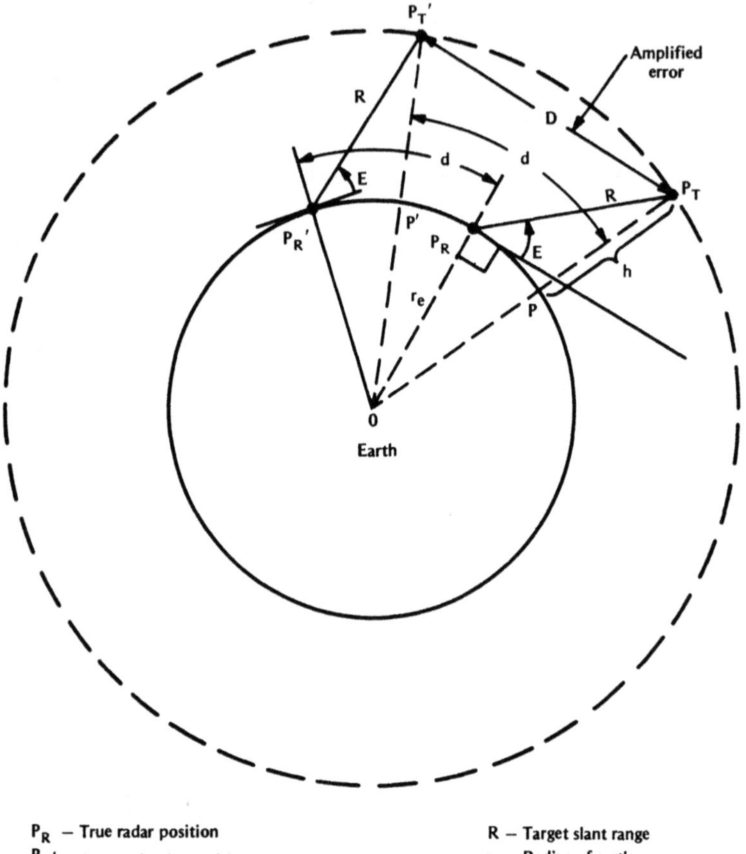

P_R – True radar position
P_R' – Assumed radar position
P_T – True target position
P_T' – Assumed target position
d – Error in radar position
D – Error in target position due to radar position error
R – Target slant range
r_e – Radius of earth
h – Target altitude above the earth
E – Target elevation angle
O – Center of earth

SOURCE: See reference 1.

Fig. 3.4. *Amplification effects of the earth's curvature on radar position error* defined by $1 + h/r_e$ where h is the target altitude and r_e is radius of the earth.

For long ranges, near the radar horizon, the effective target error due to a particular subsystem location uncertainty can be approximated by the following expression:

$$\frac{D}{d} = \left[1 + \left(\frac{R}{r_e}\right)^2\right]^{1/2}$$

where R is the slant range to the target. For short ranges:

$$\frac{D}{d} = 1 + \frac{R}{r_e} \sin E$$

where E is the target elevation angle. At ranges of 900 kilometers or less, the amplification factor can be approximated by unity.

Gravity anomalies represent another error in a tracking system but will not be discussed, since the error from this source is small compared to other system errors.

3.3.2. Total Target Error

Since all errors are randomly distributed (bias and noise), the total target error can be defined as the vector sum or rss of all subsystem rms errors related to the target at the target elevation angle. The magnitude of this vector defines the limits of a volume of uncertainty, inside which the target position cannot be determined.

Angular errors are converted to feet, allowing the total error to be expressed as a function of range. The following equation expresses the total error in terms of its components and the target range and elevation angle:

$$\sigma^2 = \sigma_R^2 + R^2[\sigma_E^2 + \sigma_A^2 \cos^2 E] \tag{3.6}$$

where

σ = total target error,
σ_R = vector sum of all rms range errors in feet-rms,
R = slant range in feet,

σ_E = vector sum of all rms elevation errors in vertical plane in radians,
σ_A = vector sum of all rms azimuth errors in horizontal plane in radians,
E = target elevation angle.

There are several advantages to total error representation:

1. The error is independent of the specific coordinate system selected. This representation permits a simple comparison of the merits of different systems on a common basis.
2. The representation avoids the complex coordinate transformation that would be necessary to arrive at the error components in a specific frame of reference. It also permits the errors of each subsystem to be totaled in the most convenient frame.
3. The representation permits simple evaluation of the "typical intercept" capability.

The main disadvantage of this error representation is that it does not permit consideration of the total error component in a specific direction.

3.4. Smoothing of Position and Velocity Data

References 1, 10, 32–36

An integral step in computing the overall tracking error is the application of digital smoothing techniques to decrease high frequency random errors. The position and noise reduction factors of various filters are discussed and represented in the final error model. Only those high-frequency errors due to the tracking and stabilization subsystems can be filtered; they are mostly random noise derived from target scintillation, glint, beacon modulation, servo jitter and receiver noise. The positioning and data stabilization long-term random errors, as well as the radar system systematic errors and biases, cannot be smoothed to any significant extent, because they are essentially composed of errors whose period is long relative to the real-time data-sampling interval used in filtering. The smoothing process itself, consists of fitting a polynomial of sufficiently high order to the measured test data on the basis of least square deviation.

Selection of a suitable filter depends not only on the noise reduction factor required, but also on the characteristics of the target and the phase of its trajectory. For example, during launch phase the computing rate is of paramount importance. So during this part of the trajectory very little lag can be tolerated through the filter, since errors increase with time lag. On the other hand, the ability to smooth data increases with increasing time lag. Figure 3.5 illustrates the behavior of the noise factor in a typical low order filter as a function of the number of data points N either side of the filter midpoint and the total number $2N+1$ of data points in the filter. The noise factor is defined as the output noise $\overset{(n)}{\sigma}(t)$ ($n=0$ for position, $n=1$ for velocity, etc.) divided by the input (instrument) noise and for effective smoothing assumes values less than one. The number of data points N each side of the filter midpoint is defined as the ratio of the filter lag t_f and interval Δt between data points. It can be seen from the figure that the noise factor increases with filter time lag t_f but decreases as the number of data points is increased.

The actual selection of filter order, sampling period, and filtering lag involves a tradeoff of three factors: (1) noise reduction (including filter time lag) (2) curvefitting errors, and (3) computer computation cycle time. These factors must be optimized for each phase of the trajectory (launch, midcourse, and terminal).

Although digital filtering is a powerful tool for reducing random noise in measured data, it does have limitations. It

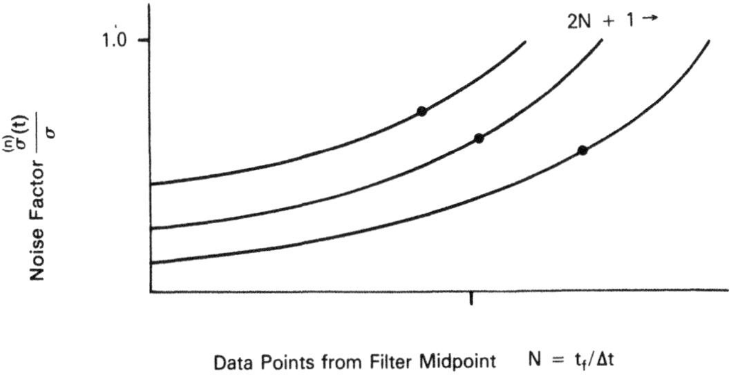

Fig. 3.5. *Trend of errors in a typical low order mid-point filter*

cannot, for example, enable a medium-accuracy system to match the accuracy of an advanced system. The noise reduction achievable can only be traded off against the uncertainties introduced by filter lag time, curve-fitting error, and the computer cycle time and complexity. Moreover, filtering is only useful in reducing random errors; it is not effective in reducing bias or low-frequency errors. Also, if the dominant noise frequencies are not well separated from the data frequencies, filtering may introduce frequency-dependent errors, whose significance should be considered in the design of the filter. In high-accuracy systems, therefore, for missions that can tolerate only small filter time lags (approximately 0.2 to 0.5 seconds), the radar and stabilization noise must be reduced considerably.

3.5. Noise Reduction

References 1, 10, 32–36

Since the tracking radars use digital computers for data processing and computation, digital filtering techniques are useful. Filter analysis presented in this section is one approach to the filtering problem. Other methods presently being implemented include feedback filters and recursive methods which operate on the last data point.

3.5.1. General Filter Equation

A commonly used digital linear filter has an output, for equally spaced data, of the form:

$$\overset{(n)}{x}(t) = \sum_{i=1}^{M} \left[\sum_{j=0}^{\nu-1} \frac{j!}{(j-n)!} w_{ji} t^{j-n} \right] d_i \tag{3.7}$$

where

d_i = input position data, $i = 1, 2, \ldots M$,

$\overset{(n)}{x}(t)$ = filter output at time t from filter midpoint.

$n = 0$ (position), $n = 1$ (velocity), . . . ,*

M = number of data points,

$\nu - 1$ = order of filter, $j = 0, 1, 2, \ldots, \nu - 1$.

*The order of the derivative of the parameter over which it appears.

Values for the coefficient enclosed between brackets in equation (3.7) can be obtained by assuming the order for the filter and computing the weights w_{ji}. The derivation of the general filter equation for position is given in the references.

The noise reduction factor is given by:

$$\left(\frac{\overset{(n)}{\sigma(t)}}{\sigma}\right)^2 = \sum_{i=1}^{M}\left[\sum_{j=0}^{\nu-1}\frac{j!}{(j-n)!}\,w_{ji}t^{j-n}\right]^2 \tag{3.8}$$

where

$\dfrac{\overset{(n)}{\sigma(t)}}{\sigma}$ = ratio of output noise to input noise.
$n = 0$ (position), $n = 1$ (velocity), . . . ,*
(Input noise is assumed equal for all measurements).

Equation (3.8) can be easily evaluated for second-, third-, and fourth-order filters. Noise reduction for such filters is shown by the dots in Fig. 3.5. These show that noise reduction occurs as more data points are taken in the sampling interval.

3.5.2. Midpoint Filter

For the midcourse or orbital portion of the trajectory, a second-order equation is quite satisfactory, since the motion is almost purely Keplerian. Filter lag time is relatively unimportant in this phase, and a midpoint filter provides good noise reduction with tolerable lag. The noise reduction equation for the second-order midpoint filter for positon (at the midpoint of the data span where $t = 0$) reduces to:

$$\frac{\sigma(0)}{\sigma} = \sqrt{\frac{3(3N^2+3N-1)}{(2N+1)(2N-1)(2N+3)}}\ \text{m/m} \tag{3.9}$$

and for velocity to:

$$\frac{\dot{\sigma}(0)}{\sigma} = \frac{1}{\Delta t\sqrt{N/3(N+1)(2N+3)}}\ \text{m/second/m} \tag{3.10}$$

*The order of the derivative of the parameter over which it appears.

where

$\frac{\sigma(0)}{\sigma}, \frac{\dot{\sigma}(0)}{\sigma}$ = ratios of output position noise and output velocity noise to input noise, respectively,

N = data points each side of midpoint of the sampling period,

Δt = data sampling time.

Noise reduction for position and velocity for midpoint smoothing as a function of the number of data points is shown in Fig. 3.6. The sampling time is assumed to be 0.1 seconds. For N large, equations (3.9) and (3.10) can be written as follows:

$$\text{For position} \quad \frac{\sigma(0)}{\sigma} = \frac{1.06}{N^{1/2}}$$

$$\text{For velocity} \quad \frac{\dot{\sigma}(0)}{\sigma} = \frac{1.21}{\Delta t N^{3/2}} \tag{3.11}$$

The noise reduction factor given by equations (3.11) is valid only at the filter midpoint, i.e., at $t = 0$. If a prediction is made

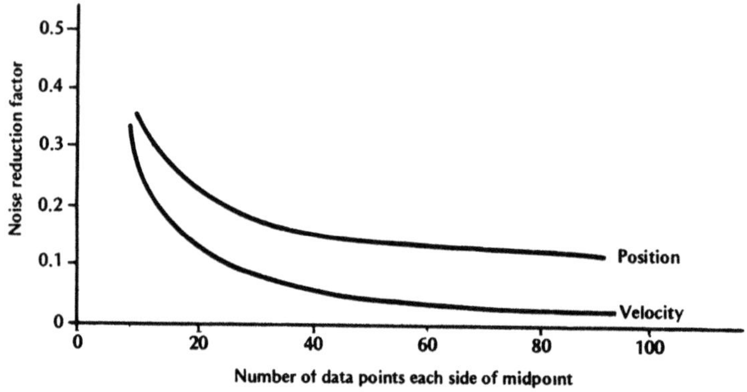

Fig. 3.6. *Output-to-input noise reduction factor plotted for a midpoint second-order filter as a function of number of data points in the sampling interval*

at some other time t the noise reduction factor for position in a second-order filter can be obtained approximately by:

$$\frac{\sigma(t)}{\sigma} = \frac{2.12}{N^{1/2}}(t/t_f)^2 \tag{3.12}$$

where

t = prediction time; $t = N\Delta t$ measured from the filter midpoint,
t_f = filter lag time; equals $\frac{1}{2}$ the tracking interval of the filter; measured from the filter midpoint,
N = the number of data points each side of the filter midpoint, and
Δt = interval between data points.

Equation (3.12) indicates that the noise reduction factor for position prediction in a second-order filter decreases as the square root of the number of data points N, each side of the filter midpoint, increases. Moreover however the noise reduction factor increases as the prediction is made into the future. Equation (3.12) holds only for the condition $t \geq t_f$.

3.5.3. Curvefitting Error

Assumption of a second-order curve is satisfactory for midcourse tracking, but may not be so for tracking during launch and reentry, where high rates of acceleration change are common. This is particularly the case during reentry, when the dynamic reaction of the vehicle with the atmosphere results in high deceleration forces. Upon entering the atmosphere, the trajectory is more accurately defined by a third- or fourth-order equation.

In general, a curvefitting error, or filter error, results from passing derivatives of higher order than those for which the filter was designed. The curvefitting error in position is bounded by:

$$e(\nu) = \left(\frac{N\Delta T}{t_c}\right)^{\nu} \cdot \frac{1}{1 - \dfrac{N\Delta t}{t_c}} \tag{3.13}$$

where

N = data points each side of midpoint of the sampling period,
Δt = data interval time,
t_c = interval of convergence,
$\nu - 1$ = order of the filter.

Equation (3.13) represents the maximum error in the prediction of missile position.

For the case of a second-order filter $\nu = 3$ so that a comparison of the curve-fitting error for position given by equation (3.13), and the noise reduction error for position given by the first of equations (3.11) indicates that the former error increases, in most applications for which $N\Delta t \ll t_c$, as the cube of $N\Delta t$ while the latter error decreases as the inverse square root of N and is independent of the data interval time Δt. The time $N\Delta t$ is sometimes referred to as the filter lag.

Equation (3.13) includes all derivatives which have been disregarded in the truncated series representing the measured parameter (range, angle, etc.). In general, errors due to derivatives two orders higher than the filter are not significant and are usually disregarded. Equation (3.13) therefore represents the curve-fitting error which is due to the derivative obtained following truncation of the series. The equation for velocity error resulting from passing derivatives of a higher order than those for which the filter was designed can be found in a similar fashion.

3.5.4. Computation Time

Another factor entering into the selection of a suitable filter is the computation, or cycle time to produce, for example, an impact point every Δt seconds where $1/\Delta t$ is the sampling rate. For trajectory prediction, six filters are required, one each for radar range, azimuth, elevation, and their derivatives. The filtering process could take up too much of the cycle time, particularly when the computer multiply time is very high.

3.6. Error Model

References 1, 10, 32–36

The error model which can be constructed for the evaluation of single-look position accuracy, smoothed position accuracy, and smoothed velocity accuracy for an overall tracking system is shown in Fig. 3.7. For the sake of completeness, error sources have been included which pertain to mobile and stabilized radar platforms. By tracing through the error model, the procedure for determining the single-look position accuracy should be clear.

Parametric error curves may be developed from the error model. These curves become useful tools for quick evaluation of the tracking errors for various missions. They represent the total subsystem errors and total combined subsystems errors as a function of slant radar range, signal-to-noise ratio, and elevation angle.

3.6.1. Error Inputs

As can be seen in Fig. 3.7, input error sources to the error model include stabilization system, target-dependent, radar-dependent, flexure, radar platform, and position and velocity errors. Each error source will next be described briefly.

3.6.1.1. Stabilization System Errors

Errors due to the stabilization system appear as vertical and heading errors having both long-term bias and short-term filterable components. The stabilization system removes the vehicle pitch, roll and yaw, from the coordinate system, in mobile systems and therefore affects tracking elevation and train errors directly. As such, the stabilization errors can be summed on an rss basis with the radar subsystem errors. Since all errors are resolved at the target elevation, the stabilization system heading error is multiplied by the cosine of the elevation angle.

3.6.1.2. Target-Dependent Errors

Target-dependent errors are caused by such phenomena as glint, scintillation, and beacon modulation and by thermal or receiver noise, and are defined as the range and angle noise. They are, generally, functions of the target range. Their magnitude depends on the target cross-sectional area for skin

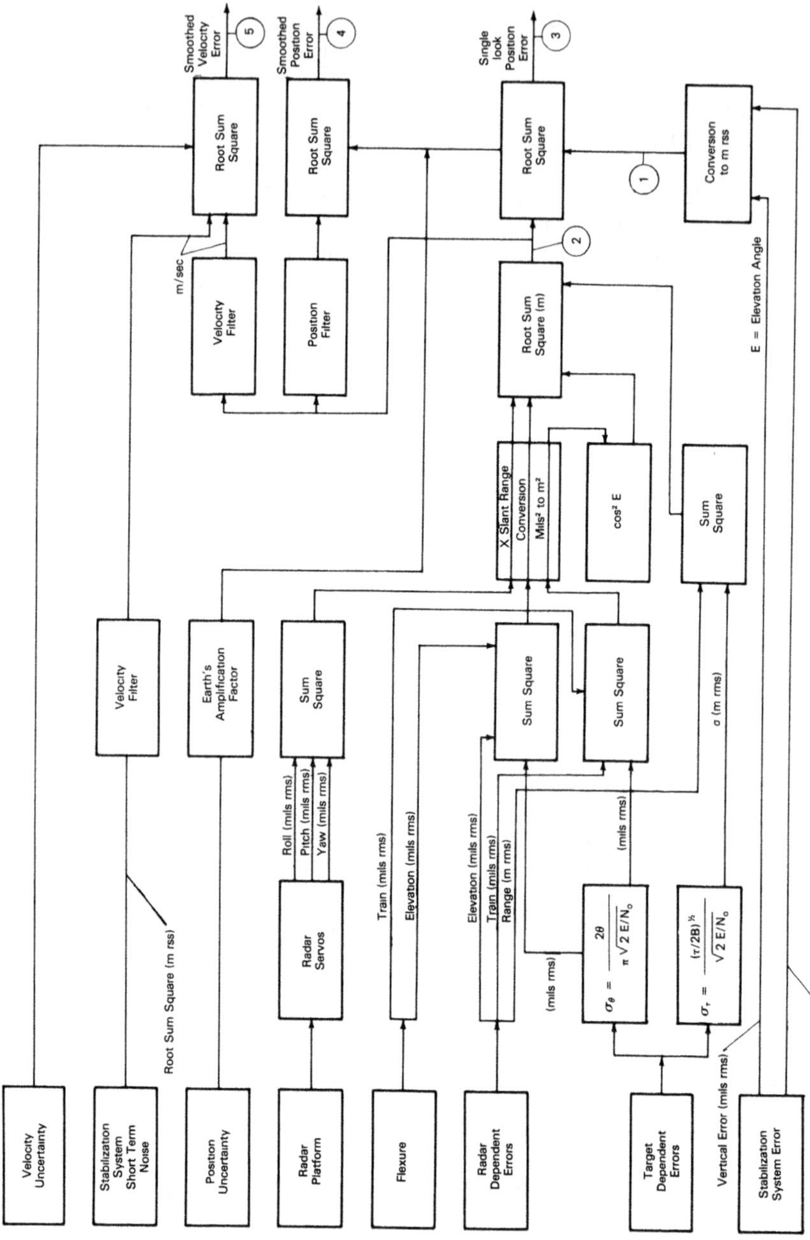

Velocity Uncertainty
Stabilization System Short Term Noise
Position Uncertainty
Radar Platform
Flexure
Radar Dependent Errors
Target Dependent Errors
Stabilization System Error
Root Sum Square (m rss)
Velocity Filter
Earth's Amplification Factor
Radar Servos
Roll (mils rms)
Pitch (mils rms)
Yaw (mils rms)
Sum Square
Train (mils rms)
Elevation (mils rms)
Elevation (mils rms)
Train (mils rms)
Range (m rms)
Sum Square
Sum Square
(mils rms)
$\sigma_\theta = \frac{2\theta}{\pi\sqrt{2\,E/N_o}}$
(mils rms)
$\sigma_\tau = \frac{(\tau/2B)^{1/2}}{\sqrt{2\,E/N_o}}$
σ (m rms)
Vertical Error (mils rms)
X Slant Range
Conversion
Mils² to m²
cos² E
Sum Square
Velocity Filter
Position Filter
Root Sum Square (m)
m/sec
Root Sum Square
Root Sum Square
Root Sum Square
Conversion to m rss
E = Elevation Angle
Smoothed Velocity Error
5
Smoothed Position Error
4
Single look Position Error
3
2
1

tracking, and the beacon power for beacon tracking. Range and angle noise due to the receiver was discussed in section 2.2.11. Range noise due to the receiver, is determined by the following equation:

$$\sigma_\tau = \frac{(\tau/2B)^{1/2}}{\sqrt{2E/N_o}} \text{ seconds} \tag{3.14}$$

where

σ_τ = range error due to thermal noise,
τ = radar pulse length or period of observation,
E/N_o = energy-to-noise ratio = $S/N \cdot B/F$,
S/N = single-look signal-to-noise power ratio,
B = noise bandwidth (receiver noise bandwidth for track-while-scan, and range-servo-noise bandwidth for tracking),
F = pulse repetition frequency.

Initially, the pulse length should be as large as possible, in order to obtain a reasonable signal-to-noise ratio. Once this is achieved, the pulse length can then be reduced to allow for greater target definition.

The range-servo-noise bandwidth, for the case of a tracking radar, is set to a minimum value to obtain maximum velocity and acceleration constants. This decreases the servo lag errors but increases the range noise error, as shown by equation (3.14). The pulse repetition frequency is adjusted to as large a value as possible in order to maintain small range and angle errors without creating target ambiguity.

Similarly, angle errors due to the receiver are determined by the standard expression:

$$\sigma_\theta = \frac{2\theta_B}{\pi\sqrt{2E/N_o}} \text{ radians} \tag{3.15}$$

where

σ_θ = angular error due to thermal noise,
θ_B = antenna beamwidth,
E/N_o = energy-to-noise ratio.

Both range and angle noise due to the receiver are high frequency random errors and so can be filtered.

It can be seen from equations (3.14) and (3.15) that the range and angular accuracy is inversely proportional to the square root of the energy-to-noise ratio E/N_o. Since the energy-to-noise is inversely proportional to the fourth power of the range, both the range and angular errors due to the receiver, are proportional to the square of the target range. In contrast, target glint is inversely proportional to the target range. Thus, glint predominates at short range, while receiver noise is the main concern at long range, when tracking targets. Only the receiver noise is represented in Fig. 3.7.

3.6.1.3. Radar-Dependent Errors

Radar-dependent range and angle noise errors are associated with the electromechanical properties of the radar itself. They include refraction-compensation errors, alignment errors, transmission errors, and tracking-servo lag errors due to target dynamics. (The last error is peculiar to tracking radar, should such a subsystem be included.) In general, radar-dependent errors are not filterable over the sampling period, although in advanced systems some of the errors (tracking-servo lags, flexure, etc.) can be either measured or computed and thus compensated for in the digital computer.

3.6.1.4. Flexure Errors

Flexure results from the twisting and bending of structures, causing angular errors between the inertial reference coordinate system and the radar coordinate system. Its effects can be minimized by an optical monitoring system or by close mounting of the inertial reference and antenna. With an optical monitor, the angular tracking errors can be measured and used to compensate the tracking data.

3.6.1.5. Platform Errors

For mobile radars, errors in the radar stabilization loop are caused by servo dynamic lag in following vehicle pitch, roll, and yaw. The lag errors are functions of the vehicle motion amplitudes related to the train and elevation axes, and are determined from the servo characteristics and the estimate of the vehicle platform motion amplitudes and frequencies.

3.6.1.6. Position and Volocity Uncertainty

An error in position is a translation bias at the target as discussed previously. However, for slant ranges under 900 kilometers the amplification of position error due to earth curvature is ignored; the error is therefore treated as an rms vector and is sum squared with the other subsystem errors. Vehicle velocity uncertainty is a bias error and is therefore summed on an rss basis.

3.6.2. Parametric Error Curves

If we sum the rss errors at points 1, 2, and 3 of the error model in Fig. 3.7, we obtain rss errors for the (1) inertial stabilization subsystem (for mobile radars), (2) radar subsystem, and (3) combined error, respectively, as a function of slant range, with the signal-to-noise ratio and target elevation as a parameter. Figure 3.8 gives radar subsystem errors for a medium-range radar that would be obtained at point 2 in the error model. These curves show that the combined error for a target at a given range decreases with increased signal-to-noise ratio and target elevation. Similar curves can be plotted for points 1 and 3. These curves make possible a quick estimate of system tracking error and equipment configuration, once a tracking mission has been defined. In addition, different radar and inertial systems can be compared directly in terms of the resultant vector error. Combining the errors at point 3 provide the single-look position accuracy in a mobile position, inclusive of position uncertainty. Smoothed position and velocity can be obtained at points 4 and 5 using conventional filtering techniques.

The general expression for position error (PE) accuracy* is given as:

*In this chapter, the terms error and accuracy are used interchangeably. Both assume the meaning of the standard deviation in statistics.

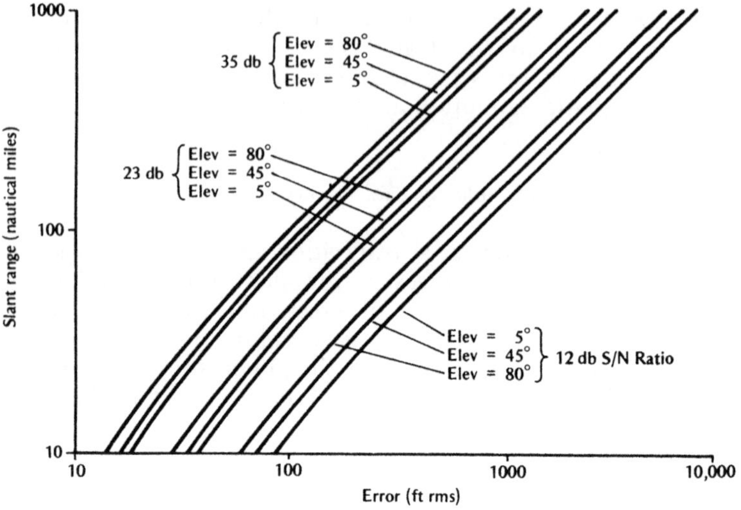

SOURCE: See reference 1.

Fig. 3.8. *Radar subsystem error, medium-range radar, error depends on the S/N ratio and the target elevation angle*

$$\sigma_p^2(0) = [(\text{radar noise})^2 + (\text{stabilization noise})^2]\left(\frac{\sigma_0}{\sigma}\right)^2 + [(\text{radar bias})^2 + (\text{stabilization bias})^2] \quad (3.16)$$

where σ_0/σ is the position noise reduction factor previously discussed. The stabilization terms apply only to mobile systems. It can be argued that if a given radar is made mobile, the result will be a less accurate system, since by equation (3.16) the addition of the stabilization noise and bias terms can be expected to increase position error.

The general expression for velocity error in a mobile system is given as:

$$\sigma_v^2(0) = (\text{radar noise})^2 + (\text{stabilization noise})^2$$
$$= [\sigma_{\dot{R}}^2 + R^2(\sigma_{\dot{E}}^2 + \sigma_{\dot{A}}^2 \cos^2 E)]\left(\frac{\dot{\sigma}_0}{\sigma}\right)^2 + [\sigma_{\dot{p}}^2 + \sigma_{\dot{r}}^2 + \sigma_{\dot{h}}^2 \cos^2 E]\left(\frac{\dot{\sigma}_0}{\sigma}\right)^2 \quad (3.17)$$

where the radar terms have been defined in connection with equation (3.6) and

$\dot{\sigma}(0)_r/\sigma$ = velocity noise reduction factor,
σ_p = pitch noise,
σ_r = roll noise,
σ_h = heading or yaw noise.

From equations (3.16) and (3.17) it is seen that radar noise and stabilization noise can be filtered to reduce position error and velocity error; however, the position error can at best be reduced to the level of the bias.

The error model in Fig. 3.7, which shows the combined error flow in vector rss form, provides a simple and direct method for analyzing the relative importance of subsystem errors in terms of total vector error. As subsystems become more-accurate and error monitoring equipment becomes more sophisticated, it may be possible to ignore some errors. Certain other errors that are now negligible may become predominant. Thus, as we become able to control random noise, bias errors may become increasingly significant.

3.7. Design Procedure

References 1–10, 28–37

In a general sense, radars are only the sensors for the overall defense system. The system detects, identifies, monitors, locates, tracks and destroys targets. Its subsystems include radars, communication links, data processing equipment, interceptors, command control functions, and their operators, i.e., both engineering and non-engineering functions.

The functions and performance requirements of the key subsystems will now be delineated. A word of caution at this point: the overall system itself must be eventually considered with regard to its response, accuracy, cost, data-handling capacity, rate, etc. It then becomes the responsibility of the major subsystem engineering designers to select or develop equipment to meet the performance requirements.

3.7.1. Mission Objectives and Radar Functions

One of the principal problems of the systems engineer involved in the preliminary stage of a military system development is the selection of the optimum design from a number of proposed system designs. Early in the development of any military system there will be a number of suggested approaches, each possessing different technical performance characteristics and costs. The value of a system for a given application will depend on the relative importance of such performance criteria as the system's ability to detect, identify, monitor, locate, track, and destroy targets as well as upon capacity, range, effectiveness, reliability, and cost considerations.

In the past, the assessment and comparison of proposed designs was often made by the systems engineer on the basis of his own judgment and experience. Thus, design considerations and specific radar parameters were generated by the systems engineer from the known performance requirements for the system. A general format for performing this operation is given in Table 3.1.

For that class of systems where broad questions of strategy are not a factor, and where the required performance in all pertinent categories can be stated explicitly, selection of a design can be made on a simple objective basis. The lowest cost system that provides the desired performance is selected. Here, cost is considered in the broad sense to include procurement, training, maintenance, and similar factors. In many military systems the value of particular performance characteristics is difficult to assess because of their strategic implications. It is with this latter class of problem that most systems analysis is concerned.

3.7.2. Design Selection

This section outlines a simple mathematical formulation that includes all the factors relevant for system selection. This formulation also illustrates the change in value of any system as the weightings of the different performance characteristics are varied.

Value is defined as the sum of all the measurable performance and cost parameters that apply to the system, each parameter having been normalized and weighted in accordance with its importance to the particular application. Such a definition of value serves as a numerical yardstick to aid in system valuation. Identifying, evaluating, and mathematically relating these parameters are the steps necessary for objective

Table 3.1. *Relation of system requirements to radar parameters*

SYSTEM REQUIREMENTS	DESIGN CONSIDERATION	RADAR PARAMETER
Performance Surveillance Detection Identification Monitoring	1. Capacity—Number and type of targets 2. Target density—resolution. 3. Target dynamics. 4. Environment and terrain. 5. Reaction time. 6. Reliability. 7. Mean repair time. 8. Adverse environment operation a. weather b. clutter c. jamming d. chaff e. nuclear	1. Volumetric coverage—maximum and minimum range, elevation and azimuth angular coverage. 2. RCS of targets. 3. Waveform for detection, velocity and range, adverse environments. 4. Energy—based on probability of detection and false alarm rate for target fluctuation model. 5. Pulse repetition rate for above range and MTI requirements. 6. Data rate. 7. Azimuth and elevation beamwidths for resolution, data rate, and power-aperture requirements. 8. Transmission frequency. 9. Peak and average power. 10. Redundancy. 11. Automatic maintenance and repair.
Acquisition Locate Track Destroy	9. Same as 1–8 plus measurement accuracy capability. 10. Acquisition logic (order of selection of targets to be tracked). 11. Track monitoring logic (method of determining if target is still in track). 12. Reacquisition logic (method of reacquiring targets that have dropped out of track).	12. Same as 1. 13. Same as 2. 14. Same as 3 plus accuracy. 15. Same as 4 plus accuracy. 16. Sampling rate and smoothing time for each tracked target. 17. Same as 7 plus accuracy. 18. Same as 8. 19. Same as 9. 20. Same as 10. 21. Same as 11.
Cost R&D Procurement Training Maintenance	13. Cost-effectiveness.	22. Same as 12–21 plus cost.
Lead time	14. Risk assessment.	23. Same as 22 plus risk.

system evaluation. A careful statement of mission objectives indicates the relative importance of particular performance factors. These factors can be normalized using appropriate mathematical functions or graphical techniques to provide normalized performance coefficients. The sum of the weighted products of the pertinent normalized coefficients is then the numerical value for the value of the system.

An equation for system value can be written in the following form:

$$V = \sum_i w_i p_i \tag{3.18}$$

where

V = system value,
w_i = weighting,
p_i = normalized performance or cost parameter.

The performance parameters p_i will be normalized so that:

$$0 \leq p_i \leq 1.$$

The weightings will be assigned so that:

$$\sum_i w_i = 1$$

Typical performance and cost factors of interest might include: maximum search and tracking range, resolution, accuracy, size, weight, target capacity, reliability, mean repair time, procurement cost, training cost, maintenance cost, etc. These parameters must be normalized to their expected values for use in the value equation. The determination of the normalization functions, i.e., the parameter variation about its expectation, is based on an interpretation of the original statement of the system performance requirement. Table 3.2 gives a detailed list of the principle performance and cost parameters of an ABM system. These are grouped in the four areas of system availability, penetration hardness, effectiveness and cost. Each class in turn is broken down further and includes many submembers.

Before proceeding, it will be well to consider the limitations of this and any other scheme for the comparative evaluation of

Table 3.2. *List of four principle ABM system factors*

p_{1j} = AVAILABILITY	p_{3j} = EFFECTIVENESS
Hardware Operational Assurance	Size
Reliability	Range of Operation
Maintainability	Resolution
Flexible C&C Network	Accuracy
Cable	Target Capacity
Ground	
Airborne	p_{4j} = COSTS
Satellite	Research
Survivability	Development
Blast Hardening	Design
Active Defense	Fabrication
Deceptive Basing	Testing
Mobile Systems	Installation
	Training
p_{2j} = PENETRATION HARDNESS	Operation
Targeting Flexibility	Maintenance
Counter-Countermeasures	

complex systems. It must be recognized that no approach will unerringly select the optimum system unless a very specific set of data is available. This data would have to cover the performance of the system and the weights to be assigned to different aspects of performance and costs. Clearly, if such reliable data is available, the system has already been designed and selection of the optimum system is a simple matter.

Generally, there are many uncertainties, particularly in the value to be assigned to aspects of performance and the assessment of strategic performance which influences the price one is willing to pay for this performance. These uncertainties cannot be removed by the manipulation of variables in equations. Selection of a particular evaluation technique will depend on the nature of the problem and the most convenient method of assessing the uncertain data. When data is in probabilistic form a probability approach is indicated.

In the foregoing approach, equation (3.18), the weightings will usually incorporate the major subjective judgments. It is in these weightings that the tradeoffs between the desired performance and the willingness to pay for it are incorporated.

Using equation (3.18), a numerical value can be computed for each proposed system design. The variation in the value of each system can then be made as the weights (subjective judgments) are varied. This procedure can be repeated in the early phases of design selection, using revised inputs as these become available from systems analysis, which is primarily concerned with determining system performance parameters and their expectations. In this manner, a rational selection of the military system which is to be built can be made.

3.7.3. System-Design Analysis

A generalized functional block diagram of an ABM radar was shown in Fig. 2.11. The main subsystems shown were a passive parallel-fed array using phase shifters, a beam-steering computer, transmitter, receiver, system computer, display and control console and input/output interface. The antenna shown generates a single steerable beam.

A system design analysis is required to establish the integrity of the design. The design problem is not usually associated with the design of individual units but rather with the interfacing of the large number of individual units into a system whole which is amenable to being controlled by the automated computer program.

The ABM radar system design consists of two main components: the facility and the prime electronic equipment. The facility consists of four critical engineering components: the structure, prime power, air conditioning and fire protection. The prime electronic equipment consists of nine critical engineering components: transmit array, receive array, beam steering, system computer, radar interface and control equipment, signal processor, monitor and controls, target identification and the computer program. These components will now be described in terms of a PAR-type radar.

3.7.3.1. Facility

The facility includes the building structure, prime power, air conditioning and fire-fighting equipment.

The steel-reinforced heavy cement building is designed to withstand to a great extent the nuclear environment in all but a direct hit and provides the rigidity required for the antenna face. The building itself resembles a truncated pyramid and is about 100 m long by about 50 m high. The depth of the building below the surface approximately equals its height. Antennas can be placed on each side of the building to provide a full 360° surveillance coverage. Fewer antennas provide lesser coverage, of course. The building provides adequate room for the radar equipment, service and shop spaces, and personnel space.

Prime power is provided by a commercial power transmission line or by a 10 Mwatt gas turbine in case of commercial power failure. The distribution system provides power to the radar equipment, lights, service bays and air conditioning.

Air conditioning is used throughout the building to maintain the temperature to within several degrees of room temperature at 75°. Equipment and building air is precooled for exhausting at room temperature to insure the absence of an infrared signature by the site. Several compressor units provide the refrigeration to operate the system. Appropriate isolation features are installed to prevent the internal building's contamination by nuclear radiation.

The fire protection system is extensive and automatic. It includes compartmentalization of the building structure, a smoke-removal system, a CO_2 system and a water system. Hose reels are located throughout the building. A fixed automatic flooding system is located in the transmitter face, the receiver face, and under the computer floor, and is supplied by a 15 ton CO_2 system from a central storage tank. In addition, water storage and high-rate water wells are available. High-rate water pumps supply water to the face deluge system and to sprinklers located throughout the building.

3.7.3.2. Prime Electronic Equipment

The prime electronic equipment includes the transmit array, receive array, beam steering, system computer, radar interface and control equipment, signal processor, monitor and controls, target identification and the computer program.

The transmit array consists of the transmitter antenna panel assemblies, the transmitter modules, the cabling and distribution system, the master-oscillator distribution group, the power modulation group, the transmitter-array calibration group, the beam-steering distribution group, and the modula-

Table 3.3. *Transmit array characteristics*

ARRAY AREA	Square 72×72 module grid
NUMBER OF MODULES	5184
BEAMWIDTH AT BORESIGHT	1.4°
PEAK POWER OUTPUT/MODULE	10 kw
ILLUMINATION	Uniform
PULSE LENGTHS	1, 5, 10, 25, 125, and 250 microseconds in various burst waveforms
CENTER FREQUENCY	442 MHz
BANDWIDTH	10 MHz

SOURCE: See reference 34.

tor control console. Table 3.3 gives typical transmit array characteristics of a test system.
The transmit antenna element is a shortened resonant dipole with a parasitic element director. The dipoles are made short by increasing the diameter and shaping the surface of the dipole arms. The resulting pear-shaped arms reduce the mutual coupling and allow placement on 0.55 centers. Each transmitter module is about 18 by 21 by 80 cm and weighs 23 kg.

The receive array consists of the antenna panel assemblies, the receiver modules, the calibration beam-steering and local oscillator, distribution assemblies, the calibration equipment and the beam-forming network. Table 3.4 gives typical receive array characteristics of a test system.

Table 3.4. *Receive array characteristics*

ARRAY AREA	152 elements in diameter
NUMBER OF RECEIVER MODULES	19,500 crossed dipoles 4,660 active dipoles
BEAMWIDTH AT BORESIGHT	Cluster of 9 (3×3) squinted beams on 0.4° centers with 1 db crossover between beams

SOURCE: See reference 34.

Beam steering is accomplished by phase shifting a single frequency at power levels which can be controlled by diodes driven by digital logic. Power amplification is done after the control. The same phase-shifter outputs are employed in both the transmit and receive arrays. The only difference results from the fact that only seventy-two of the 152 signals are required for transmit and from the fact that the signal must be amplified to the higher level for transmit.

The system computer consists of multiple IBM system 360/65I central processors. Each processor has access to all peripheral equipment consisting of nine disk drives, twelve tape units, four displays, a printer and a card reader/punch, data adapters to interface with the radar interface and control equipment and the external data link. The central processors communicate with each other such that upon failure of the on-line system the standby unit takes over. The Model 360/65I has 131,072 32-bit words of core memory and has a basic add time of 14 microseconds.

The radar interface and control equipment consists of a computer buffer, memory, the radar controller, the radar data buffer and peripheral equipment and is used to provide the interface between the computer and the radar equipment.

The signal processor consists of analog and digital equipment necessary to convert the IF signals from each of the nine receive beams into digital information suitable for use by the radar computer. This equipment also includes the system frequency generator which provides the stable frequencies for the transmitter, receiver, beam-steering and signal-processing channels and also includes the signal simulator used for calibration and closed-loop testing of the radar. Table 3.5 provides a summary of waveforms employed by the radar and gives a brief indication of their functions. Three separate processors are employed for processing the various waveforms: search, track and coherent.

The system monitor and control is functionally divided into the areas of operations and maintenance and serves the purpose of monitoring and controlling the transmit and receive arrays.

The target identification consists of analog and digital equipment necessary to obtain the target signature and its comparison with signatures of known objects. This equipment is necessarily multichannel to permit the identification of a large number of simultaneous targets.

Table 3.5. *Summary of waveforms employed by ABM radar*

DESIGNATION	DESCRIPTION	PRIMARY FUNCTION
SEARCH CHIRP PULSE	250 μsec FM	Long-range search and track acquisition
SEARCH SIMPLE PULSE	10 μsec	Short-range search and track acquisition
TRACK CHIRP PULSE	250 μsec FM	Long-range track
TRACK SIMPLE PULSE	1 μsec	Short-range track and short-range SLBM search
COHERENT EXTENDED RANGE TRACK	40 pulse burst of 125 μsec pulses; burst duration 1 sec	Extended-range track and signature
COHERENT INTERMEDIATE DOPPLER	40 pulse burst of 25 μsec pulses; burst duration 0.2 sec	Intermediate-range doppler and signature
COHERENT PRECISION RANGE	40 pulse burst of 5 μsec pulses, stepped frequency; burst duration 1.2 milliseconds	Precision-range tracking
ARRAY CALIBRATION WAVEFORM	60 μsec	Transmitter and Receiver Module calibration

SOURCE: See reference 34.

The computer program consists of tactical routines plus radar testing, maintenance, diagnostic and operating system routines. The various programs are assigned to computers both on-line and off-line. The multiprogramming structure permits external interrupts or timing stimuli to interrupt whatever processing is in progress. The interrupt is examined for its operational priority and if it is found to be of higher priority than the task currently in progress, control of the central processor is preempted. This sequence may occur successively at many priority levels resulting in situations where many tasks are simultaneously in various stages of completion.

3.7.3.3. *Design Features*

A principle feature of the design of ABM radars is the redundancy of equipment in order to achieve high availability. Also, similar equipments used to provide backup upon exhaustion of redundant equipment.

Another important feature in the design is the automatic maintenance and calibration of individual equipments followed by overall system performance checks. Overall system checks are also employed during normal operation while independent maintenance and calibration are allocated time by the computer program on a time-available basis to make detailed checks. All timing pulses in the system and all frequencies are derived from common sources, thus simplifying the task of maintaining synchronization throughout the system.

Other features in the system are the system modularity to permit ready exchange of parts and spares in maintenance or upgrading tasks, counter-countermeasures equipment to permit system operation in jamming environments, sequential detection, dual-receive polarization, optimized power allocation based on target range and cross-section and growth flexibility.

3.7.4. Performance Analysis

Suppose we are required to search the hemisphere out to 2,000 km for 0.1 m^2 targets, completing the search in 50 seconds. Suppose further that the radar should operate at 0.5 GHz (UHF-band) and provide a nominal 0.5° angular and 150 m (1 microsecond) range resolutions against a maximum of

25 simultaneous targets, tracking each target at ranges less than 1,500 km and completing the track at 500 km. In this example, the total time available is computed using equation (3.1) and is found to be 150 seconds or 50 seconds for search and 100 seconds for tracking. We will now briefly discuss the search, tracking, resolution, accuracy, smoothing, target capacity and countermeasures environment performance of the radar in this example.

3.7.4.1. Search

Clearly the dominant requirement for the radar operation is the complete surveillance of the maximum search range at the sampling interval of once every 50 seconds, i.e., the quadrant must be patrolled at the rate of once per 50 seconds.

The total number of beam positions can be computed approximately by:

$$b = \frac{\Omega}{3\omega} = 7{,}500 \tag{3.19}$$

where

b = number of beam positions in quadrant,
Ω = solid angle of quadrant, and
ω = solid angle of beam.

The factor 1/3 which appears in equation (3.19) is due to the fact that the beam broadens when scanned away from its boresight position normal to the antenna face. Thus somewhat fewer beam positions are needed than would be the case for a constant beamwidth. In fact, the beam varies from its minimum value of about 0.5° at boresight to about 0.6° at the horizon and about 1° at zenith in the vertical plane and to about 0.7° at the limits of its horizontal scan in the horizontal plane.

As a consequence of the number of beam positions indicated by equation (3.19) and the required surveillance time, a single beam can dwell for only $\frac{50}{7{,}500}$ or about 6.7 milliseconds in a given beam position. A portion of this dwell-time is used for switching the beam, the remaining time being available for the transmission and reception of pulses. Thus, if the beam switching time is 0.1 milliseconds, then only 6.6 milliseconds is available for transmission and reception. During this time echoes will appear within about 1,000 km. It appears therefore that twice the available dwell-time would be needed. Consider now the case where the beam transmits in one beam position and then goes back to the previous beam position to receive. For example, if the beam transmits for 0.6 milliseconds in one beam position and then goes back to the previous beam position to complete 6 milliseconds of dwell for receiving echoes it will receive these from the range between 1,000–2,000 km. Clearly, some such program for the beam will be needed to obtain the correct echo range. In this simple example the pulse width is 0.6 milliseconds and the interpulse period is 6 milliseconds. In some such manner, all 7,500 beam positions will be scanned in about 50 seconds, the required surveillance time. Time sharing the single beam for both transmit and receive functions and beam reversing avoids the problem of implementing parallel beams. This latter approach would considerably increase the number of antenna elements and the related equipment for beam steering and controlling the beams. Beam paralleling is needed, however, for receiving echoes simultaneously and thereby for reducing the radar response search and tracking times.

The average transmitted power in a quadrant can be obtained using the following equation:

$$P_T = \frac{(4\pi)^3 kT\, R^4 (E/N_o) L}{G^2 \lambda^2 \sigma \tau} \tag{3.20}$$

The following assumptions will be used to compute the value of equation (3.20):

$(4\pi)^3 = 2000$,
kT = the product of the Boltzman's constant and the absolute antenna temperature; $kT = 4 \times 10^{-21}$ watts/Hz,
R = Surveillance range; $R = 2 \times 10^6$ m,
E/N_o = Signal energy-to-noise power spectral density ratio; For reliable detection at maximum range $E/N_o = 100$; 20 dB
L = Miscellaneous losses; $L = 10$ dB,
g = Antenna gain; $G = 4\pi/\omega$; $G = 1.66 \times 10^5$; 52.2 dB,
λ = wavelength; $\lambda = 0.7$ m,
σ = radar cross-section; $\sigma = 0.1$ m^2,
τ = interpulse period; $\tau = 6.7$ milliseconds.

Using these assumptions, the average transmitted power per quadrant is found to be 14.15 kw or about 56.6 kw for all four quadrants. The peak power is found using the following equation:

$$P_{\text{peak}} = \frac{P_T \times \tau}{PW} \tag{3.21}$$

where

P_T = average power; 14.15 kw,
τ = interpulse period; 6.7 milliseconds,
PW = pulsewidth; 0.6 milliseconds.

Thus, the peak power per quadrant is 158 kw. The peak power per quadrant increases to 158 Mwatts if the pulsewidth is reduced to 0.6 microseconds. It is interesting to note that the power requirement in this example is quite modest and can easily be handled by a conventional reflector antenna. However, the array can provide the high beam switching rates which cannot be obtained using the reflector.

In the search operation, the basic switching rate would be about 150 Hz. At this rate the beam dwells in a given beam position for a total of about 0.6 milliseconds for transmitting and then switches to the previous beam position where it dwells for a total of about 6 milliseconds for receiving pulses. The beam follows this transmission-reception cycle and eventually returns to a beam position after a delay of about 50 seconds. In this manner the quadrant can be searched in 50 seconds with pulses having a maximum of 0.6 milliseconds in length and at the peak power level of 158 kw or at 10^3 times this level if 10^{-3} times shorter pulses are generated. In any case the power is spread over the several thousand or more radiating elements in the transmitting array so that each element can be expected to handle only a modest small amount of power. The radar would continue its search operation scanning the quadrant at 50 second intervals and, upon detection of a target, would allocate some or all of its power to the tracking function.

3.7.4.2. Tracking

Since the total tracking time is $T = 100$ seconds and since the maximum number of targets that must be tracked is $n = 25$, the available time for tracking a target is $\tau_d = 100/25 = 4$ seconds. In a straight tracking mode the radar allocates all its power to the tracking function and applies 4 seconds of continuous dwell-time to the target and then moves on to the next target. In a track-while-scan mode the radar allocates a portion of its power to the tracking function and, for example, might scan the target about 90 times—the ratio of the available track time allocated to the basic dwell period of 6.7 milliseconds. Overall, the target is illuminated for less time in the track-while-scan mode than in the straight tracking mode by virtue of the dwell-time loss. Since the lowest PRF is 150 Hz a minimum of 150 Hz × 4 seconds—600 hits will be made on the target over a relatively short span of its trajectory using the straight tracking mode. This compares with the 90 hits over a relatively long span of trajectory using the track-while-scan mode. Generally, the track-while-scan mode implies no urgency on the part of the radar.

The tracking logic must now be decided; specifically, acquisition logic, track monitoring logic, and reacquisition logic.

Acquisition logic is the order of selection of targets for tracking. This may be on a first-come, first-served basis, or it may be based on some other criterion such as speed of approach. Track monitoring logic is the method of determining if a target is still in track. In a straight tracking mode, where the radar follows a single target exclusively and continuously for four seconds, track monitoring may be obtained using the received signal itself as an indicator. In a track-while-scan mode, where the radar continuously scans and performs its normal surveillance operation and provides target data to the computer which then develops the target track, track monitoring must be a computer routine.

Reacquisition logic is the method of reacquiring targets that have dropped out of track. In a straight tracking mode, the computer commands the radar into a limited search pattern in the expected target direction. In a track-while-scan mode the radar continues its basic search rhythm and the computer correlates successive target positions.

Depending on the tracking logic adopted, the sampling rate and smoothing time is determined. In a straight tracking mode the sampling rate is at the system PRF. In a track-while-scan mode the sampling rate is at the system scan rate, with one or more hits obtained in a single sample. The latter rate is a much slower rate and generally results in fewer measurements in a given time interval. The straight tracking mode is therefore a fast mode and herein lies the tradeoff between the two modes—speed of obtaining data versus the facility of handling many targets in a track-while-scan mode. The straight tracking mode involves the incorporation of a tracking servo loop within the radar. It is generally considered when the available time to make measurements runs out, as would be the case in a terminal engagement when only seconds separate the incoming swarm of radar targets from impact.

3.7.4.3. Resolution

The radar is required to have resolution of 0.5° in angle and 150 m in range. The angular resolution is obtained using 0.5° beams. At 1,000 kilometers, targets must be 8.7 km apart in order to be resolved in angle. This separation decreases linearly with range. Since an array has been selected, the range resolution may be obtained in the straightforward method by using a one microsecond equivalent pulse. For example, the use of a 0.6 millisecond pulse would require a pulse compression type waveform with a compression ratio of 600:1. Since no velocity resolution is required, the waveform can be incoherent, consisting of one microsecond equivalent pulses at the minimum pulse repetition frequency of 150 Hz. This selection of waveform would require only noncoherent processing.

3.7.4.4. Accuracy

Accuracy is the total deviation between the actual target location in space and that location which is reported by the radar processed output. Accuracy includes bias errors such as geodetic uncertainties and local surveys, and atmospheric propagation errors.

Precision, on the other hand, represents the repeatable and measurable radar error variance and unknown bias errors. It is primarily a measure of circuit stability, thermal noise, and instrumentation errors. Instrumentation errors include antenna pointing (electrical and mechanical), range and angular signal processing, data conversion, and analogue-to-digital quantization errors encountered between the antenna, digital computer, and beam steering unit that cannot be compensated or removed with bias corrections.

The major sources of error are itemized in Table 3.6. Detailed discussions on these error contributions can be found in the sections referenced in the table.

It is apparent from the discussion of the error model in section 3.6 that errors are obtained independently and then combined. Since errors which are most significant are waveform and signal-to-noise ratio dependent (discussed in section 2.2.11), it is desirable to provide a single waveform for all operating modes (search, acquisition, track) unless the terms of power consumption and performance degradation are excessive. Under these latter circumstances separate waveforms, sampling rates, and beamwidths are employed in the search, acquisition, and track functions. Usually, all the above-mentioned procedures are performed several times before the final operational radar parameters are determined.

Equations (2.37) may be employed to compute the single-look accuracy for range, velocity, and angular measurements. In our example radar problem, for the one microsecond equivalent pulse system:

$$\tau = 10^{-6}\ \text{seconds; } 150\ \text{m},$$

$$B = \frac{1}{\tau} = 10^{6}\ \text{hertz},$$

$$\frac{E}{N_o} = 100.$$

The single-look measurement accuracy for the equivalent pulse system is therefore:

$$\begin{aligned} \sigma_\tau &= 7.5\ \text{meters}, \\ \sigma_\nu &= 1{,}950\ \text{meters/second,*} \\ \sigma_\mu &= 0.0004\ \text{degrees}. \end{aligned} \tag{3.22}$$

The noncoherent system cannot determine velocity from a single look. However, the velocity can be determined from two observations separated by time T by taking the rms value of

*This measurement accuracy applies only if the system is made coherent.

Table 3.6. *Major sources of error in a radar system*

ERROR	TYPE	REFERENCE SECTION
TARGET DEPENDENT	RANDOM	3.6.1
RANGE NOISE	RANDOM	2.2.11
VELOCITY NOISE	RANDOM	
COHERENT SIGNAL		2.2.11
INCOHERENT SIGNAL		3.7.4.4
ANGLE NOISE	RANDOM	2.2.11
RADAR DEPENDENT	RANDOM/SYSTEMATIC	3.6.1
REFRACTION	SYSTEMATIC	Appendix
ALIGNMENT	SYSTEMATIC	
TRANSMISSION	SYSTEMATIC	
TRACKING SERVO LAG	SYSTEMATIC	
ILLUMINATION	RANDOM/SYSTEMATIC	
QUANTIZING	COMPUTER	
INSTRUMENTATION	COMPUTER	
CONVERSION	COMPUTER	
TRUNCATION (CURVEFITTING)	RANDOM/SYSTEMATIC	3.5.3
MULTIPATH		

FLEXURE	RANDOM/SYSTEMATIC	3.6.1
ANTENNA POINTING		
TEMPERATURE		
WIND		
SOLAR		
GRAVITY		
MANUFACTURING TOLERANCES		
RADAR PLATFORM	SYSTEMATIC	3.6.1
SERVO DYNAMIC LAG		
STABILIZATION SYSTEM	RANDOM	3.6
POSITION AND VELOCITY UNCERTAINTY	SYSTEMATIC	3.6.1

the two single-look accuracies. The velocity measurement accuracy for the noncoherent system is therefore given by:

$$\sigma_\nu = \frac{\sqrt{2}}{T}\sigma_\tau \tag{3.23}$$

It can be seen from equation (3.23) that an improvement in the velocity measurement accuracy can be obtained by increasing the time T between observations.

In the straight tracking mode T can be as small as 1/150 second (the time between successive pulses) and as great as 2 seconds (the total tracking time), giving measurement accuracies of 1,570 meters per second and 5.25 meters per second.

In the track-while-scan mode T can be as small as 1 second (the time between successive scans) and as great as 50 seconds (the total tracking time), giving measurement accuracies of 10.5 meters per second and 0.21 meters per second.

In both cases, the velocity accuracy improves in time from the initiation of tracking. It can be seen that the noncoherent system will provide a velocity measurement accuracy of 5.25 meters per second or 0.21 meters per second, depending on the tracking mode, by the time the target is at 490 kilometers range. Additional velocity accuracy at earlier time may be obtained by employing a coherent radar system or by smoothing the data.

For a coherent system which requires a velocity resolution of $\nu = 100$ meters per second, the pulsewidth must increase in accordance with the second of equations (2.35), to provide the needed velocity resolution. The bandwidth must remain fixed to assure the needed range resolution of 150 m. Thus, for the coherent system:

$$\tau = \frac{\lambda}{2}\frac{\sqrt{3}}{\pi\nu} = 275 \times 10^{-6}\text{ seconds},$$

$$B = 10^6\text{ hertz},$$

$$\frac{E}{N_o} = 100.$$

The single-look measurement accuracy for the coherent system is therefore:

$$\sigma_\tau = 130 \text{ meters},$$
$$\sigma_\nu = 7.3 \text{ meters/second}, \quad (3.24)$$
$$\sigma_\mu = 0.0004 \text{ degrees}.$$

If, therefore, velocity resolution is required, the waveform must be coherent and consist of 275 microsecond pulses at the pulse repetition frequency of 150 hertz and most likely of the pulse-compression variety with a compression ratio of 275/1, thus providing both range and doppler capability.

These results show that the velocity obtained over time T in the noncoherent system can be obtained in a single look in the coherent system. It is, however, obtained at the expense of degrading the single-look range accuracy and, of course, by using a more sophisticated type of waveform.

Coherent radar operation may be classified in a manner similar to the straight track mode of operation, in that both provide data at the faster rate. The implementation of either or both operations in a given application will of course increase the sophistication, complexity, and cost of the radar system and should be considered when time is at a premium.

Whether noncoherent or coherent, the radar data can be smoothed to reduce the noise. Smoothing is obtained, however, at the expense of computation complexity, cost, and time.

Increasing the complexity and cost of the radar reduces the time for obtaining accurate data measurements. The same accuracy can be obtained at the expense of time by either reducing the radar complexity or by assuming more complexity in the computer for smoothing the data. Given a needed level of final measurement accuracy to be obtained in a given time, the designer must be able to evaluate the design, effectiveness, and cost of a variety of system designs which can meet the measurement accuracy criteria.

3.7.4.5. Smoothing

Smoothing is employed to increase the accuracy of predicting the missile's present or future position, velocity, acceleration, etc. Two basic means are used to increase the accuracy of prediction: higher order filters and more data points. Both means however are accomplished at the expense of more computer capacity and data processing. For the example at hand, a second-order midpoint filter will be assumed. In the straight tracking mode a total of 600 data points become available in the 4-second tracking interval, at the rate of 150 data points per second. In the track-while-scan mode a total of 20 data points become available in the 10-second data interval, at the rate of 1 data point every 5 seconds.

The noise reduction factor for position in a second order midpoint filter can be obtained using equations (3.11). Thus for $N = 10, 300$:

$$\begin{aligned} \sigma(0)/\sigma &= 0.33, \text{ 10 point filter} \\ \sigma(0)/\sigma &= 0.06 \quad \text{300 point filter} \end{aligned} \tag{3.25}$$

where σ is the single-look measurement accuracy, given by equations (3.22) and (3.24)—the case for the present example.

Equations (3.25) state that the position of a target can be obtained to about 33 percent of the radar position measurement accuracy using the track-while-scan mode and to about 6 percent of the radar position measurement accuracy using the straight tracking mode. Clearly in the present example the straight tracking mode affords the most significant smoothing of the data. It should be recalled however that the straight tracking mode collects data points at a rate higher than the track-while-scan mode.

The noise reduction factor for position prediction in a 2nd order midpoint filter can be obtained using equation (3.12). Thus for $N = 10, 300$:

$$\begin{aligned} \sigma(t)/\sigma &= 0.92\ (t/t_f)^2, \quad \text{10 point filter} \\ \sigma(t)/\sigma &= 0.176\ (t/t_f)^2, \quad \text{300 point filter} \end{aligned} \tag{3.26}$$

where

t = prediction time; measured from the filter midpoint, and
t_f = filter lag time; measured from the filter midpoint; equals $\frac{1}{2}$ the filter tracking interval; $t \geq t_f$.

Assuming that the target's ballistic coefficient is high, i.e., above 5,000 kg/m^2, and that the target speed is about 10,000 m/sec., the prediction time and filter lag for filters located at the midpoint of the tracking interval are:

$$t = 700 \text{ km}/10 \text{ km/hr} = 70 \text{ seconds},$$

$$t_f = 25 \text{ seconds, 10 point filter},$$

$$t_f = 1 \text{ second, 300 point filter}.$$

Substitution of these values for the prediction and filter lag times into equations (3.26) produces the noise reduction factor for position prediction for the case of our example. Thus:

$$\begin{aligned} \sigma(t)/\sigma &= 7.2, \quad \text{10 point filter} \\ \sigma(t)/\sigma &= 870. \quad \text{300 point filter} \end{aligned} \tag{3.27}$$

where σ is the single-look measurement accuracy, given by equations (3.22) and (3.24) for the case of our example.

Equations (3.27) state the fact that the impact location of the target missile cannot be predicted within the radar measurement accuracy. Clearly, however, the track-while-scan tracking mode affords the best prediction for the missile impact location. This is due to the fact that the prediction time is a lesser multiple of the filter lag time (in the track-while-scan mode) and this overcomes the initial advantage of significant smoothing of the data obtained using the straight tracking mode.

Review of equations (3.22) and (3.24) indicates that the position measurement determines the measurement accuracy; 7.5 m for the non-coherent system and 130 m for the coherent system. As a consequence, using the first of equations (3.22), the impact point location of the target missile can be determined quite accurately using the track-while-scan tracking mode. The prediction can be accurate to about 56 m using the non-coherent system and to 940 m using the coherent radar. Coherency therefore has been obtained at the expense of impact location prediction accuracy. Of course, the straight tracking mode is less accurate for the prediction of the missile's impact point in the present example. This latter mode can be used quite successfully however in terminal tracking when the radar has no time to track-while-scanning and must obtain its data on the target in the least possible time.

It is interesting at this time to also compute the truncation error which results from using a finite order filter. If we assume that the missile time of flight is $T_f = 1{,}800$ sec., from launch to impact, and the missile orbit period to be 4,400 sec., for the particular case of our example, the interval of convergence can be computed as $t_c = 2{,}200 + 900 - 70 = 3{,}030$ sec. Thus, using equation (3.13), the ratio of the prediction to convergence time is $70/3{,}030 = 0.025$ which can be used to find the maximum possible truncation error for the second order filter ($\nu = 3$). Thus, the truncation error is found to be about 2×10^{-5} normalized to about 15 m at the midpoint of the filter (target tracking interval). In this example, the truncation error is very small.

3.7.4.6. Target Capacity

The target capacity is defined as the number of targets that can be tracked simultaneously and was discussed briefly in connection with equation (3.1). The number of targets n that can be tracked simultaneously by a single radar is given by:

$$n = \frac{1}{\tau_s/T + \tau_d/T} \qquad (3.28)$$

where

τ_s = beam switch time,
τ_d = beam dwell-time, and
T = total tracking time available.

As discussed previously the beam dwell-time τ_d and the total tracking time T which is available to track targets are not controlled by the defense but are to a great extent controlled by the offense. The beam switch time τ_s is determined by the available technology. However, the defense may have some option in setting these parameters somewhat more to its advantage.

Basically the total tracking time T is a linear function of the target tracking interval ΔR and inverse function of the target speed v. Thus:

$$T = \frac{\Delta R}{v} \qquad (3.29)$$

It can be seen from equation (3.29) that a doubling of the target tracking interval ΔR holds some potential for doubling the target capacity. However, this can be accomplished only at the expense of increasing the radar power level and most likely decreasing its impact-point prediction accuracy since a longer prediction time will be required at the longer range. Somewhat compensating this effect is the availability of more time for smoothing the data and therefore obtaining better noise reduction factors.

The dwell-time τ_d can be decreased only at the expense of again increasing the radar power level and most likely increasing the beam switch time τ_s.

The general conclusion is that the target capacity can be increased through the employment of high average power by the radar. Arrays therefore offer singular opportunities for the transmission of large amounts of power.

The computer capacity which is needed to store and process the data is on the order of:

$$C_{\max} = (\nu M)^2 nb \tag{3.30}$$

where

$C_{\max}$ = maximum computer capacity,
$\nu - 1$ = order of the tracking filter,
M = number of data points in the tracking interval,
n = number of targets, and
b = number of bits in a radar word.

Equation (3.30) represents the order-of-magnitude estimate for computer capacity based on a straightforward solution of the required equations of motion. A lesser capacity is required if the expedient of algorithms is used for the solution of equations. Thus the maximum capacity which is required of a computer for tracking 25 targets using a second-order filter with a total of 10 data points in the tracking interval is, assuming a 50-bit computer word, about 1 million bits.

3.7.4.7. *Noise Jamming Environment*

The performance of a radar in a noise jamming environment can be assessed in terms of the degradation of the radar's measurement accuracy. The measurement accuracy of a radar in a jamming environment is given by:

$$\sigma = k\ (J_o/E)^{1/2} \tag{3.31}$$

where

σ = measurement accuracy,
J_o = jamming spectral noise density; in watts/Hz,
E = signal energy; in watt-sec., and
k = constant of proportionality; in units of the measured parameter (range, angle, velocity, etc.).

Equation (3.31) is the statement for the well-known measurement accuracy of a radar in a jamming environment. When the jamming ceases $J_o = N_o$ and the thermal noise N_o determines the accuracy. The jamming noise is considered to be always greater than the receiver noise, i.e., $J_o > N_o$.

If a jammer increases the receiver noise level by a factor of 20 db, the measurement accuracy of the radar decreases by a factor 10. Thus if a radar predicts the impact location of a target missile with an accuracy of 56 m the presence of 20 db of noise jamming in its receiver will reduce this accuracy to 560 m, which may be unacceptable to the defense. Therefore, in a noise jamming environment, the defense may require high average power as a counter-countermeasure to the noise jamming. Again arrays offer single opportunities for the transmission of large amounts of power. Thus the effect of noise jamming on the measurement accuracy can be minimized by increasing the average power level. Alternatively, the defense may increase the dwell-time τ_d of the radar beam, and increasing the signal energy E in this manner, or it may require higher noise reduction factors and data filtering rates in its computers.

References

1. K. E. Peress and A. Schwartz, "Tracking Error Analysis of a Range Ship", *Sperry Engineering Review,* Fall 1962. The accuracy with which a shipboard tracking system can determine the position and velocity of a space vehicle, missile, or aircraft is derived. Discusses formulation of tracking mission requirements, target

model and tracking conditions, analytical procedures for combining the error contributions, and the system error model. Explains the presentation of results as parametric curves.
2. A. S. Locke, et al., *Guidance*, ed., G. Merril, Van Nostrand, 1957. An excellent in-depth technical treatise on guided missiles and related topics.
3. R. B. Dow, *Fundamentals of Advanced Missiles*, Wiley, 1958. A Useful engineering reference covering many subjects on a first principle basis. Recommended in its entirety for follow-on to the material of this chapter.
4. M. I. Skolnik, ed., *Radar Handbook* McGraw-Hill, 1969. An in-depth technical handbook on radar and related topics.
5. M. I. Skolnik, *Introduction to Radar Systems*, McGraw-Hill, 1962. A detailed reference on radar systems.
6. R. Kyle, "Phased-Array Antenna Fundamentals", *Countermeasures*, March 1978.
7. D. L. Ashcroft and W. T. Whistler, "Capabilities and Limitations of Array Radars", Second Winter Convention on Military Electronics, Los Angeles, Feb. 1–3, 1961. This early paper is a summary of the decision-making process for selection of array radar.
8. John L. Allen, "Array Radars, A Survey of Their Potential and Their Limitations", *Microwave Journal*, May 1962. Qualitatively explores the potentials and limitations of array radars, with emphasis on ground-based, long-range radar applications. Reviews (1) fundamental array types; (2) radar parameters most basic to target detection, resolution, and measurement; and (3) capacity of array configurations to perform these functions.
9. D. Lehman, "Multibeam Array Controls RPVs", *Microwave Systems News*, October 1978.
10. Constant, J. N., *Introduction to Defense Radar Systems Engineering*, Spartan/Hayden Books, 1972. Presents topics on defense radar systems. Chapter 4 gives a radar system error model and Chapter 10 gives array antennas.
11. F. E. Terman, *Radio Engineers Handbook*, McGraw-Hill, 1943. Reference book summarizes basics of radio and electronics. The discussion of the propagation of radio waves is given in Chapter 10.
12. S. S. Attwood, *Radio Wave Propagation*, Academic Press, 1949.
13. W. M. Hall, "General Radar Equation", *Space/Aeronautics R&D Handbook, IEEE* (Institute of Electrical and Electronic Engineers) 1962–1963. Derives the general radar equation and special forms of this equation are used to illustrate various radar problems and systems.
14. D. K. Barton, *Radar System Analysis*, Prentice-Hall, 1965. A detailed reference on radar systems. Particular attention is directed to Chapters 1, 4, 5, 6 for the theory of radar detection and to Chapter

2 for the theory of radar measurements. Chapter 13 provides a discussion for the processing of radar data. Chapters 10 and 13 give error analysis of angle trackers and processing systems.
15. T. C. Bartlee, *Digital Computer Fundamentals*, McGraw-Hill, 1966. A general reference for modern digital computers.
16. B. Widrow, *et al.*, "Adaptive Antenna Systems", *Proceedings of the IEEE*, December 1967.
17. C. E. Cook and M. Bernfield, *Radar Signals*, Academic Press, 1967. A detailed study of the theory of radar signals.
18. M. H. Carpentier, *Radars: New Concepts*, Gordon and Breach Science Publishers, 1968. Provides a comprehensive theoretical basis for modern radar. Particular attention is directed to the discussion of ideal receivers, radar performance ambiguity functions, and clutter elimination.
19. F. E. Nathanson, *Radar Design Principles*, McGraw-Hill, 1969. Gives extensive discussion of radar target characteristics, atmospheric effects, and the backscattering properties of sea and land.
20. W. M. Brown and C. J. Palermo, *Random Processes, Communications and Radar*, McGraw-Hill, 1969. Introductory text. Particular attention is directed to the brief discussion of detection in Chapter 6.
21. W. S. Burdic, *Radar Signal Analysis*, Prentice-Hall, 1968.
22. J. V. Di Franco and W. L. Rubin, *Radar Detection*, Prentice-Hall, 1970. Gives a powerful and highly mathematical treatise on detection theory.
23. "Detection Theory", *Proceedings IEEE*, Vol. 58, No. 5, May, 1970. A detailed reference and summary of detection theory. Gives basic outlines, recent theoretical developments and areas of application.
24. Special Issue, "Radar Technology" *Proceedings IEEE*, June 1974.
25. "Trends in Modern Radar System Design", Session 25, ELECTRO 76, Boston, 11–14 May 1976.
26. E. Torero, "What went wrong III?: The bypass of 3-D Radar", *IEEE Spectrum*, October 1976.
27. F. Moncrief, "Radar 77", *Microwave Systems News*, October 1977. London's Radar 77 follows the two and one-half year sequence established at Radar 75 in Washington. Papers at this years conference emphasize current radar trends, with phased arrays, bistatic or multi-static systems, and computer applications sharing the limelight.
28. D. Howard, "Predicting Target Caused Errors in Radar Measurements", *Electronic Warfare*, February 1978.
29. B. Miller, "Rising Costs Stimulate the Search for better Ground Surveillance Radar", *Electronic Warfare*, February 1978.
30. "The Nuclear Environment—What's it like?", *Electronic Warfare*, March 1978.
31. "Radar Requirements on the Upswing", *Microwave System News*, March 1978.
32. P. J. Kahrilas, "Design of Electronic Scanning Radar Systems",

Proceedings of the IEEE, Nov. 1968. Presents a format for the design of electronic scanning radar systems, emphasizing theoretical and practical design considerations unique to electronic scanning. An introductory discussion of the role of electronic scanning technology in meeting modern radar requirements is followed by a brief description of the generic types of electronic scan. Lists system parameters affecting mission performance. Delineates corresponding radar functions particularly affected by electronically scanned radar, with a discussion of the merits of the selection. Major decisions are (1) single or multiple simultaneous beams; and (2) narrow, intermediate, or wide bandwidth performance capability. Discusses automatic multitarget acquisition logic, computer controlled iterative tracking equations, interpolative-null tracking, error budgets, and errors unique to electronic scanning. The design procedure is illustrated with practical examples.

33. Special Issue on Electronic Scanning, *Proceedings of the IEEE*, Nov. 1968. An up-to-date summary of array theory, mutual coupling theory, components, array elements, and array implementation. The article by P. Kahrilas is particularly useful from a systems viewpoint.
34. J. Reed, "The AN/FPS-85 Radar System", *Proc. IEEE*, March, 1969. The AN/FPS-85 is a large operational UHF array radar. This paper gives a description of the radar's operation, some insights into the system design, pictures of actual hardware, a summary of system testing and a project summary.
35. P. J. Kahrilas, "HAPDAR—An Operational Phased Array Radar", *Proc. IEEE*. Vol. 56, No. 11, 1968. The HAPDAR (Hard Point Demonstration Array Radar) described is a multifunction phased array radar in operation at the White Sands Missile Range. The principle design feature is the TACOL (Thinned Aperture Computed Lens) array. The array and system design concepts are discussed and a detailed physical, functional and analytical description of the radar is presented. Recently obtained operational test results on array patterns and multitarget tracking performance are discussed and compared to design specifications.
36. R. Leach, "Evaluating the Quality of Prediction for a Position Predicting or Tracking System", *ARS Journal*, November 1962. Discusses how radar data should be processed to give a best prediction of position and assessment of the prediction.
37. C. Young, "Monograms simplify phased Array design". *The Electronic Engineer*, November 1969.

4

ABM System Survivability Analysis

This chapter will examine those factors which are required to determine the probability of survival of an ABM system. Survivability is defined in terms of two factors: first, the threat estimate and, second, the ABM system vulnerabilities with respect to electromagnetic, physical, and heterogeneous effects. Electromagnetic effects include all types of radar countermeasures; physical effects include blast damage, thermal and nuclear radiation; heterogeneous effects include chemical, biological, and radiological warfare, and sabotage. Several techniques for the analysis of ABM system survivability will be considered. In general, survivability analysis is made on the basis of a postulated threat and given ABM system capabilities, to arrive at a prediction for the ABM system survivability.

4.1. Survivability with respect to ECM

References 1–8

The survivability of the ABM system to ECM can be discussed in terms of the ECM threat, ABM system vulnerabilities, and analysis of the ECM environment.

To begin with, a number of parameters must be defined in order to analyze the effects of ECM on the ABM system. These parameters relate mainly to specification of the threat and the determination of ABM system vulnerabilities. Criteria therefore of effective ECM must be established for each radar and communication link in the ABM system. These criteria are in terms of word unintelligibility and probability of word error as functions of the signal-to-noise ratio in each channel. A radar word is information which represents the measurement of the target range, range rate, azimuth and elevation position, and radar cross-section. A communications word is information which represents the coded or uncoded message in the communication link. Although word error is used as the ECM criterion, it usually will be necessary in a complete ECM analysis to relate word error to message intelligibility. The jamming signal-to-desired signal ratio which produces the various signal-to-noise ratios can be derived from the receiver characteristics.

To arrive at the jamming-to-signal, J/S, ratio at the input to the receiver, both the desired and jamming signal levels must be calculated. The desired signal level that occurs at the receiver input is determined from a knowledge of the transmitter and receiver, their locations, antenna characteristics, operating frequencies, propagation characteristics, and power levels. The jammer signal level is calculated from a knowledge of the jammer and its location and characteristics, which frequently must be postulated.

4.1.1. ECM Threat

References 9–11

In order to postulate the ECM threat two main sources of information may be used; intelligence information and friendly ECM capabilities. The ECM parameters that must be postulated are: ECM location, ECM antenna characteristics, ECM technique, and ECM power, frequency, and frequency stability.

4.1.1.1. ECM Location

High-powered, long-range active jammers may be located anywhere in enemy territory. Shipboard active jammers may be located at any point accessible to ships or submarines. Airborne or spaceborne active or passive jammers may be located at any point accessible to aircraft, ballistic missiles, or satellites. Small, low-power, portable active jammers may be handcarried or parachuted to optimum locations. Satellite-borne jammers should not be minimized in the threat evaluation.

4.1.1.2. ECM Antenna Characteristics

For the case of active jammers, the antenna bearing both in azimuth and elevation, antenna gain, antenna height, and antenna polarization must be considered. Clearly such characteristics become a function of the frequency of operation, the space which is available to install the antenna, and in ballistic applications, the severe reentry environment.

4.1.1.3. ECM Techniques

With regard to radar systems, there are three powerful techniques for the discrimination of targets; trajectory prediction, signature analysis, and atmospheric filtering. The prediction of a given target trajectory can be made quite accurately, thus eliminating non-threatening targets and ECM-induced signals. Signature analysis is the deductive process by which the target characteristics, size, weight, shape, and motion are determined. The pattern of radar reflections from a target as it moves relative to the ground radars is unique and is defined to be its signature. Many patterns are now available for signatures from different targets. These patterns have been obtained both theoretically and from actual practice from flight test data and from the measurement of models on test ranges. Comparison of a measured pattern with known patterns can be made, thus eliminating non-matching patterns which would presumably be due to ECM. Atmospheric filtering of most lightweight ECM devices will be made during late reentry, thus eliminating all but the heavy objects as threats.

Each particular ECM technique must be analyzed for its particular effect on the radar or communications receiver. Both active and passive ECM techniques must be thoroughly evaluated, i.e., spot, barrage, and sweep jamming, spot and volume chaff, high- and low-altitude decoys, RV shaping, absorbing materials, and wake visibility.

4.1.1.4. ECM Power, Frequency, and Frequency Stability

The parameters of power, actively or passively generated, frequency, and frequency stability, in conjunction with the particular ECM technique which is employed, determine the level, duration, and fluctuations of power which arrives at the receiver at frequencies within its pass band.

When all the aforementioned ECM parameters have been specified, the resulting type of propagation and path loss can be computed to arrive at the receiver from each type ECM device. Nuclear attenuation effects may be included with path losses both for the ECM and legitimate signals.

4.1.2. ECM Vulnerabilities

References 12–26

In order to evaluate ECM vulnerabilities, it is first necessary to establish criteria by which effective ECM may be specified quantitatively. This must be done for both radar and communications links in the ABM system.

4.1.2.1. Confusion ECM

If noise is added to an ordinary radar or communications channel, the receiver output decreases in intelligibility as the noise-to-signal ratio is increased. ECM effectiveness is evaluated by measuring the percentage of unintelligible signals or words resulting from a given signal-to-noise ratio. This can be accomplished by using the results of Chapter 2. Figure 4.1 is a plot of the probability of detecting a single sinusoidal pulse in white Gaussian noise, as a function of the signal-to-noise ratio and for a given low probability of false alarm.

It can be seen in Fig. 4.1 that the effect of increasing the noise in the receiver is to decrease the probability of detection of the signal itself. Similar curves can be obtained once the character of the noise and character of the signal are known

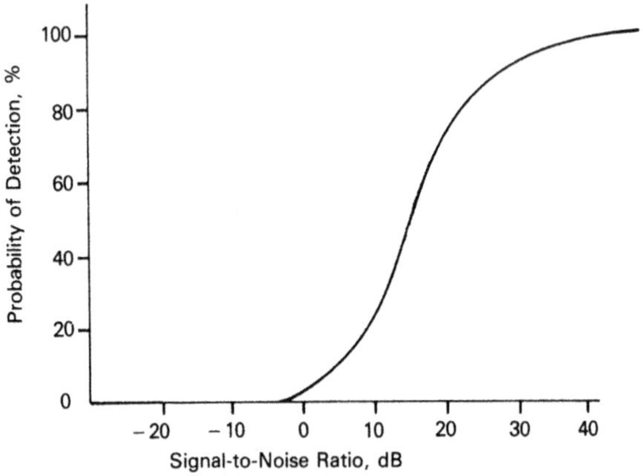

Fig. 4.1. *Probability of detection of a sinusoidal signal in white Gaussian noise (probability of false alarm 10^{-6})*

and this can be accomplished for various threshold settings in the receiver, i.e., at given probability of false alarm. Furthermore, for most threshold settings, the probability of detection of signals falls off dramatically when the signal-to-noise ratio falls below about 10–20 dB.

4.1.2.2. *Deception ECM*

Not all ECM techniques add noise to the receiver being jammed. Notable exceptions are the class of active jammers known as repeaters, as well as decoys and RCS reduction. Repeaters and decoys generally simulate legitimate signals while RCS reduction techniques tend to modify legitimate signal characteristics. As a consequence, criteria other than that of noise must be established to describe effective ECM performance. However, for such ECM techniques, no clear-cut quantitative descriptor of ECM performance can be specified. Each case must be investigated separately on its own merit.

The probability of detection of a non-noise or deception ECM is a strong function of the false target altitude. This is also true to a certain extent for the noise type or confusion ECM, since the signal-to-noise ratio at the receiver can be related to the ECM device altitude. As a consequence, a cri-

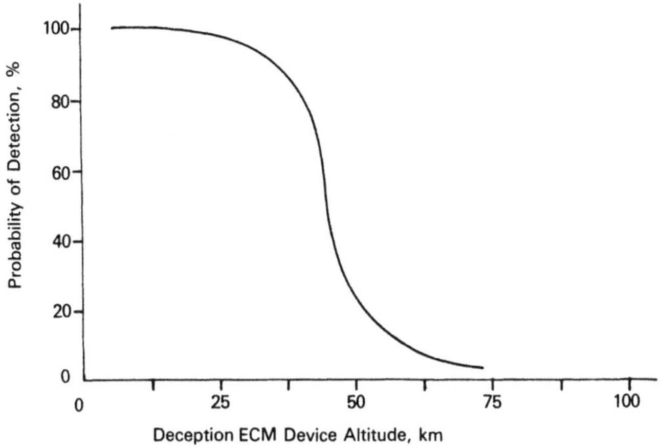

Fig. 4.2. *Probability of detection of deception ECM*

terion for ECM effectiveness can be established for deception ECM based on the probability of detection of the deception by a certain altitude for a large number of similar but not identical engagements. Figure 4.2 is a plot of the probability of detecting a given deception ECM as a function of altitude. Effective deception ECM can be defined as having a low probability of detection above some given altitude, i.e., the radar decision altitude.

4.1.2.3. Probability of Survival

The probability of survival of a defended target in an ECM environment is the compliment of its probability of kill, i.e.:

$$P_s + P_k = 1 \tag{4.1}$$

where P_s, P_k are the probabilities of target survival and kill.

The probability of kill in equation (4.1) assumes that the target is undefended and the probability of its penetration P_p by a warhead is assured, i.e., $P_p = 1$. Otherwise, P_k in equation (4.1) must be replaced by the combined probability of penetration and kill P_pP_k. The probability of penetrating a target is the compliment to its probability of being detected and destroyed P_d. Mathematically:

$$P_p + P_d = 1 \tag{4.2}$$

Thus, the probability of survival of a defended target is found by replacing P_k by P_pP_k and solving for P_s in equation (4.1) which, in view of equation (4.2), gives the result:

$$P_s = 1 - (1 - P_d)P_k \tag{4.3}$$

which indicates that the probability of survival P_s of a defended target being attacked by a warhead is a function of the defense capability to detect and destroy the warhead and the offense capability to escape detection and destruction and then to kill the target. The ECM environment which is provided by the offense has the objective of degrading the defense probability of detecting and destroying the warhead.

4.1.3. ECM Analysis

Once criteria have been established and related to the signal-to-noise ratio at the receiver output for the case of confusion type ECM, and to the ECM device altitude for the deception type ECM, the ECM analysis can begin.

4.1.3.1. Confusion ECM

The equation which most appropriately describes the radar performance in the presence of active noise jamming is given by:

$$\frac{E}{J_0} = \frac{P_T G_T G_R \sigma \tau R_j^2}{4\pi J_T G_j G_a L R^4} \tag{4.4}$$

where

E/J_0 = ratio of received energy-to-jamming power per Hz,
P_T = average transmitter power; in watts,
G_T = gain of the transmitter antenna,
G_R = gain of the receiver antenna,
σ = radar cross-section of target; in meters squared,
τ = time of observation; in seconds,
R_j = range to the jammer; in meters,
J_T = jammer transmitter power per Hz,
G_j = gain of the jammer in direction of radar,
G_a = gain of the radar receiving antenna in the direction of the jammer; When the receiving antenna is pointed at the jammer $G_a = G_R$,
L = ohmic and similar losses in the antenna, transmission line; $L = 1$ when no losses are present in the antenna and transmission line; a typical loss is 10, or its equivalent 10 dB,
R = range to the target; in meters; If the jammer is on-board the target $R_j = R$.

The effects of propagation have been neglected in equation (4.4) but these can be easily accounted for. Propagation effects should include nuclear bursts along the paths connecting the receiver and target as well as those connecting the receiver and jammer.

The ratio of received energy-to-received jamming energy

(power per Hz), E/J_0, given by equation (4.4) is related to the signal-to-jamming power ratio, S/J, at the receiver by:

$$\frac{E}{J_0} = \frac{S}{J}(B\tau)FD \tag{4.5}$$

where

E/J_0 = ratio of received energy-to-received jamming power Hz,
S/J = signal-to-jamming power ratio at receiver input,
$B\tau$ = Product of receiver bandwidth B and observation time τ,
F = noise figure of receiver,
D = degradation factor; represents the ratio of the signal power required in the actual system to the signal power that would be required in the matched filter system; is therefore a function of the receiver design.

Equations (4.4) and (4.5) can be utilized to compute the signal-to-jamming ratio for a variety of parameters for the target and jammer and the known parameters of the radar. This can be accomplished for a number of threshold settings in the receiver. The results will be similar to that given in Fig. 4.1 and, in general, will give the probability of detection of the signal as a function of the signal-to-jamming power ratio at the receiver input.

The equation which most appropriately describes the radar performance in the presence of volume chaff is given by:

$$\frac{E}{C_0} = \frac{\sigma V_c}{\sigma_d N V} \tag{4.6}$$

where

E/C_0 = ratio of received energy-to-received chaff energy; target and chaff are co-located,
σ = radar cross-section of target; in meters squared,
σ_d = radar cross-section of a dipole; $\sigma_d = 0.15\lambda^2$ where λ is the wavelength in meters,
N = the total number of dipoles in the chaff cloud,
V_c = volume of chaff cloud,
V = volume of the radar resolution cell; $V = R^2\theta\phi p$ where R = range to target, θ,ϕ = azimuth and elevation beam-width of radar beam, p = pulse-width of radar pulse.

The radar cross-section of a conical reentry vehicle, as viewed from nose-on aspects, is itself proportional to λ^2 so that equation (4.6) can be further simplified. In fact, for most frequencies, the cross-section of a cone-sphere for nose-on aspects is less by a factor 2 than the cross-section of a dipole which also varies as λ^2. For this particular case, equation (4.6) is simplified as:

$$\frac{E}{C_0} = \frac{V_c}{2NV} \tag{4.7}$$

The ratio of received signal energy-to-received chaff energy, given by either equations (4.6) or (4.7), is related to the signal-to-chaff ratio S/C at the receiver input by:

$$\frac{E}{C_0} = \frac{S}{C}(B\tau)FD \tag{4.8}$$

where all parameters have been previously identified in connection with the discussion of equation (4.5).

Equations (4.6), (4.7), and (4.8) can be utilized to compute signal-to-chaff ratios for a variety of parameters for the target and the chaff, and the known parameters of the radar. This can be accomplished for a number of threshold settings in the receiver to determine the probability of detection of the target. The results will be, as was also the case for jamming noise, similar to those given in Fig. 4.1 and, in general, will give the probability of detection of the signal as a function of the signal-to-chaff power ratio at the receiver input.

The signal-to-jamming power ratio or the signal-to-chaff ratio, as the case may be, can also be expressed as a function of the range or altitude between the radar and the target. In this manner, the probabilility of detection of a target in the confusion type ECM environment can be compared directly with the radar decision altitude. For its survival, the radar must detect the target with high probability of detection prior to the target's arrival at the radar decision altitude.

4.1.3.2. *Deception ECM*

As mentioned previously, no single equation appropriately describes the radar performance in a non-noise or deception-type ECM environment. Three powerful techniques, however, can be used for the discrimination of targets; trajectory prediction, signature analysis, and atmospheric filtering.

The prediction of a target trajectory can be made by taking a number of position (range, azimuth, and elevation) measurements over some span of the target trajectory and then using these data to predict the target's future position. If the predicted target impact point is non-threatening the target can be disregarded as being either of no interest or deceptive. If $2N+1$ measurements are made, each separated by a time interval Δt, then the total time consumed in making measurements is $2N\Delta t$.

The accuracy of predicting the warhead impact point, following a series of position measurements, is given by the following equation:

$$\sigma(t) = k_m(N)t^m\sigma, \qquad t = \frac{R}{v} = \frac{h}{v \sin \epsilon} \tag{4.9}$$

where

$k_M(N)$ = a constant which is determined by m and N,
m = order of the position filter
N = measurements taken each side of the center of the measurement span,
t = time reckoned from the midpoint of the measurement span to target impact,
R = target range,
h = target altitude,
ϵ = target elevation,
v = target speed,
σ = radar error.

Equation (4.9) is a simplified expression which determines the standard deviation for predicting the warhead impact point.

Figure 4.3 illustrates a trajectory which impacts at point A. Shown is the span $2N\Delta t$ over which measurements are made, the time t between the center of the span of measurements, the impact point A, and the warhead altitude and elevation, as measured by the radar. Also shown is the true target location at point B which is separated from the impact point A by distance d. Drawn about each point are error circles with their respective standard deviations. The standard deviation of the impact

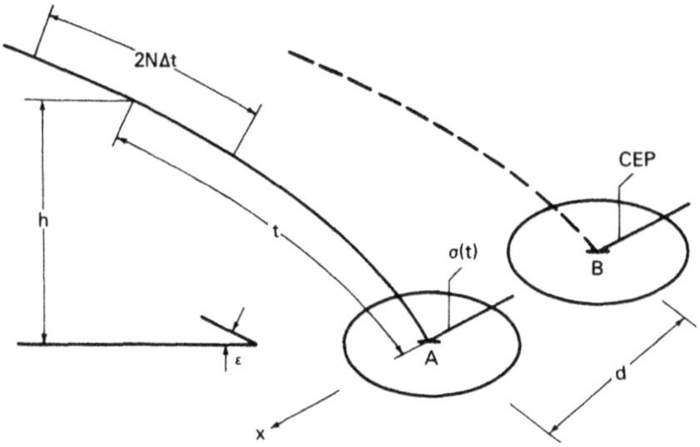

Fig. 4.3. *Trajectory prediction technique*

point at A is $\sigma(t)$ as determined by equation (4.9) while the standard deviation of the target at B is the CEP, the missile circular error probability.

The probability of detecting or discriminating between the two trajectories A and B, the non-threatening trajectory directed to the impact point A and a corresponding threatening trajectory directed to the actual target B, is determined to be the combined distribution of two Gaussian distributions, one with regard to the accuracy of prediction of the impact point at A and the other with regard to the missile distribution at the target point B. Thus, using the target point B as the center of coordinates, the probability of detecting or discriminating between the two trajectories is given by:

$$P_d = \frac{1}{2\pi(\mathrm{CEP})\sigma(t)} \exp -\frac{1}{2}\left[\frac{(x-d)^2}{\sigma^2(t)} + \frac{x^2}{(\mathrm{CEP})^2}\right] \tag{4.10}$$

In view of equation (4.9), the probability of detection given by equation (4.10) can be expressed as a function of the ECM deception altitude h for a variety of radar and jamming parameters and smoothing techniques. The results of plotting equation (4.10) will be similar to that shown in Fig. 4.2 and, in general, will give the probability of detection or discrimination of the trajectory A from a threatening trajectory B, as a function of the deception device altitude. Of course, to be effective using this technique, the defense must obtain a high value for the probability of detection prior to its decision altitude.

The discrimination of a target on the basis of signature analysis can be accomplished by taking a number of measurements over some span of the warhead trajectory and then using the data to identify its unique pattern. If $2N+1$ measurements are made, each separated by a time interval Δt, then the total time consumed in making measurements is $2N\Delta t$. Two general approaches have been implemented to obtain the necessary patterns for comparison; theoretical and empirical, with the latter including the measurement of actual targets. The two approaches obviously complement each other, each producing a body of data against which the patterns of unknown objects can be compared. The empirical method produces a thick catalogue of recorded patterns, while theoretical studies lead to a relatively smaller number of mathematical expressions which can describe possible patterns. The advantage of reducing patterns to mathematical expressions is that rather straightforward computer techniques can be used to make the signature analysis automatic. Simplifying assumptions must however be made about the relationship between the reflective properties of legitimate warheads and these assumptions normally diminish the coverage of possibilities.

The probability of discrimination between an unknown pattern $p(t)$ comprising $2N+1$ data points and a known pattern $p_0(t)$ is essentially, for a large number of data points, the cross correlation integral between the two functions:

$$p_d(t) = \int_0^{2N+1} p(\tau)p_0(t+\tau)\,d\tau \tag{4.11}$$

where

$p_d(t)$ = probability density of detection or discrimination between two patterns or signals at time t reckoned from the midpoint of the measurement interval $2N+1$, and

$p(t), p_0(t)$ = pattern or signal time functions of the measured and stored patterns, respectively.

The integral given by equation (4.11) is a well-known integral in radar theory and for similar patterns results in a maximum when the correlation time is zero, i.e., when the two patterns match at $t=0$. It essentially defines the response of a filter or special receiver which can be designed for pattern recognition. The probability density $p_d(t)$ drops off quickly either side of its maximum at $t = 0$.

The probability of detection or discrimination between patterns can therefore be found from:

$$P_d = \int_0^t p_d(t)\, dt \tag{4.12}$$

Since the time t is nearly proportional to the warhead altitude h, the probability of detection given by equation (4.12) can be expressed as a function of altitude. The result of plotting equation (4.12) will be similar to that shown in Fig. 4.2 and, in general, will give the probability of detection or discrimination between a known and unknown pattern. Again, to be effective using this technique, the defense must obtain a high value for the probability of detection prior to its decision altitude.

The technique of filtering lightweight objects by the atmosphere can be used to eliminate many active and passive ECM devices. To be successful in using this technique, however, the defense system must be designed for quick response. This technique forms the basis for effective terminal defense and indeed has been implemented in the U.S. Safeguard ABM system by using the fast Sprint missiles. Quick response becomes necessary to launch the fast interceptors following the filtering of lightweight objects by the atmosphere and the subsequent discrimination of the heavier objects. Most confusion and low-quality deception devices will be filtered by the atmosphere, thus leaving the defense unencumbured to discriminate the presence of high-quality decoys, i.e., high-quality deception devices. When using this technique the warheads are being discriminated at very low altitude and short range, for which it becomes necessary that the defense system have a quick response and fast interceptors.

The atmosphere is also useful in another way, namely it provides an altogether separate technique of signature analysis based on the recognition of a reentering object's wake. The hot ionized envelope that surrounds a reentering object considerably modifies its radar cross-section over that which is obtained when the object is viewed as a bare body. Much work has been done to predict the influence of the reentry plasma sheath on the bare body cross-section and its overall visibility. Given a certain reentry vehicle configuration and entry speed it is possible to calculate roughly the electron density and the dielectric constant of the plasma sheath and wake which surround the reentering vehicle and thus to predict the onset of the reentry wake phenomenon and its progress with descending altitude. The shape and extent of the wake is unique for each reentry vehicle. The onset, width, length, diameter, and other visible characteristics of an object's wake can be compared with wake characteristics for known objects, thus forming the basis for discrimination of high-quality decoys and deception ECM.

Survivability enhancement with respect to ECM is generally referred to as electronic counter-countermeasures (ECCM) and these will be discussed in Chapter 5. Suffice it to say that much work has been done in this field both for radar and communications ECCM.

4.2. Survivability with respect to Nuclear Bomb Attacks

References 27–31, 36–42, 45

Since a bomb or missile aimed at a target is subject to random aiming errors, the criterion that is used to specify survivability is the probability of survival of the target. For defense design purposes, the desired probabilities of target survival can be specified for defense system functions and components. These desired probabilities will depend upon such factors as technical requirements, critical time periods, and defense priorities. Once the required probabilities have been specified and a threat is postulated, the defense system may be designed with capabilities that will produce its required degree of survivability.

4.2.1. Nuclear Threat

References 1–3

The threat to a defense system relative to nuclear bomb attacks may be evaluated in three areas: weapon parameters, weapon effects, and weapon targeting, the latter of which includes strategies and tactics.

4.2.1.1. Weapon Parameters

The weapon parameters of concern in the estimation of the

Table 4.1. *Effect of weapon parameters on survival probability of a single 1.4 kg/cm² (20 psi) target*

NUMBER OF WEAPONS	YIELD	CEP	P_s
1	1 MT	5 km (3 mi)	87%
1	10 MT	5 km (3 mi)	55%
1	1 MT	2.5 km (1.5 mi)	55%
4	1 MT	5 km (3 mi)	55%

SOURCE: See reference 1.

threat are the numbers of weapons (modified by weapon reliability), yield, and accuracy. These parameters change continuously with technological advances, and must be frequently updated on the basis of recent information from two important sources. The first source is intelligence information, while the second is relative information of friendly weapons, with the assumption that the enemy technology and hardware is no better.

The estimate of the number of available weapons determines how many weapons may be expended per target, while the yield (in equivalent tons of TNT) and accuracy (in circular error probability) determine the killing power of a single weapon. These weapon parameters have varying effects on the kill probability of a target. Table 4.1 shows the probabilities of survival of a 1.4 kg/cm² (20 psi) target as a function of the three weapon parameters of interest.

It can be seen that for the parameter ranges of interest, an increase in accuracy by a factor of 2 is roughly equivalent to increasing the yield by 10 or, alternatively, using four weapons instead of one.

4.2.1.2. *Weapon Effects*

The second area of threat relative to nuclear attacks is the area of nuclear bomb effects. The nuclear weapon effects of concern in the threat evaluation for the defense system are blast, thermal radiation, nuclear radiation, and effects on propagation. The distribution of the effects of a nuclear explosion are shown in Fig. 4.4.

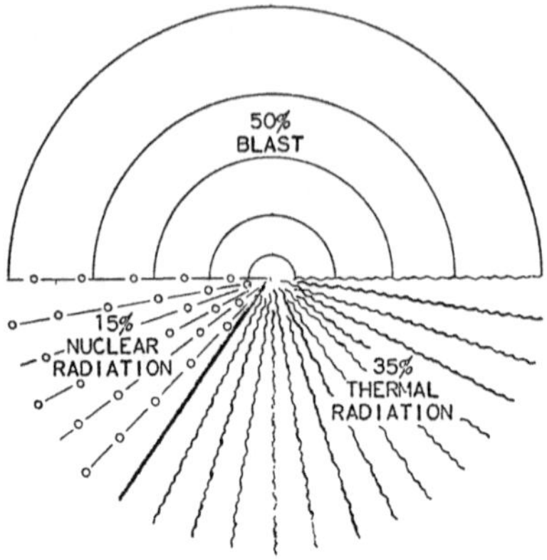

SOURCE: See reference 1.

Fig. 4.4. *Distribution of effects of a nuclear (fission) explosion.*

Approximately 50 percent of the energy released by a nuclear explosion escapes in the form of blast and shock. Blast is generally the most destructive effect of an atomic explosion with respect to the ABM system as well as with respect to a general target. It is specified in terms of overpressure which is the pressure above the normal atmospheric pressure occurring at any given distance from the explosion.

An additional 15 percent of the energy yield of a nuclear explosion is released in the form of nuclear radiation, some of which occurs immediately, within one minute of the detonation, and some over longer periods of time, reckoned in months or years after the detonation. There are several types of radiation that are of concern and several different units used in the quantitative specification of these types. Although radiation does degrade the performance of many electronic components and, particularly, semiconductor devices at ranges where blast is no longer critical, the chief concern with respect to the ABM system is with respect to the fallout on operating personnel.

The remaining 35 percent of the energy yield of a nuclear explosion appears as thermal radiation, which travels in straight lines. Thus any object such as a wall or dust cover will act to shield equipment and personnel. Thermal radiation is

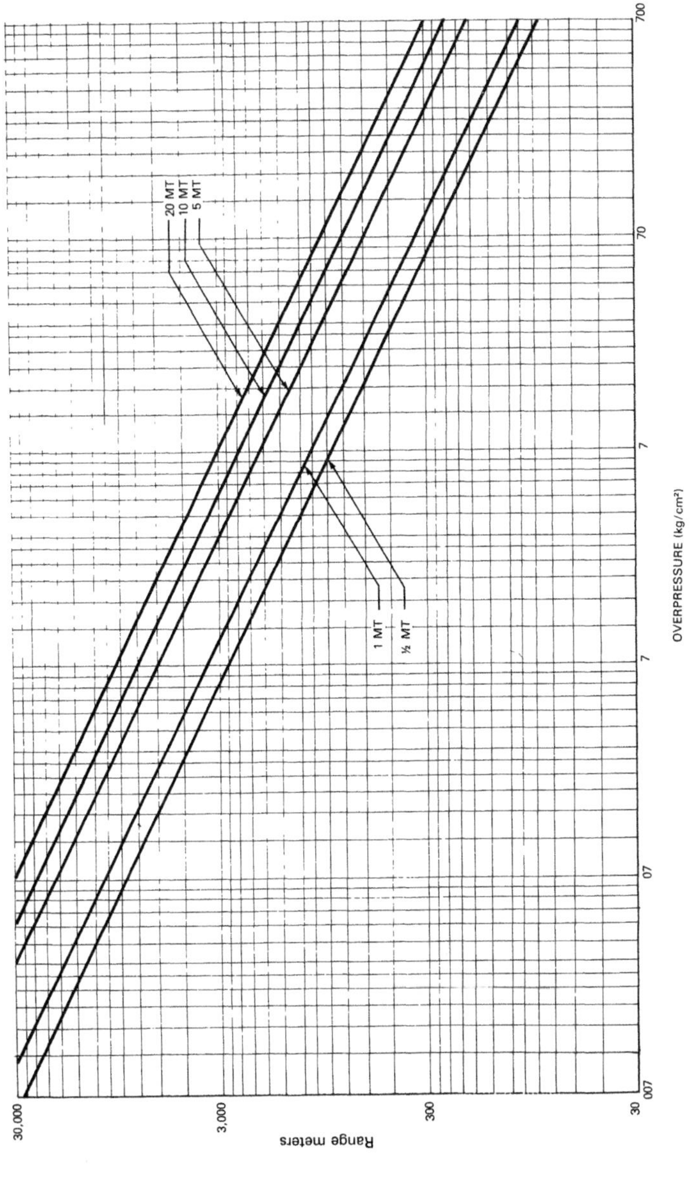

SOURCE: See reference 29.

Fig. 4.5. *Overpressure versus distance from ground zero*

measured in calories per square centimeter as a function of distance from the explosion.

It has been found that, at a given distance, blast is the most severe of the various effects with respect to the destruction of above-ground equipment such as antenna arrays and missile launchers, and it is this effect which usually is considered in survivability analysis. Figure 4.5 shows blast over pressure as a function of distance from the point of explosion.

An additional effect of nuclear explosions is to cause localized changes in propagation characteristics of electromagnetic waves. High-altitude bursts in the Pacific tests conducted by the USA also confirm the fact that nuclear detonations can cause considerable attenuation of radio signals. This particular effect of nuclear bursts must be taken into account in survivability analyses by modification of the propagation calculations for the radar and communications links that are susceptible. This is generally done in conjunction with jamming analysis.

4.2.1.3. Weapon Targeting

The third area that must be considered in evaluation of the threat relative to nuclear attacks is that of targeting. A suitable targeting model must consider reasonable attacks that are based on known enemy capabilities, strategies, and tactics for various types of scenarios and various time periods in these scenarios. The targeting models must specify those portions of the defense system that will be primary targets and those which will be bonus targets, as well as the number and yield of bombs which may be expended on each target. A variety of targeting models and strategies can be specified for the survivability analysis of an ABM system.

4.2.2. Nuclear Vulnerabilities
References 29–31

Defense system vulnerabilities with respect to nuclear bomb attacks are considered for blast, nuclear radiation and thermal radiation, with most of the emphasis placed on the capabilities of withstanding blast, since, as indicated previously, this is the effect that is of primary concern for the survival analysis of the defense system. System capabilities for attaining high degrees of survivability are limited by such factors as economics, time, technology, and enemy weapon capabilities.

Vulnerability to blast-produced overpressures is called hardness. It is expressed as the kg/cm^2 (psi) of overpressure that will cause malfunction of the component or failure of the exposed structure. The hardness ratings of primary concern in the survivability study for the ABM system are those for antennas, missile launchers, and above-ground cables and structures which house electronic equipment. Although the term hardness is not usually applied to operating personnel, it is necessary also to consider how they are affected by blast.

The hardness of most existing antenna designs is less than $0.35\ kg/cm^2$ (5 psi), resulting in rather vulnerable installations. Where it is necessary to utilize existing designs, it is possible to harden these antennas to about $0.7\ kg/cm^2$ (10 psi). In order to increase the hardness still more, new designs are needed, and many approaches to this problem have been investigated.

Overhead cables and wire lines are considered soft, with hardness values less than $0.35\ kg/cm^2$ (5 psi). The usual method of hardening is to bury them. Most building structures are also soft, having overpressure capabilities of less than $0.35\ kg/cm^2$ (5 psi). Surface buildings can be hardened to some extent, but high hardness values are achieved primarily by underground construction. The vulnerabilities of metal and stone structures to nuclear and thermal radiation are such that the structures will not be affected at distances where they will collapse due to overpressure. For this reason, the survivability analysis is made primarily on the basis of overpressure. Designs for hardened equipment and structures, in general, will possess inherently an increased resistance to all types of radiation, and the overpressure normally will remain the critical effect. A design requirement of about $21\ kg/cm^2$ (300 psi) is needed to protect ABM system components from all but direct hits.

4.2.3. Nuclear Effects Analysis

When the threats against the ABM system and its vulnerabilities have been established, the survivability analysis can be undertaken. Several basic types of analysis can be made and a complex analysis will consist of a repetitive series of calculations of the basic types. Computers usually are employed to perform the computations economically.

4.2.3.1. Bonus Point Target

The first basic problem is that of calculating the survival probability of a critical point such as an antenna or missile launcher when a bomb is aimed at some other close-by point. The geometry of the situation is shown in Fig. 4.6 and this is called the bonus-pòint target problem. The first step in the analysis is to draw a lethal radius (LR) about the critical point. The LR is a function of the bomb yield and the target hardness, and is equal to the distance at which the overpressure resulting from the bomb will equal the hardness rating of the critical point. It will be assumed that if the bomb falls inside the circle of lethal radius the critical point will be destroyed. This is the so-called "cookie-cutter" approach and is a common approach used in vulnerability studies. The aiming error of the bomb is equally probable in any direction away from the target, and thus follows a Rayleigh distribution with a given standard deviation σ. The circular error probability (CEP) is defined as the median or 50 percent probability radius, and is equal to 1.7774σ. The equation for the probability density p for this distribution is given by:

$$p = \frac{1}{2\pi\sigma^2} \exp\left[-\frac{x^2 + y^2}{2\sigma^2}\right] \tag{4.13}$$

The probability that the critical point is destroyed by a bomb aimed at the target is obtained by integrating the probability density function p over the lethal area centered at the critical

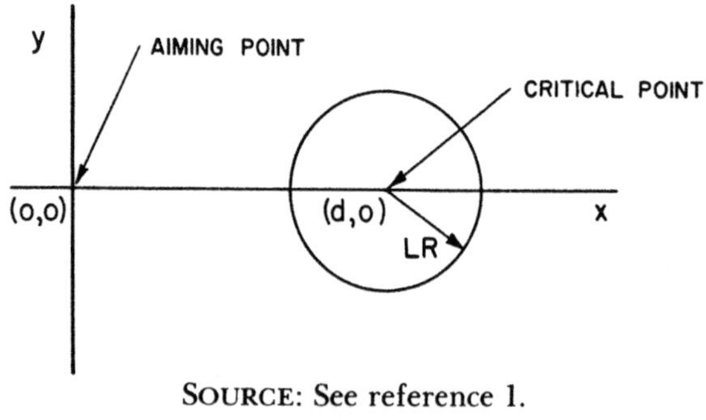

SOURCE: See reference 1.

Fig. 4.6. *Point target problem*

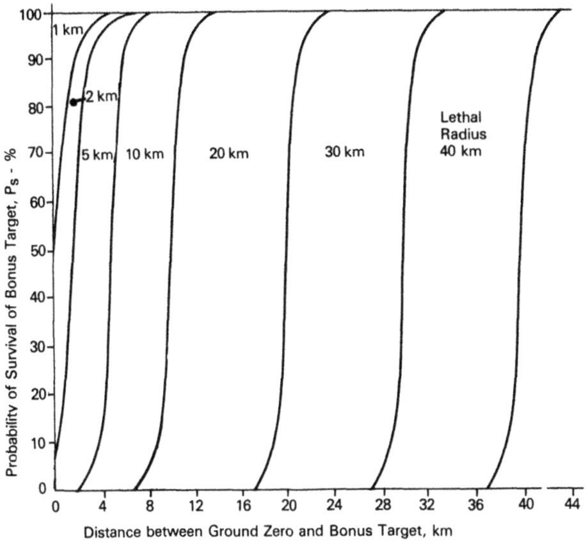

Fig. 4.7. *Probability of survival of bonus target as a function of the distance between ground zero and the bonus target; CEP = 1 km*

point. The result is given by:

$$P_k = \frac{1}{2\pi\sigma^2} \iint\limits_{\substack{\text{all } x,y \text{ in} \\ \text{lethal area}}} \exp\left[-\frac{x^2+y^2}{2\sigma^2}\right] dx\, dy \qquad (4.14)$$

Equation (4.14) represents the kill probability of the bonus-point target. Its complement therefore is the probability of survival of the bonus-point target. This latter probability is plotted in Fig. 4.7 as a function of the distance d between ground zero and the bonus target for a CEP of 1 km and various lethal radii.

4.2.3.2. Direct Point Target

The direct point problem occurs when the bomb is aimed at the critical point, and is a special case of the bonus-point target problem for $d = 0$. In this case, the general equation for the kill

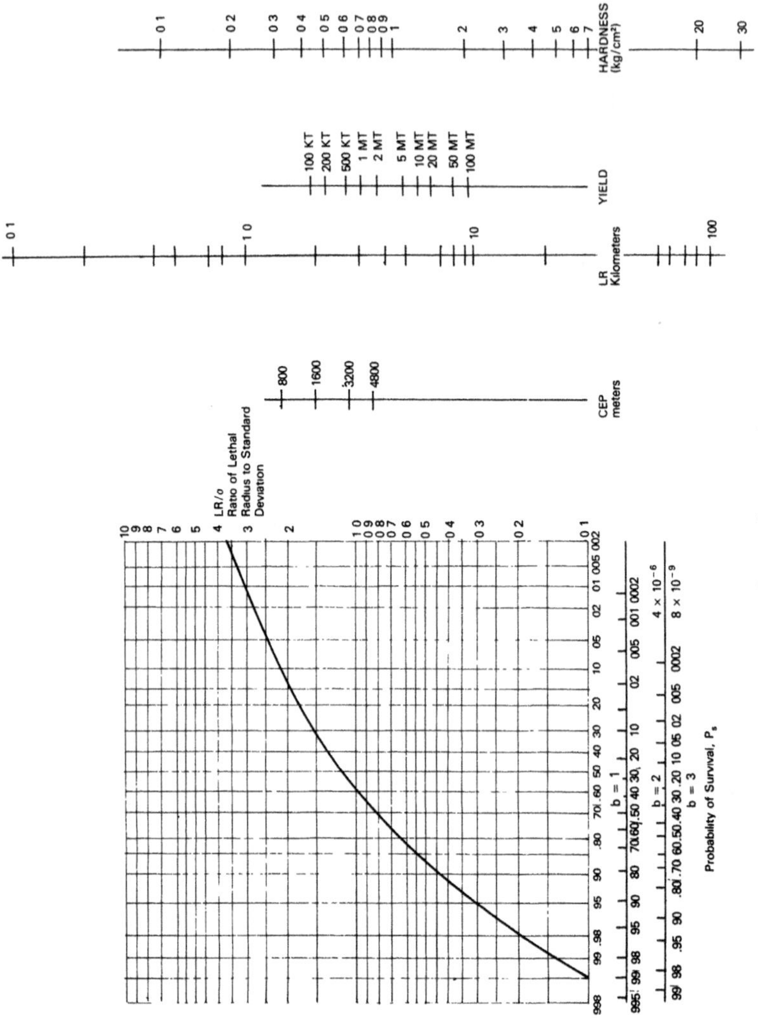

SOURCE: See reference 1.

Fig. 4.8. Single-site survival probability

probability reduces to the form:

$$P_k = 1 - e^{-0.69(\mathrm{LR/CEP})^2} \tag{4.15}$$

Figure 4.8 is a plot of equation (4.15) in the form of a nomogram. In this figure, the single-site survival probability can be obtained knowing the site hardness, the weapon yield, and the weapon CEP. Thus, a 7 kg/cm^2 (100 psi) target and 1 MT weapon with an 800 m CEP results in a lethal radius LR = 1,100 m and ratio, therefore, of lethal radius to standard deviation of LR/CEP = 1.4 for which the probability of survival in this example is 40 percent.

4.2.3.3. *Replicated-Dispersed Targets*

Another common problem which can be solved using the basic equation (4.14) is the calculation of the survival probability of replicated-dispersed targets such as a missile or interceptor farm. Figure 4.9 is a plot of the survival probability of four replicated-dispersed targets in the form of a nomogram, for various values of the target complex radius to the standard deviation of the weapon R/σ. The chief assumption made for the construction of this figure is that only one of the replicated-dispersed points needs to survive in order for the site to survive. The complex radius R is reckoned from the centroid of points in the replicated-dispersed target.

4.2.3.4. *Line Target*

Another basic problem is the calculation of the survival probability for a line target such as a cable, communication, or road. Figure 4.10 illustrates the line target problem. Since the aiming errors again have a Rayleigh distribution, equation 4.14 again applies. However, the integration instead of being over a circular area is now over an area defined by the limits $-\infty$ and $+\infty$ in the x direction and a and b in the y direction. This integral can be expressed in the form of an error integral, or error function, erf x. The equation is given by:

$$P_k = \tfrac{1}{2}(\mathrm{erf}(0.85b/\mathrm{CEP}) - \mathrm{erf}(0.85a/\mathrm{CEP})) \tag{4.16}$$

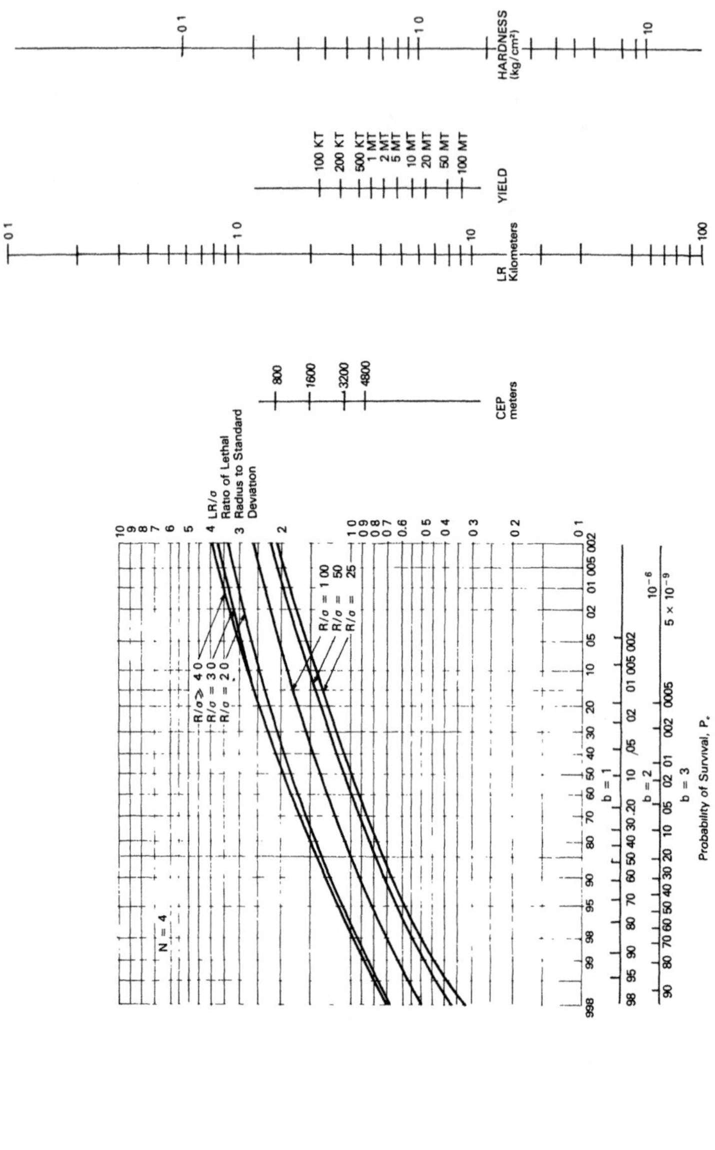
HARDNESS (kg/cm²)
YIELD
100 KT
200 KT
500 KT
1 MT
2 MT
5 MT
10 MT
20 MT
50 MT
100 MT
LR Kilometers
CEP meters
800
1600
3200
4800
LR/σ Ratio of Lethal Radius to Standard Deviation
N = 4
b = 1
b = 2
b = 3
Probability of Survival, Pₛ

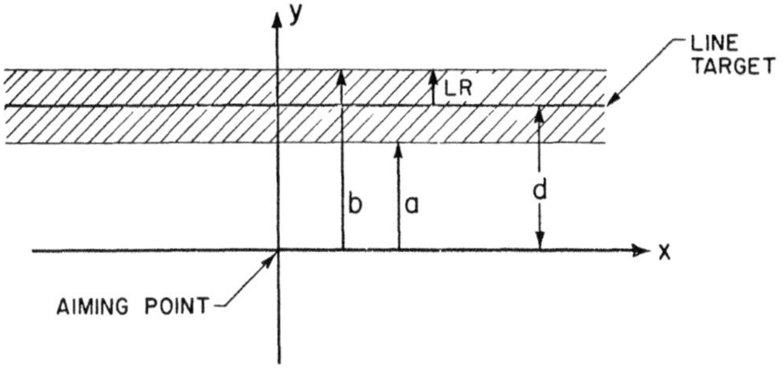

SOURCE: See reference 1.

Fig. 4.10. *Line target problem*

A special case occurs when the line itself is the direct target. In this case $a = -b$ and equation (4.16) for this case is given by:

$$P_k = \text{erf}(0.85b/\text{CEP}) \tag{4.17}$$

Equations (4.16) and (4.17) can be evaluated using standard tables for the error integral. Figure 4.11 is a plot of equation (4.17) in the form of a nomogram.

4.2.3.5. *Combinations of Simple Targets*

More complex problems involving combinations of the basic problems can now be evaluated by combining probabilities for the basic problems. Several examples of problems which include combinations of basic problems are shown in Fig. 4.12.

Another case which is even more difficult to analyze is the case where the target is free to move in an area, e.g., submarines in the ocean or mobile targets on land. The analysis of such cases is generally obtained in terms of an equivalent hardness and this is used to calculate the lethal radius in the usual manner.

The target complex of an ABM system includes radars and their antennas, communications, road and access ways, interceptor farms, and strategic missile silos. All of the foregoing analyses may be invoked in the larger problem of determining the survivability of such complex targets with respect to nuclear bomb damage.

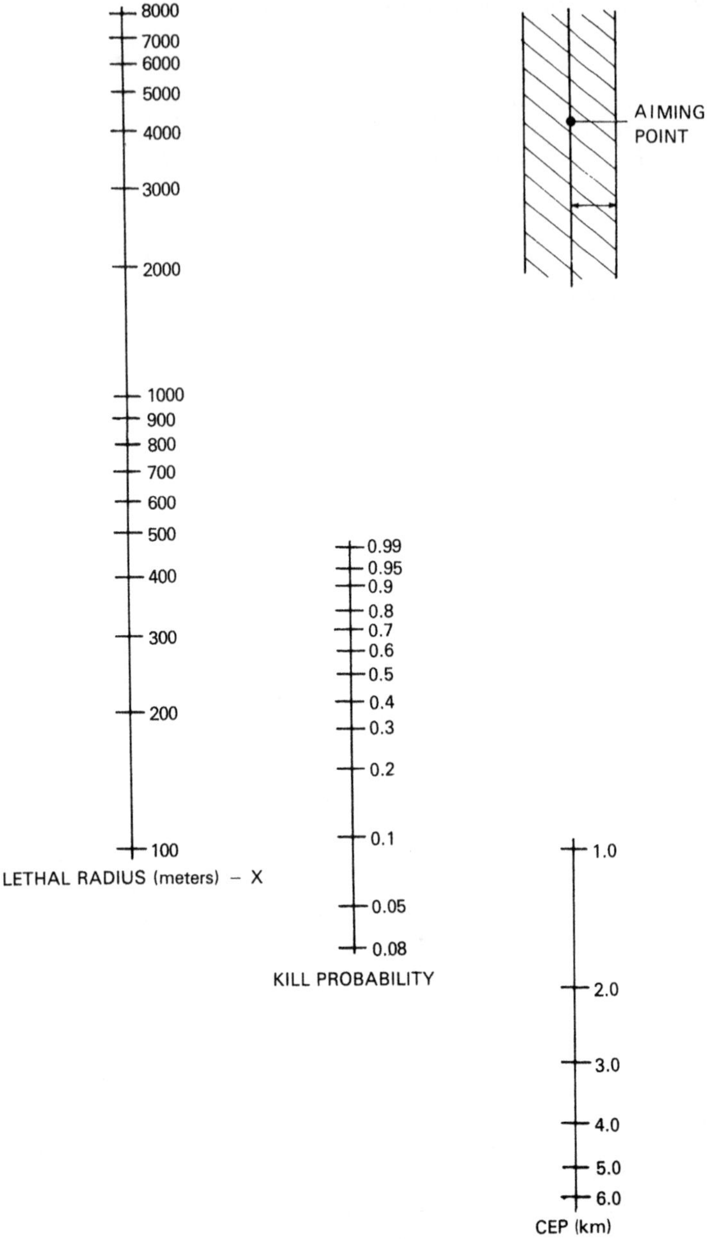

SOURCE: See reference 1.

Fig. 4.11. *Kill probability for link with aiming point in center of link*

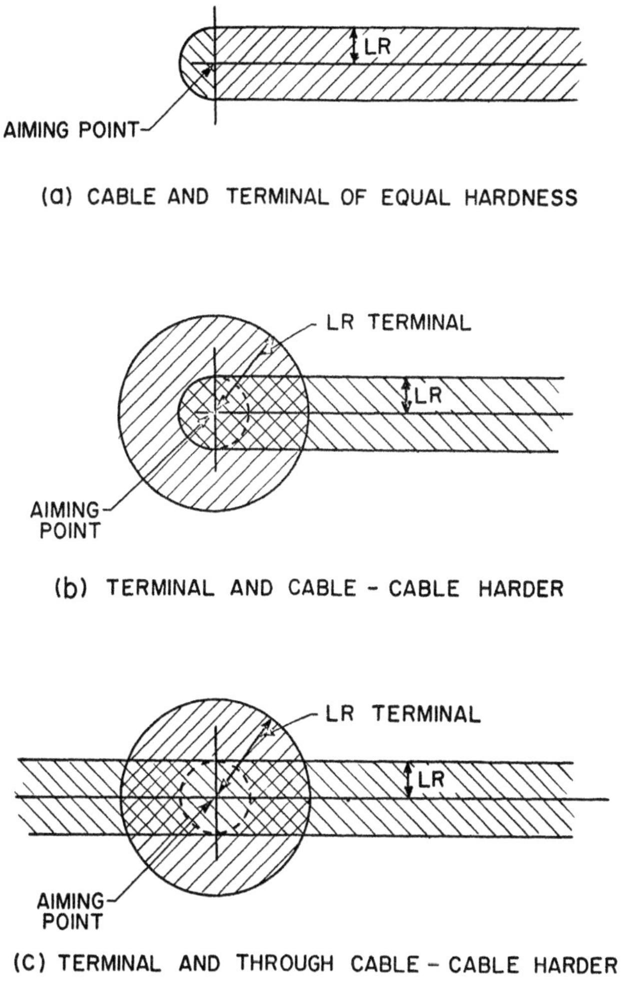

SOURCE: See reference 1.

Fig. 4.12. *Combination problems*

4.3. Survivability with respect to Chemical, Biological, Radiological Warfare, and Sabotage (CBR&S)

References 32–35, 43–47

The survivability of the ABM system with respect to CBR&S warfare can also be discussed in terms of the threat, vulnerabilities, and analysis of the CBR&S environment.

As is the case with ECM and nuclear bombs, a number of parameters which relate to the specification of the threat and the determination of ABM vulnerabilities must first be defined in order to analyze the effects of CBR&S on the ABM system. Criteria must therefore be established for effective CBR&S in terms of personnel and equipment being affected by the CBR&S threat.

4.3.1. CBR&S Threat

References 32–34

With the exception of some component degradations due to radiation, mentioned in connection with the discussion of nuclear bomb effects, CBR&S warfare by its nature can be considered primarily as a threat against the personnel required to operate the ABM system rather than against the equipment portions of this system. Sabotage is an exception and may be considered primarily as a threat against the equipment portions of the ABM system.

Although chemical and biological weapons are linked together in the popular literature, seven categories of such weapons or techniques can be distinguished: lethal, incapacitating, biological, chemical, anti-plant, radiological, and sabotage weapons. The distinction between between lethal and incapacitating agents is not altogether clearcut since a continuous spectrum of agents exists from the highly lethal to the generally non-lethal.

4.3.1.1. *Lethal Biological Weapons*

References 32–34

Lethal biological weapons operate by disseminating clouds of disease germs over the target area or upwind from it. The germs would then be inhaled by the target population.

Table 4.2. *Lethal biological agents by category and effects*

AGENTS	DISEASES	INCUBATION PERIOD (DAYS)	EFFECT OF TREATMENT	CONTAGIOUSNESS
VIRUSES	EASTERN EQUINE ENCEPHALITIS	5 TO 15	NONE	BY VECTOR
	TICK-BORNE ENCEPHALITIS	7 TO 14	NONE	BY VECTOR
	YELLOW FEVER	3 TO 6	NONE	BY VECTOR
RICKETTSIA	ROCKY MOUNTAIN SPOTTED FEVER	3 TO 10	GOOD	BY VECTOR
	EPIDEMIC TYPHUS	6 TO 15	GOOD	BY VECTOR
BACTERIA	ANTHRAX	1 TO 5	MODERATE	LOW
	CHOLERA	1 TO 5	GOOD	HIGH
	PNEUMONIC PLAGUE	2 TO 5	MODERATE	HIGH
	TULAREMIA	1 TO 10	GOOD	LOW
	TYPHOID	7 TO 21	GOOD	HIGH

SOURCE: See reference 32.

Table 4.2 gives a list of lethal biological agents by category and effect. In the last column of the table contagion by vector means that transmission by certain species of mosquitos or other insects is indicated.

Among the lethal biological agents that might be considered for military use are the viruses of Eastern equine encephalitis and yellow fever, the rickettsia causing Rocky Mountain spotted fever, and the bacteria causing anthrax, plague, cholera, glanders, and melioidesis. As example, the lethality of the disease anthrax is devasting if the bacteria are sprayed in the air in an aerosol. The inhalation of approximately 50,000 spores of the bacterium B-anthracis weighing less than a millionth of a gram is believed to be enough to cause a 50 percent chance of contracting pulmonary anthrax. Symptoms would first appear about a day after the attack. The onset might be mistaken for a common cold, but this would be followed by severe coughing, cyanosis, respiratory failure, and death. Untreated pulmonary anthrax is almost always fatal.

Lethal biological weapons have the potential of killing human populations over large areas even on a par with nuclear weapons. The threat is more formidable since the technology for such weapons is considerably less complex than the comparable technology for producing nuclear weapons. The substitution of lethal biological weapons for nuclear warheads is a distinct possibility.

4.3.1.2. Incapacitating Biological Weapons

Incapacitating biological weapons are disseminated in the same way as are lethal biological agents. The clouds of disease germs are spread over the target area or upwind from it. The germs are then inhaled by the target population.

Table 4.3 gives a list of incapacitating biological agents by category and effect. As with the previous table, contagion by vector means transmission by certain species of mosquitos or other insects.

Among the incapacitating biological agents that might be considered for military use are the viruses of chikungunya fever, dengue fever, and Venezuelan equine encephalitis. The latter causes severe headache, nausea, and prostration but has a case fatality rate in natural epidemics of 0.5 percent or lower. Other incapacitating biological agents are rickettsiae causing Q

Table 4.3. *Incapitating biological agents by category and effect*

AGENTS	DISEASES	INCUBATION PERIOD (DAYS)	EFFECT OF TREATMENT	CONTAGIOUSNESS
VIRUSES	CHIKUNGUNYA FEVER	2 TO 6	NONE	BY VECTOR
	DENGUE FEVER	5 TO 8	NONE	BY VECTOR
	VENEZUELAN EQUINE ENCEPHALITIS	2 TO 5	NONE	BY VECTOR
RICKETTSIA	Q FEVER	10 TO 21	GOOD	LOW
BACTERIA	BRUCELLOSIS	7 TO 21	MODERATE	NONE
FUNGI	COCCIDIOIDOMYCOSIS	7 TO 21	POOR	NONE

SOURCE: See reference 32.

fever, bacteria causing brucellosis, and fungi producing coccidioidomycosis.

The use of incapacitating biological agents might be considered in certain unlikely and extreme situations. Perhaps the most attractive scenario for their use is the case where a large friendly force is entrapped by enemy troops deployed over an extensive area and intermingled with civilians. In this situation, the employment of incapacitating biological weapons in forward and rear areas might impede the enemy advance long enough to allow reinforcement or evacuation of the friendly troops when the alternatives might be to use nuclear weapons or surrender.

4.3.1.3. Lethal Chemical Weapons

Modern lethal chemical weapons employ the nerve gases first developed but not used by Germany during World War II. These agents are hundreds of times more poisonous than the poison gases of World War I; they kill when they are inhaled or when they are deposited as liquid droplets on the skin. The term nerve gas derives from the fact that these agents operate by interfering with the transmission of nerve impulses across synapses. They do so by inactivating the enzyme chloinesterase, which normally functions to terminate the transmission of a nerve impulse. In the presence of a nerve agent, nerve impulses continue without control, causing a breakdown of respiration and other functions. Death caused by nerve gas poisoning results from asphyxiation. It is preceded by blurring of the vision, intense salivation, and convulsions.

Table 4.4 gives a list of both lethal and incapacitating chemical agents by category and some of their major characteristics. The table gives chemical warfare agents included in the U.S. stockpile. This stockpile includes two kinds of nerve agent. One is GB, a code name for isopropylmethylphosphonofluoridate. It is also known as Sarin and was produced in limited amounts by Germany during World War II. Sarin is a volatile liquid that evaporates at room temperature to a colorless and odorless gas. Weapons containing Sarin release it as a spray, which then evaporates to create a respiratory hazard for unprotected personnel. The lethal exposure for man is estimated to be approximately 10 milligram-minutes per cubic meter. This means, for example, that a man would accumulate a lethal dose in ten

Table 4.4. *Lethal and incapacitating chemical agents by category and major characteristics*

COMMON NAME	SARIN	VX	MUSTARD	BZ	CS AND CS-2
MILITARY CLASSIFICATION	LETHAL AGENT (NERVE GAS)	LETHAL AGENT (NERVE GAS)	LETHAL AND INCAPACITATING AGENT	INCAPACITATING AGENT (PSYCHOCHEMICAL)	HARASSING AGENT
PHYSICAL STATE	LIQUID	LIQUID	LIQUID	SOLID	SOLID
FORM AS DISSEMINATED	VAPOR, AEROSOL OR SPRAY	AEROSOL OR SPRAY	SPRAY	AEROSOL OR DUST	AEROSOL OR DUST
DURATION OF CONTAMINATION	HOURS OR DAYS	DAYS OR WEEKS	DAYS OR WEEKS		CS—MINUTES CS-2—WEEKS

SOURCE: See reference 32.

minutes if the concentration of Sarin in the air were 10 milligrams per cubic meter. The other kind of nerve agent in the U.S. stockpile is VX. The chemical formula of VX is still secret, although it has been suggested that the agent is ethyl S-dimethylaminoethylmethylphosphonothiolate. It is a member of a class of compounds first prepared in the mid-1950s in the search for improved insecticides. Sarin was also the outcome of insecticide research. Also a liquid, VX is several times more toxic than Sarin and much less volatile, being lethal either when inhaled or when deposited on the skin. VX kills in a matter of minutes, and by contaminating the ground and objects on which it is deposited it can make an entire area hazardous for many days. The lethal dose of VX applied to the skin has been estimated from 2 to 10 milligrams, depending on the site of application. Since contact with even a small droplet of VX can be fatal, adequate protection requires the wearing of a special suit as well as a gas mask.

Although U.S. policy now treats toxins in the same way as biological weapons, they also have been classified technically as chemicals because they do not reproduce. Toxins are poisonous substances produced by living organisms including plants, animals, and bacteria. Examples are ricin from castor bean, tetrodotoxin from the globefish, and botulin from the bacterium Clostridium botulinum. Some toxins, such as botulin, are highly lethal to man; others such as the staphylococcus enterotoxin, the substance responsible for staphylococcal food poisoning, are usually only temporarily incapacitating. Even though toxins are not capable of reproduction and therefore cannot cause epidemics, they do induce many of the symptoms associated with infection by disease organisms. Indeed, the principle pathological symptoms of several bacteriological diseases are thought to be caused by toxins produced within the human body by the living micro-organisms. Thus, toxic weapons, both in terms of the means of their production and the symptoms they cause, are closely related to the biological ones. For use as weapons, toxins would be dispersed in much the same way as biological and chemical weapons. The chief military argument for having toxins is that, because of their great potency, the weight of toxic munitions needed to cover a given area would be lower than the corresponding requirement for standard chemical munitions.

4.3.1.4. *Incapacitating Chemical Weapons*

Modern incapacitating chemical weapons are of two types, one with effects lasting considerably beyond the period of exposure and one with brief effects. An example of the former type is the U.S. agent BZ. This is a psychochemical, the chemical identity of which is still secret. It has been speculated that it belongs to the family known as benzilates. BZ is a solid that can be dispersed as an aerosol to be inhaled by enemy personnel. It affects both physical and mental processes, causing blurred vision, disorientation, and confusion. Its incapacitating effects can last for several days.

The principle short-term incapacitant now in military use is CS, which is the designation for orthochlorobenzalmalonitrile. This compound was first synthesized in the USA in the 1920s. After World War II it was developed by the British government as a riot-control agent and named after its American discoverers, Ben Corson and Roger Staughton of Middlebury College. When employed for military purposes, it is more accurately described as a harassing agent. The first CS munitions operated by vaporating the agent from a pyrotechnic mixture. The CS then condenses to form an aerosol of micron-size particles. A newer form, designated CS-2, is used in both bulk-disseminating devices and bursting bombs; it is a powder consisting of micron-sized particles treated with silica gel and silicone compound to improve its flow properties and persistence in the field. CS-2 can be effective in the field for several weeks. It is reintroduced into the air by the wind and by the movement of people and vehicles. The effects of CS depend on the particle size of the aerosol. Particles larger than about 50 microns exert their predominant effect on the eyes, whereas smaller particles are more effective as lung irritants. For military use, CS-2 is milled fine enough to achieve the latter effect. Exposure to either form of CS causes intense pain in the eyes and upper respiratory tract, progressing to the deep recesses of the lungs and giving rise to feelings of suffocation and acute anxiety. In humid weather, moderately heavy skin exposure can cause severe blistering that requires many days for healing. If the exposure is not excessive, the symptoms usually pass within a few minutes after the exposure ends. The lethal dose for man, as estimated from animal experimentation, is very much higher than that required to cause intense irritation. Nevertheless, heavy or prolonged exposure, such as might be expected in confined spaces or in close proximity to a munition emitting the agent, could cause serious lung damage and death, particularly among infants and the infirm. No long term after-effects of moderate exposure to CS have been demonstrated.

4.3.1.5. Anti-Plant Agents

Anti-plant agents are chemical weapons directed principally against the target plant environment. These agents were first developed for military purposes during World War II and subsequently came into wide use for weed control. Three principal anti-plant agents or mixtures have been in military service. They are designated Orange, White, and Blue. Agent Orange is an equal mixture of the *n*-butyl esters of 2,4-dichlorophenoxyacetic acid (2,4-D) and 2,4,5-trichlorophenoxy-acetic acid (2,4,5-T). The agent is mainly directed against forest vegetation. It is applied undiluted at three gallons per acre, approximately twenty-five times the average amount for domestic application. Within a week or more after its application the leaves fall from most jungle trees. Some concern has been voiced that 2,4,5-T may cause human birth defects. Agent White is a 4:1 mixture of the triisopropanolamine salts of 2,4-D and 4-amino-3,4,6-trichloropicolinic acid. The latter component is known as picloram. The agent is sprayed from a water solution at a rate of approximately 3.42 kg (7.6 lbs) of total herbicide per acre. It is generally used for the same purposes as agent Orange, although its somewhat lower volatility makes it preferable for operations where drifting would be a hazard, as in the vicinity of friendly installations. The extreme resistance of picloram to biodegradation has been a factor in limiting its employment. Agent Blue is a water solution of sodium dimethylarsinate, applied at the rate of 4 kg (9 lbs) per acre and is used mainly against rice crops.

4.3.1.6. Radiological Weapons

Approximately 50 percent of the energy released by a nuclear explosion escapes in the form of radiation. Of this energy, approximately 30 percent is released in the form of nuclear radiation while the remaining 70 percent is released as thermal radiation. As discussed previously, radiation degrades the performance of many electronic components, particularly semiconductor devices, at ranges where the bomb blast is no longer critical. The chief concern, however, with radiations is with their long-term effect on the personnel required to operate equipment. Regarding nuclear radiation effects, there are several types of radiations of concern some of which occur immediately, within one minute of a bomb's detonation, and some over considerably longer periods of time, months or even years.

The nuclear bomb is considered to be the prime example of a radiological weapon. Other less spectacular weapons however are also feasible. These are based on the location of radiation sources in areas used by target personnel where proximity to the source or prolonged exposure may be either lethal or incapacitating. The dissemination of low-level radiological weapons may be made in a way quite similar to that for disseminating biological and chemical agents.

4.3.1.7. Sabotage

Sabotage is an act by enemy personnel, few in numbers, against critical elements or components of the ABM system. A few strategically-placed saboteur agents in a highly organized ABM system could systematically reduce its effectiveness without being detected. In a way which simulates ECM, the saboteur teams must first infiltrate or penetrate the system which is to be destroyed. Accordingly, a probability of penetration and probability of kill can be defined for sabotage. The target-system vulnerability to sabotage can therefore be expressed in terms of a probability of survival.

It appears intuitive that the probability of survival of the target system would be a strong function of the number of saboteur teams and their points of access to the ABM system. To be successful, the saboteur teams would have to penetrate to the most critical points, nodes and lines, of the ABM system and, furthermore, have the technical capability to destroy the system once infiltration has been accomplished. Saboteur teams would most likely remain dormant until actual hostilities broke out.

4.3.2. CBR&S Vulnerabilities

References 32–34

ABM system vulnerabilities to CBR&S warfare are chiefly a function of countermeasures taken for the protection of personnel and equipment. Such countermeasures may include effective filtration systems for hard installations, effective gas masks and effective protective clothing for individuals, inoculations, and availability of medical supplies. With regard to sabotage, countermeasures may include screening of personnel employed in critical system operations and the implementation of guard and security measures.

There are moderately effective vaccines and antibiotics against some biological weapons, but none are available against others. Moreover, such protections might be overwhelmed by a massive attack, and for some biological agents, antibiotics can be circumvented by the use of drug-resistant strains. Protection can be given by gas masks or air-filtered shelters if there is early warning of an attack, but no satisfactory early-warning device has yet been developed. In any case, a program for supplying the general civilian population with masks and shelters and maintaining discipline for their use would require a major and sustained economic and political effort without achieving a corresponding reliable protection against biological attack. It is also not to be expected that biological warfare agents would be deliberately chosen to be contagious; that would maximize the risk of spreading disease far beyond the intended target, and possibly to the territory of the attacker and his allies. Nevertheless, the unnatural conditions inherent in military operations create the possibility that widespread epidemics would be unintentionally started. There is also the hazard, difficult to evaluate, that the bacteria or viruses used in an attack, or even used in a field test of the biological weapon, could subsequently emerge from exposed populations of humans, rodents, birds, or other animals, with increased persistence, virulence, and contagiousness for man. Thus, lethal biological weapons present a devastating threat of killing human populations over large areas. The relative ease with which such weapons can pass to third-party nations if the technology becomes established along with their uncontrollable performance have persuaded the USA to renounce them altogether. The possession of such weapons by the USA, it is felt, would add little, if anything, to the U.S. strategic deterrent capability, whereas their proliferation would present a major threat to all mankind.

Any decision to launch an incapacitating biological attack must take into account the fact that substantial numbers of civilians, particularly infants and the infirm, will almost inevitably be killed, even if the case fatality rate is only a few tenths of a percent. In addition, there is the possibility that the fatality rate under military conditions might be much higher than the rate estimated from natural occurrences of the disease and from various kinds of experimental data. To be distinguished from effects caused by the first use of biological weapons is the possible escalation and enlargement of their use in retaliation or in subsequent strikes. A principle long-term cost of using incapacitating biological weapons, or even maintaining them, would be the stimulation of the proliferation of germ weapons, including lethal ones, and the corresponding increase in the probability of occurrence of accidents. The facilities for developing, producing, and delivering incapacitating biological agents are essentially the same as those required for lethal germs. The risks of using incapacitating biological weapons appear to outweight the possible gains.

In the case of chemical weapons, since Sarin affects mainly the respiratory system, the use of gas masks provides good protection. Modern gas masks are capable of reducing the concentration of all known war gases by a factor of about 10^5. In addition, there are chemical antidotes for nerve agents that can provide protection if the dose of the agent is not very great and the antidote is administered promptly. On the other hand, even a small drop of VX can be fatal, so that adequate protection requires the wearing of a special suit as well as a gas mask. Protective suits and masks are extremely cumbersome. They are mechanically awkward, and the buildup of heat is a serious problem. Fighting efficiency would be seriously reduced by the wearing of full protective equipment and also by the strict observation of various special precautions necessary for survival in a lethal chemical environment. The use of lethal chemicals against an enemy without protective equipment would be devastating. Against an enemy possessing suits and masks and able to enforce their use of its troops, the use of such chemicals would result in a nuclear-type stalemate or parity.

Given adequate warning, a highly disciplined population could be defended against a nerve gas attack by a combination of gas masks, protective shelters, and antidote therapy. Although strategic chemical weapons would add nothing significant in terms of capability to the arsenals of the nuclear powers, the proliferation of such weapons among third nations would constitute a serious hazard. Even though it is unlikely that a third nation could deliver a chemical attack over wide enemy areas, it could nevertheless penetrate a limited number of cities using conventional airborne or shipborne delivery means.

For use as weapons, toxins could be dispersed as aerosols in much the same way as biological and chemical weapons. Because toxins are not absorbed effectively through the skin, gas masks would provide protection, as would shelters fitted with special air filters. Protection can also be afforded by prior immunization with specific toxoid. Each toxoid, however, is effective only against a particular kind of toxin, and for some toxins the margin of protection is not enough to be of practical significance.

4.3.3. CBR&S Effects Analysis
References 32–34

Analysis of the AMB system with respect to CBR&S warfare utilizes the same methods as those discussed generally under survivability of nuclear bomb attack, the only difference being that other effects are substituted for overpressure and different vulnerability parameters are substituted for hardness. One additional modification for the case of chemical and biological agents is that the wind velocity in the target area must be included in the specification of the contours of the particular effect being considered. Mathematically, this involves a modification of the lethal area by the wind velocity vector. This very probably will cause integration difficulties, but graphical solutions are available for the solution of odd-shaped lethal areas.

Chemical and biological weapons have serious shortcomings from a military viewpoint. Their effects are not as predictable as those of other weapons. They might get out of control. Alternative and already available weapons are preferable, and the acquisition of a biological weapons capability would be in addition to, not a substitute for, preexisting military expenditures and programs. Military officers and political leaders are strongly disinclined to use biological weapons, partly because of

Table 4.5. *Comparison of nuclear, chemical, and biological weapons*

	TYPE OF WEAPON		
AGENTS	1 MT NUCLEAR	15 tons CHEMICAL	10 tons BIOLOGICAL
AREA AFFECTED	UP TO 300 SQUARE KILOMETERS	UP TO 60 SQUARE KILOMETERS	UP TO 100,000 SQUARE KILOMETERS
TIME TO EFFECTIVENESS	SECONDS	MINUTES	DAYS
DAMAGE TO STRUCTURES	WIDESPREAD DESTRUCTION	NONE	NONE
OTHER EFFECTS	PROLONGED RADIOACTIVITY IN AREA OF 2,500 SQUARE KILOMETERS	CONTAMINATION FOR DAYS OR WEEKS	POSSIBLE EPIDEMIC OR NEW FOCI OF DISEASE
NORMAL USE OF AREA	3 TO 6 MONTHS	LIMITED FOR DAYS OR WEEKS	VARIABLE
EFFECT ON MAN	90 PERCENT DEATHS	50 PERCENT DEATHS	50 PERCENT MORBIDITY

SOURCE: See reference 32.

unfamiliarity and partly because of the moral revulsion and apprehension that are undeniably associated with these weapons. Nevertheless, these weapons possess a wide area of coverage. A single strategic missile with a biological warhead could devastate an area perhaps 100 times as large as that devastated by a conventional nuclear weapon.

Table 4.5 compares three weapons for their damaging effects. The weapons are a 1 MT nuclear bomb, 15 tons of a chemical nerve agent, and 10 tons of a biological agent. In each case, the table assumes delivery of the weapon by a single missile on an unprotected population. It can be seen in the table that the 1 MT warhead would affect an area up to 300 square kilometers. On the other hand, it would require perhaps twenty to thirty missiles to dump chemical and biological agents in the required amounts to affect up to 60 square kilometers, in the case of chemicals, and 10^5 square kilometers in the case of germ gas. Thus, a single strategic missile with a biological warhead could be ten to twenty times more effective than the same missile carrying a nuclear warhead. This possibility must be acknowledged in strategic warfare.

Sabotage is difficult to analyze, since most of the variables cannot be defined by natural or physical laws. The variables of interest are centered necessarily about the individuals involved in both the sabotage effort and in the security control which protects the system. Some studies of this problem have been made, particularly with respect to missile systems. In general, these studies attempt to arrive at a probability of successful sabotage as a function of a number of assumptions made for the specific system. For the ABM system analysis, similar assumptions must be established for sabotage attacks against the various system installations, and then probabilities calculated for inclusion in overall survival probabilities.

4.3.4. International Agreements Evolution
Reference 35

The most important international effort to prohibit the use of chemical and biological weapons is the Geneva Protocol of 1925. It prohibits (1) the use in war of asphyxiating, poisonous, or other gases, and of all analogous liquids, materials or devices, and (2) the use of bacteriological methods of warfare. The protocol does not prohibit research, development, testing, or production of gas or germ weapons. It does not prohibit the use of such weapons in reprisal against their first use by the enemy. It does not prohibit the use of riot-control gases or other agents for domestic police purposes. It does not prohibit the use in war of non-toxic smokes used for concealment or of flamethrowers, napalm or other incendiary weapons.

The peace treaties of World War I treated gas warfare as being already prohibited and specifically forbade the manufacture and importation of war gases by Germany and her wartime allies. The 1922 Washington Treaty on Submarines and Noxious Gases prohibiting gas warfare was ratified by the USA, Great Britain, Italy, and Japan but never came into force because France, whose ratification was required, objected to its provisions on submarines. The 1925 Geneva Protocol for the Prohibition of the Use in War of Asphyxiating, Poisonous, or Other Gases, and of Bacteriological Methods of Warfare was signed by representatives of thirty-eight nations. This protocol however was not ratified by the USA mainly for lack of public support and opposition by the chemical industry and other special interests.

By 1939 the Geneva Protocol had been ratified by forty-four nations, including all major European powers. At the outbreak of World War II, England, France, and Germany exchanged assurances that they would abide by the protocol. In 1943, the U.S. President declared that gas warfare was outlawed by the general opinion of civilized mankind and that the USA would under no circumstances resort to the use of gas weapons unless they were first used by the enemy. As a result, gas was not used in World War II. Since World War II, the USA has on numerous occasions declared its support for the no-first-use principle of the Geneva Protocol, but did not take the step of ratifying the protocol itself. In 1966 the U.N. General Assembly called for strict observance by all states of the principles and objectives of the Protocol and invited all nations that had not done so to accede to the Protocol.

4.3.4.1. Present International Agreements

A total of 106 nations are now parties to the Geneva Protocol; thirty-eight of them have ratified it since the 1966 U.N. resolution. The parties include all North Atlantic Treaty Organization (NATO) members, including the USA, all members of the

Warsaw Pact, including the USSR, and all nuclear powers, including the People's Republic of China. All major industrial nations are now parties.

The 1925 Protocol bans the first use but not the retaliatory use of gas and bacteriological weapons. Going beyond the 1925 Protocol, the 1972 Biological Weapons Convention banned the development, production or other acquisition of any biological weapon. However, as of now, there is no prohibition on the production or stockpiling of chemical weapons, including nerve gases. The USA and presumably the other powers have large stocks of extremely potent nerve gases. The USA and other signatories of the 1972 Convention committed themselves at that time to eventually eliminating all chemical weapons.

Whether or not the Geneva Protocol prohibits the use of riot-control gases and herbicides is a subject of some dispute. Since 1965, questions have been increasingly raised regarding the employment by the USA of such agents, CS and herbicides, in Vietnam. In 1969 the U.N. passed a resolution holding that the Geneva Protocol prohibits the use in war of all chemical agents directed at men, animals, or plants. The resolution was passed by a vote of 80 to 3, with 36 abstentions. Portugal, Australia, and the USA voted against the resolution. The USA has argued in the past that the Geneva Protocol does not, for humanitarian purposes, prohibit the use in combat against an enemy of agents commonly used to control domestic riots or of chemicals used to control unwanted vegetation.

4.3.4.2. USA Policy

The USA has renounced all forms of biological weapons and the first use of most chemical weapons, including riot control gases and herbicides. However, in the past, the USA, unlike the rest of the world, has held that riot control gases and herbicides are not agents of chemical warfare.

In 1969 President Nixon proclaimed that the USA will never be the first to use lethal or incapacitating chemical weapons and will not use biological weapons under any circumstances, even in retaliation. He also announced that he would submit to the Senate for its ratification the 1925 Geneva Protocol prohibiting the use of chemical and biological weapons in warfare. In 1970 the President extended the policy for biological weapons to cover toxins as well. Under the new policy the USA will destroy existing stocks of germ and toxin weapons and will no longer engage in their development, production, or stockpiling. The USA biological program will be restricted to research for strictly-defined defensive purposes, such as techniques for immunization. Accordingly, plans now in force dismantle military biological and toxin installations and the existing sizable facilities are being transferred to non-defense agencies of government for peaceful uses.

With regard to chemical weapons, the USA policy is less sweeping, in that it reserves the right to continue the development, production, and stockpiling of lethal and incapacitating chemical weapons. It pledges, however, never to use them first. The same pledge is embodied in the treaty commitment by the 106 states now party to the Geneva Protocol.

In January, 1975 the USA acceded to the Geneva Protocol of 1925 banning any first use of gas and bacteriological weapons, and to the Biological Weapons Convention of 1972, banning the development, production, or possession of bacteriological and toxin weapons. The USA policy is presumably based on the review of existing chemical and biological weapons, weapons that might conceivably be developed, the possible military value of each type of weapon to other weapons available, the possible consequences of proliferation, strategic questions concerning the use of such weapons for threat or deterrence, the political and moral impact of their development, and possible use, the potential hazards for civilian populations, possible defenses against the various weapons, and the policies and activities of other nations.

4.4. Summary

The survivability considerations presented in this chapter lead to the following general conclusions.

First, the determination of the survivability of an ABM system is a complex problem. It involves the consideration of numerous factors of threat, vulnerability, and analysis that in themselves are complex. Yet each factor must be individually defined, specified, and analyzed before any collective determination of the probability of system survival can be attempted.

Second, the determination of survivability is in a constant state of flux because of changing offensive capabilities and corresponding changes in defensive strategy against these capabilities. The effect of such changes in the survivability of the ABM system must be continually updated, modified, and evaluated and the results applied to the ABM system.

Third, there appears to be a need for establishing survivability criteria and a methodology for automating the analysis and for quickly obtaining results for variations of its assumptions and parameters.

Fourth, there appears to be a need for classifying existing disciplines of science and for developing new and improved tools and concepts for use in survival analysis.

Fifth, the nature of survival analysis extends beyond the survival of an ABM system following an outbreak of war to the survival of the civilian population in any circumstance. Clearly, the policy a nation must follow to survive includes non-military as well as military considerations. Survivability analysis therefore appears to warrant an emancipation for its purely military scope and use as a distinct and important tool in non-military evaluations of pending policy decisions.

References

1. O. G. Becker and H. E. Struck, "Aircom Survivability Considerations", 2nd Winter Convention on Military Electronics, IRE, Los Angeles, Feb. 1–3, 1961. Examines the parameters and develops techniques to determine the probability of survival of the U.S. Air Force worldwide communications system.
2. J. Constant, *Introduction to Defense Radar Systems Engineering*, Spartan Books, 1972. Presents topics on ballistic radar systems. Contains a treatment of ECM.
3. G. B. Panero, *Survivability of Air Defense Systems,* Mitre Corp., November, 1959. There is an extensive literature covering survivability and associated subjects, and this reference and the following three references are part of this literature, some of which is classified. Although the references provided are by no means complete with respect to all of the currently available material; they do list sources of data and results that may be useful to the reader.
4. J. Medhurst et al., "Distortion in Frequency Division Multiplex FM Systems Due to an Interfering Carrier", *IEEE Journal,* May, 1958.
5. Cook Research Laboratories, "Study of the Susceptability of Communications Equipment to Jamming", Report 54 WCLN 2052-41, Contract AF33(616)-115.
6. S. Reiger, "Error Probabilities of Binary Data Transmission Systems

in the Presence of Random Noise", IRE Convention Record, Part 8, 1953.
7. F. Stumpers, "Theory of Frequency Modulation Noise", *IRE Proc.*, September, 1948.
8. S. Rice, "Statistical Properties of a Sine Wave Plus Random Noise", *Bell System Tech. Jour.*, Vol. 27, January, 1948.
9. M. Skolnik, *Introduction to Radar Systems*, McGraw-Hill, 1962. ECM and ECCM are discussed in Section 12.10.
10. Eds., "The silent war, an EW primer", *Countermeasures*, March, 1975. Few basic articles exist on Electronic Warfare (EW). The general tendency for the EW community is to mask its efforts in secrecy even to the point of not discussing information requiring no security classification. Consequently, individuals entering industry, workers in peripheral areas, and non-technical managers are relatively uninformed about EW even though they serve as program and hardware development managers. This article provides a general understanding and evaluation of the EW role, along with a glossary of associated terms.
11. J. Kamstra, "The History of ECM", *Countermeasures*, May/July, 1975.
12. J. Klass, "Avionic War Aims at Deceit, Confusion", *Aviation Week*, November 25, 1957.
13. J. Holahan, "Tools and Techniques of Electronics Warfare", *Space/Aeronautics*, April, 1960.
14. R. J. Schlesinger et al., *Principles of Electronic Warfare*, Prentice-Hall, 1961. A text on electronic warfare. Perhaps the only one available.
15. News item, "U.S. Weighs Israeli Request for Latest ECM Equipment", *Electronic Design*, November 8, 1973. Describes the performance of deceptive versus noise jammers.
16. H. David, "Mideast War is Forcing Changes in American Military Electronics", *Electronics Design*, December 6, 1973. Decribes the performance of deceptive versus noise jammers.
17. Ed., *International Countermeasures Handbook*, EW Communications, Inc., 3921 E. Bayshore Rd., Palo Alto, California, 94303, May, 1975.
18. Ed., *Countermeasures Sources*, Hamilton Burr Publishing Co., Box X, Cupertino, California, 95014, October, 1975.
19. L. N. Ridenour, ed., *Radar System Engineering*, Vols. 1 and 13, M.I.T. Rad. Lab. Series, McGraw-Hill, 1947. Literature on the exact techniques and capabilities of signature analysis and wake backscatter prediction exist, but is classified. The present reference and next six references provide a general discussion of signature analysis and wake prediction.
20. S. Macko, *Satellite Tracking*, John F. Rider, 1962.
21. D. K. Barton, "Sputnik II as Observed by C-band Radar" *IRE Convention Record*, Part 5, March, 1959. Describes the first usage of signature analysis to determine the configuration of a space object.

22. C. Brindley, "Target Recognition", *Space/Aeronautics,* June, 1965. Describes the state of art for signature analysis and for the identification of satellites and reentry bodies, in the exoatmosphere.
23. P. S. Lykoudis, "A Review of Hypersonic Wake Studies", *AIAA Journal,* Vol. 4, No. 4, April, 1966. Describes the state of art for wake studies.
24. F. L. Fernandez and E. S. Levinsky, "Air Ionization in the Hypersonic Laminar Wake of Sharp Cones", *AIAA Journal,* Vol. 2, No. 10, October, 1964. Computes the electron density concentration in the wake of a slender reentry body.
25. L. Lees, "Hypersonic Wakes and Trails", *AIAA Journal,* Vol. 2, No. 3, March, 1964. Describes the distribution of observables in hypersonic wakes.
26. R. L. Garwin and H. A. Bethe, "Anti-Ballistic Missile Systems", *Scientific American,* March, 1968. The authors argue that offensive tactics and cheap penetration aids could nullify the effectiveness of a light ABM system and any other visualized so far.
27. H. J. Miser, "Bombing Accuracy and Damage from Atomic Bombs: An Introduction", OATM No. 25, August 15, 1951.
28. Planning Research Corp., "Study of the Vulnerability of Dispersed Replicated Points", No. PRC-306-1, April 4, 1960.
29. S. Glasstone, ed., *The Effects of Nuclear Weapons,* U.S. Atomic Energy Commision, June, 1962.
30. C. Hammer, "Vulnerability of Communication Networks", Memo for ITTCS, September 19, 1960.
31. Rand Corp., "The Protection of Communications Terminal Facilities and Landline Routes", September 12, 1958.
32. M. S. Meselson, "Chemical and Biological Weapons", *Scientific American,* May, 1970. Explains the nature of chemical and biological weapons and their military implications, the international effort that has been made to prevent their use, and the present situation. The USA has renounced all forms of biological weapons and the first use of most chemical weapons. The issue of whether or not to include irritant gases and anti-plant agents in the prohibition remains open.
33. *Chemical and Biological Weapons,* UN Report, 1969. Examines the essential facts about chemical and biological weapons.
34. *Chemical and Biological Weapons,* World Health Organization Report, 1969. Examines the potential effects of chemical and biological weapons on civilian populations.
35. Eds., "The Geneva Protocol Accepted", Science and the Citizen, *Scientific American,* March, 1975.
36. D. Kaye, "Military turning to CMOS/SOS for radiation hardened circuits", *Electronic Design.* CMOS/SOS is inherently resistant to radiation.
37. "HARM on target", *Microwave Systems News,* April/May, 1975. HARM (High Speed Anti-Radiation Missile), an air-launched mis-

sile designated AGM-88A, will replace both SHRIKE (AGM-45) and STANDARD ARM (AGM-78).
38. S. Drell and F. von Hippel, "Limited Nuclear War", *Scientific American*, November, 1976, and in Letters February, 1977. The USA may be committing itself to preparing for a war limited to attacks on military bases, with relatively few civilian casualties. Would the casualties really be few, and could the war stay limited?
39. J. Giles, "Simulating the Nuclear Electro-Magnetic Pulse", *Countermeasures*, August, 1977. Protecting new and existing strategic weapons and communications systems against the effects of the electromagnetic pulse (EMP) produced by a nuclear detonation is receiving major emphasis by the Department of Defense. New systems including aircraft, sattellites, ships, as well as elements of the vast Command, Control and Communications (C^3) networks, are being designed and built to EMP survivability specifications.
40. R. Davis, "Industry Pushes Radiation Hardened Microcircuits", *Electronic Warfare*, March, 1978. Despite the maturity of MOS ICs, ICBMs rely on proven bipolar technology for penetrating intense radiation environments. CMOS-SOS is being proposed and investigated for next-generation strategic missiles. I^2L also looks promising. Reliability is key.
41. "The Nuclear Environment—What's it Like?", *Electronic Warfare*, March, 1978. The atmospheric burst environment of a nuclear weapon is a function of several parameters: yield, type of device, altitude, weather conditions and atmospheric density or the altitude at which detonated. All nuclear weapons produce four lethal effects when the warhead is detonated: blast, heat radiation, prompt radiation and fallout.
42. F. Kaplan, "Enhanced Radiation Weapons", *Scientific American*, May, 1978. Although President Carter has deferred production of "the neutron bomb", it is still an alternative of U.S. policy. It remains a weapon of doubtful utility and could result in an all-out nuclear exchange.
43. S. Sesser, "Critics Charge Army Is Continuing Research on Biological Warfare", *Wall Street Journal*, 2 April, 1971.
44. "U.S. finally signs poison gas ban pact", UPI, 23 January, 1975.
45. M. Shohat, "F/O Links Subject of Hardening Study," *ME/C*, September, 1978. Work has begun on the development of fiber-optic (F/O) data links capable of operating in a nuclear radiation environment.
46. "Soviet disaster believed linked to germ warfare agents", AP news item, 19 March, 1980.
47. M. Meselson and J. Robinson, "Chemical Warfare and Chemical Disarmament", Scientific American, April, 1980. Unless the current negotiations in Geneva succeed in prohibiting the development, production and stockpiling of chemical weapons, a new and more dangerous phase of the arms race may be in store.

5

Radar Counter-Countermeasures

In order for a surveillance or tracking radar to operate in a general-type countermeasures environment, no single radar countermeasure threat can be assumed dogmatically. The radar must be flexible in order to effectively combat the many known countermeasures techniques.

Radar flexibility can be accomplished judiciously by incorporating automatic features and by providing suitable over-ride displays and circuits which enable trained operators to make proper decisions in particular countermeasures environments.

5.1. Effects of Jamming

References 1–10, 42, 43, 53–56, 66–88, 91–96, 101–108

The effect of jamming a radar is the appearance of extraneous responses in the radar displays and in the radar computers. Such extraneous responses may be few in number and may resemble real targets, or there may be numerous responses but usually not resembling real targets. In either case, the radar degrades its usefulness as part of the ABM system since it must become preoccupied with filtering or discriminating unwanted responses. The various techniques that electronically interfere with the radar performance are called electronic countermeasures (ECM) and are divided into two classes, depending on whether they are intended primarily for confusion or for deception of the radar. The various techniques that are employed to combat ECM are called electronic counter-countermeasures (ECCM).

5.1.1. Confusion ECM

The purpose of confusion ECM is to mask or hide real targets by cluttering the radar display and radar computer with numerous responses. Usually, such responses resemble ground or sea clutter and, in heavy jamming, cover more area or volume in space and consequently on radar displays and computers. The latter case is the case for effective ECM which is characterized by a complete radar loss of visibility, i.e., effective blinding of the radar for all practical purposes. Both active and passive ECM devices are capable of creating confusion at the radar. Examples of confusion ECM are noise jammers and chaff. Nuclear bursts can also create confusion by masking targets beyond the burst. The general criterion for radar performance in a confusion ECM environment is the signal-to-jamming ratio.

5.1.2. Deception ECM

The purpose of deception ECM is to present the radar with false signals which appear to be real targets. If the radar cannot discriminate the false targets by its decision altitude, they must be considered threatening in which case the deception ECM is effective, i.e., it induces the radar to take incorrect action. If, on the other hand, many false targets can be made to appear to the radar, it might not be able to process them all, in which case again the ECM is effective. Examples of deception ECM are repeater jammers, spot chaff, and decoys. The general criterion for radar performance in a deception ECM environment is the deception ECM device altitude at which the radar can accomplish target discrimination. Such discrimination can be made by the radar and its computer on the basis of trajectory prediction, signature analysis, and atmospheric filtering.

5.2. ECM Design

References 9, 10, 11–14, 72–97, 103, 106–108

The problem of radar design to combat the effects of jamming can be considered using the general criteria for radar performance in both a confusion and deception ECM environments. Both the signal-to-jamming ratio and the deception ECM device altitude at which the radar can accomplish target discrimination must be kept high to insure target visibility and discrimination by the radar decision altitude.

5.2.1. General Objectives

Any radar can be effectively jammed by a determined enemy who is willing to pay the price. Conversely ECCM are based on making the costs of penetrating the defense prohibitive.

As a general premise, the value-exchange ratio favors the defense when the following equation holds:

$$\frac{\Delta V_D}{V_D} > \frac{\Delta V_0}{V_0} \tag{5.1}$$

where

V_D, V_0 = the value of the defense and offense systems, respectively; usually expressed in terms of such factors as availability, penetration, effectiveness, and cost, and

ΔV_D, ΔV_0 = incremental values of the defense and offense systems obtained by the introduction of ECCM and ECM, respectively; also expressed in terms of the increments in system availability, penetration, effectiveness, and cost.

The introduction of ECCM by the defense is based on making the value-exchange ratio less favorable or even prohibitive for the offense, i.e., substantially increasing the divergence of inequality (5.1). Inequality (5.1), therefore, broadly determines the criteria for estimating the value of a given ECCM device. It can also be used in comparing different ECCM, as well as ECM, devices.

It should be noted that cost is but a single factor in the consideration of the value-exchange ratio. To properly evaluate the value of a given ECCM, both initial and incremental values for both the offense and defense must be available. In practice the ratio $\Delta V_0/V_0$ for the offense may not be reliably available, with consequence that it can only be estimated on the basis of limited intelligence data and from the knowledge of comparable system factors for the friendly offense. The latter data, however, may or may not be pertinent to obtaining an estimate for the enemy offense system characteristics. Nevertheless, the criterion given by inequality (5.1) can be used to calculate the incremental effect of introducing ECCM under a wide range of reasonable assumptions for the enemy offense system characteristics.

5.2.2. Design Philosophy

As a general rule, good radar design practice reduces the vulnerability of the radar to ECM. By good design is meant (1) maximizing the signal-to-jamming ratio, and (2) maximizing the altitude at which target discrimination can be made.

5.2.2.1. Maximize Signal-to-Noise Ratio
References 11, 12

The task of maximizing the signal-to-jamming ratio is most readily discussed in terms of the radar range equation. The form of this equation which most appropriately describes the radar performance in the presence of active noise jamming is given by:

$$\frac{E}{J_o} = \frac{P_T G_T G_R \sigma \tau R_j^2}{4\pi J_T G_j G_a L L_p R^4} \tag{5.2}$$

where

E/J_o = ratio of received energy-to-jamming power per Hz,
P_T = average transmitter power; in watts,
G_T = gain of the transmitter antenna,
G_R = gain of the receiver antenna,
σ = radar cross-section of target; in m^2,
τ = time of observation; in seconds,
R_j = range to the jammer; in meters,
J_T = jammer transmitter power per Hz; in watts/Hz,
G_j = gain of the jammer in direction of radar,
G_a = gain of the radar receiving antenna in the direction of the jammer; when the receiving antenna is pointed at the jammer $G_a = G_R$,
L = ohmic and similar losses in the antenna and transmission line; $L = 1$ when no losses are present in the system; a typical loss is 10, or its equivalent 10 dB,
L_{sp} = propagation loss; $L_p = 1$ when no loss is present; typically varies up to 10 for many propagation conditions; can be much larger within a plasma,

R = range to the target; in meters; if the jammer is on board the target $R_J = R$.

The form of the radar equation which most appropriately describes the radar performance in the presence of radar chaff is:

$$\frac{E}{C_o} = \frac{\sigma V_c}{\sigma_d N_d V} \tag{5.3}$$

where

E/C_0 = ratio of received signal energy-to-received chaff energy; target and chaff are co-located,
σ = radar cross-section of target; in m^2,
σ_d = radar cross-section of a dipole; $\sigma_d = 0.15\,\lambda^2$ where λ is the wavelength in meters,
N_d = the total number of dipoles in the chaff cloud,
V_c = volume of chaff cloud; in m^3,
V = volume of the radar resolution cell; in m^3; $V = R^2.\theta.\phi.p$ where R = range to the target, θ, ϕ, = azimuth and elevation beamwidth of the radar beam, p = pulse-width of the radar pulse.

Equations (5.2) and (5.3) can be utilized to determine the various possibilities which exist to maximize the signal-to-noise ratio at the output of the radar receiver. Thus, for targets with constant radar cross-section and ECM with fixed parameters, the indicated techniques are to increase the transmitted power, transmitting and receiving antenna gains, time of observation, and range to the jammer (for offset jammers only), while at the same time decreasing the gain of the radar receiving antenna in the direction of the jammer, i.e., decreasing the receiving antenna sidelobes. Of course, for the case of mainbeam on-board jammers, $G_a = G_R$ and $R_J = R$. For the case of chaff jamming the indicated technique is to increase the radar operating frequency and decrease the volume of the radar resolution cell. The latter effect is obtained by increasing the radar transmitting and receiving antenna gains. If the radar cross-section of the target on the other hand is not a constant but is itself proportional to λ^2, then the indicated techniques are slightly different. For this case, the product of receiving antenna gain and target radar cross-section is a constant so that increasing the receiving antenna gain does not work against an active noise jammer. All other indicated techniques, however, remain valid. For this case of chaff jamming, increasing the frequency of the radar will not work and decreasing the volume of the radar resolution cell is the only indicated technique for maximizing the signal-to-noise ratio.

In the present discussion, the signal-to-noise ratio which appears at the input to the radar receiver is related to the energy-to-noise density ratio by:

$$\frac{E}{N_o} = \frac{S}{N}(B\tau) \tag{5.4}$$

where

E/N_o = energy-to-noise power density ratio which appears at the input of the radar receiver,
S/N = signal-to-noise power ratio which appears at the input of the radar receiver,
$B\tau$ = product of the receiver bandwidth B and the duration of the observation time τ.

The energy-to-noise power density ratio which appears at the output of the radar receiver can be obtained from:

$$\frac{E}{N_o} = \frac{S}{N}\frac{B\tau}{FD} \tag{5.5}$$

where

F = noise figure of the receiver; a number greater than unity; for a noiseless receiver $F = 1$,
D = degradation factor in a mismatched receiver; represents the ratio of the signal power required in the actual receiver to the signal power that would be required in the matched receiver; is therefore a function of the receiver design; is a number greater than unity; for a matched receiver $D = 1$.

Equations (5.4) and (5.5) become identical for the noiseless and matched receiver. In general, equation (5.5) can be used to determine the probability of detection of a given target for a given threshold setting in the receiver. Clearly, apart from reducing the receiver noise, a means to maximize the signal-to-noise ratio is to operate the receiver as a matched filter.

5.2.2.2. *Maximize Discrimination Altitude*
Reference 13

The accuracy of a measurement or, more precisely, of a series of measurements over a given period of time can be improved considerably over the accuracy provided by a single measurement through the use of appropriate filtering techniques. The application of digital smoothing techniques can be used to reduce the noise of individual measurements. The smoothing process consists of fitting a polynomial of sufficiently high order to the measured data on the basis of least square deviation.

The selection of a suitable filter depends not only on the noise reduction factor required but also on the characteristics of the target and the phase of its trajectory. As example, high order filters are generally required during target reentry. In general, the actual selection of the filter order, the sampling rate or rate of measurement, and the filtering lag (defined as the time to process the data) involve a tradeoff of three factors; noise reduction, curve-fitting errors, and computer computation cycle time. These factors must be considered and optimized for each point of the trajectory span over which measurements are being made.

The capacity which is needed in a moving arc filter of order $\nu - 1$ with M measurements and b bits per measurement is given approximately by:

$$C = N(\nu M b)^2 \quad \text{bits} \tag{5.6}$$

where

C = computer capacity,
N = number of targets under simultaneous measurement,
$\nu - 1$ = order of the filter,
M = number of measurements in the target trajectory span, and
b = number of bits in a computer measurement; a computer measurement is defined as a computer word and consists of the combined measurements of target range, range-rate, azimuth, elevation, radar cross-section, etc.

Equation (5.6) indicates the square dependency of the computer capacity on the filter order parameter ν, the number of measurements M, and the bits per measurement b. As the sampling rate increases, the number of measurements M will increase in the fixed trajectory span being observed. Thus, for a second order filter, $\nu = 3$. If $N = 1$, $b = 100$ bits, and $M =$ 1,000 measurements, assumed to be taken for a typical reentry trajectory over a 100,000 m altitude interval, the value of equation (5.6) is about 10^{11} from which it can be seen that even for a single target measurement the computer capacity becomes quite large. This example illustrates in a qualitative way the heavy requirement for computer capacity and speed if maximizing the discrimination altitude is vigorously pursued by the defense. Adaptive techniques have been developed which reduce the requirement for computer capacity. Such techniques remove the square dependency of the computer capacity and postulate a logarithmic or linear dependency on the parameters ν, M, and b.

Maximizing the discrimination altitude requires the making of all necessary computations prior to the target's arrival at the radar's decision altitude. The quantity and speed at which such computations can be made determines the altitude at which the radar can discriminate the true nature of the target. On the other hand, making measurements at long range or at high altitudes requires higher order filters and higher sampling rates in order to preserve the accuracy of measurements, i.e., to reduce the noise. Such tactics increase the values of the parameters ν and M and require larger computer capacities for making the required computations and/or require higher computer speeds.

5.3. ECCM Techniques

References: 1–52, 57–65, 68, 72–79
82, 88–90, 97–100, 103, 106–108

A variety of techniques have been developed to combat enemy electronic countermeasures operations. These techniques are generally directed at avoiding or suppressing the noise, enhancing or discriminating the signal, optimizing radar and computer parameters, suppressing jamming parameters, and devising means for preventing catastrophic failure of radars.

Such classification of ECCM techniques is not entirely permanent since many techniques are interrelated and can be classified in perhaps several different ways according as to whether, for instance, they combat specific ones or classes of active or passive ECM devices, or whether they combat confusion- or deception-type ECM, whether they operate on the noise alone or on the signal alone, or on both the signal and the noise, etc. Nevertheless, such ECCM techniques tend to fall into several broad classifications. The remaining portion of this chapter will discuss the various classifications of ECCM techniques giving a brief summary of their use and effectiveness. Many powerful ECCM techniques are classified and therefore only those techniques available in the open literature are given below.

5.3.1. Noise Avoidance

The first technique that comes to mind is to simply avoid the enemy ECM. This involves decoupling the radar and jamming devices through frequency shifting, beamwidth shifting, and range-gate shifting. Each shift implies offsetting the radar frequency, beamwidth, and range gate from the jammer frequency, direction, and range. Frequency agility and diversity are forms of frequency shifting while spatial diversity is a form of beamwidth shifting. Range diversity is a form of range-gate shifting.

5.3.1.1. Frequency Agility

The radar frequency should be either able to be tuned or randomly located in the frequency band to avoid narrowband spot jamming. This is one of the most effective countermeasures techniques to combat active spot jammers. If the radar is tunable over a wide frequency band or if its frequency is randomly varied over a wide frequency band at a frequency rate too high for the jammer to follow, the jammer will be forced into a barrage mode to cover the wide radar frequency band, thus greatly reducing its effectiveness. Thus, for example, shifting the radar bandwidth B over a frequency band $10B$ should result in a 10 dB reduction in the jammer effectiveness.

5.3.1.2. Frequency Diversity

Locating several radars in many widely separated frequency bands is called frequency diversity and forces the enemy to carry more jammers than he may be able to allocate conveniently. If a jammer effectively jams one radar, the remaining radar or radars are left to complete the engagement. Frequency diversity can also be implemented using a single radar. Typically, bandwidths of 10 percent of the carrier frequency can be obtained for both frequency agility and single-radar frequency diversity. Since typical signal bandwidths are on the order of 0.1 percent of the carrier frequency, spot jammers can be degraded by at least 20 dB in such operations.

5.3.1.3. Spatial Diversity

The location of several radars at widely separated sites is one indicated technique for avoiding nuclear bursts. If such a burst jams one radar the remaining radar or radars are available to complete the engagement. Also, spatial diversity seriously degrades the effectiveness of high-gain jammers. If such a jammer is pointed against a single radar, it then presents low gain to the remaining radars. In general, spatial diversity forces the enemy to carry more jammers and, as a consequence, uses up more of the enemy payload.

5.3.1.4. Range Diversity

If the jammer is passive, i.e., chaff or decoys, then decoupling between the ECM device and the radar can be obtained through range diversity. In this technique, the radar range gate is centered only about the target of immediate interest and to the exclusion of the ECM at other ranges. Range diversity will not work if the target is at the same range with the ECM, as is the case for a target buried in chaff, or in active ECM or nuclear environments since, for such environments, the entire radar range in the direction of the ECM device is being jammed.

5.3.2. Noise Suppression

Reference 15

The presence of noise in a radar receiver interferes with the ability of the receiver to detect the wanted signal. Most radar receivers operate on the principle of threshold detection in which a voltage threshold is established at the receiver output. This threshold is established at a level several times the average or mean level of the noise. As long as the mean noise level remains below the threshold, only occasional violations of the threshold by the noise will occur. These are called false alarms and their probabilities for practical threshold settings are quite small. If, on the other hand, the mean noise level increases without a corresponding increase in the threshold level then higher rates of false alarms occur, i.e., the probability of false alarm increases. As a result, the probability of signal detection decreases, i.e., the receiver cannot detect the wanted signal as well as before.

Several techniques have been developed to suppress the effects of noise in radar receivers. Unfortunately, these techniques do not always minimize the noise but almost invariably minimize the signal as well. Some techniques are useful against noise with special characteristics while others are useful against noise arriving at the receiver from a given direction in space or from some given range. Nevertheless, these techniques are useful in an ECM environment.

5.3.2.1. Noise Cancellation

Reference 16

If the interfering signals are broadband they can be eliminated or reduced by subtracting the video output of the radar receiver from the output of an auxiliary receiver which is tuned to a frequency slightly different from that of the radar. In the presence of noise alone each receiver produces an equal output, so that subtraction of one output from the other produces zero net output for the system. If a signal is present in the radar receiver then the net output represents the signal alone, the noise components from each receiver being exactly cancelled. When the outputs of the two receivers are subtracted from one another, ideally only the radar signal remains plus the receiver noise from both receivers. The subtraction is performed in the video portions of the receivers following the second detector. This type ECCM is effective in reducing large impulsive interference but does so at the expense of reducing the normally present signal-to-noise ratio, since the receiver bandwidth is increased by the bandwidth of the auxiliary receiver by a factor 2, i.e., at the expense of a 3 dB drop in the radar performance. This type of noise cancellation can also be applied against a wideband barrage jammer but is slightly less effective against a spot jammer. Noise cancellation can also be applied against chaff if the chaff response itself is broadband.

5.3.2.2. Noise Silencing
References 17, 18

Another way of eliminating broadband off-frequency impulsive and pulse-like interfering signals is to precede the normal receiver by a wideband amplifier and a limiter. Such interfering signals may be intense enough to impair normal reception. The wideband amplifier is designed to include most of the spectrum of the interfering signal. If both the desired signal and the interference are not present simultaneously, and if the limiter is set above the amplitude of the desired signal, then the interference is reduced by the action of the limiter. This ECCM technique is particularly effective against short-duration noise spikes created by active- or passive-type jammers. The wideband amplifier preserves the short duration of the noise spikes, which are then clipped by the limiter to remove a considerable portion of their energy. If the noise spikes were not limited, they would appear as false alarms at the receiver output or, if sufficiently intense, would shock-excite the narrowband IF amplifier and produce an output pulse much wider in duration than the input impulse or noise spike, in effect masking the presence of desired signals. If limiting of the noise spike was not present the interference would effectively mask the presence of desired signals for time durations much longer than the width of the spike alone. The limiter therefore does not allow the noise to influence the receiver for a time longer than the duration of a noise spike when the noise spike has high intensity. When the noise spike is of a lesser intensity, the limiter simply reduces the receiver's probability of a false alarm. In fact, by limiting noise in general, a beneficial effect results in that a constant false alarm rate can be obtained in the receiver.

5.3.2.3. Hard Limiting
References 19–24

A special case of interest in noise silencing techniques is the technique of hard limiting. With a hard limiter preceding a matched filter, all legitimate signals have the same amplitude when detected in the receiver. This technique insures that the range sidelobes of a sin x/x shaped pulse with a large amplitude are not detected as separate targets displaced in range. Although limiting in general is used advantageously as a perfect gain control technique, as well as a noise silencer, it has some disadvantage in a chaff environment. If the chaff returns are of the same nature as those for the legitimate target and if a chaff noise spike exceeds the target return in amplitude and overlaps the target, then effective suppression or masking of the target signal can result. This effect was discussed in the previous section on noise silencing and therein it was indicated that relief from this type of impulsive or noise-spike ECM may be obtained using a wideband amplifier preceeding the limiter. The combination therefore of a wideband amplifier and hard limiter preceding the normal receiver provides an effective ECCM against many types of active and passive jamming.

5.3.2.4. Constant False-Alarm Rate (CFAR)
Reference 25

If the radar receiver employs threshold detection, in which signals are detected if they exceed a preset threshold, then the presence of noise jamming can increase the rate of false alarms to a considerable extent. The effect of additional false alarms is generally to confuse the radar operator or computer. A trained operator might turn down the gain of the receiver during the presence of jamming noise, or else he might be able to ignore the noise if it emanates from sectors other than the direction of the target. As the rate of false alarms increase, the automatic processing circuits tend to overload. The ECCM technique of maintaining a constant false alarm rate in the presence of a noise environment is essentially that of obtaining fast automatic gain control (AGC).

One form of CFAR uses the average noise level to provide an automatic gain control, much like an operator adjusting a manual gain control. As the noise level increases the receiver gain falls off. Automatic CFAR therefore reacts faster and is superior to an operator keeping the false-alarm rate constant. A possible disadvantage of AGC based on the average noise level is the delay encountered between the detection of the noise level and resetting of the threshold. As mentioned previously, another possible means of implementing CFAR is to employ a hard limiter either in connection with noise silencing at IF or at the output of the receiver at video.

CFAR maintains the false-alarm rate constant but does so at the expense of reducing the probability of target detection. If the threshold level is increased to maintain the desired constant false-alarm rate it may prevent weak signals from being detected at the higher threshold setting. Therefore, CFAR does not provide immunity to jamming; it makes the receiver less sensitive but permits its operation in an otherwise effective jamming environment. It permits the engagement of the stronger legitimate targets at the expense of weak targets.

5.3.2.5. Automatic Gain Control (AGC)
References 26–28

The function of AGC is to maintain the dc level of the receiver output constant while input signals vary over a wide range. One of the purposes of AGC in any receiver is to prevent saturation of the receiver by large signals. Another purpose of AGC is to maintain constant false alarms in the receiver. Saturation of the receiver results in the masking of targets or receiver shutdown while an increased false-alarm rate decreases the probability of detecting weak targets. This latter type AGC is referred to as constant false alarm (CFAR) and has been discussed previously.

There are various reasons input signals vary; the inverse fourth power law for targets, the inverse square law for jammers, conical scan modulation, amplitude fluctuations in the radar cross-section of targets, directional jamming antennas, and repeater jammers. All such signals tend to either saturate the radar receiver at short range or to increase the rate of false alarms at long range.

The required dynamic range of the AGC will depend on the variation in range over which targets and jammers are to be observed and the variations expected in the radar cross-section, and the jammer peak power. The dynamic range in typical jamming operations may well be on the order of 100 dB, or perhaps more. Since the practical maximum gain variation which may be obtained with a single IF stage is on the order of 40 dB, two, three, or even more stages of the IF amplifier must be gain-controlled to accomodate the total dynamic range. Since the first IF stage influences the receiver noise figure and since the last IF stage in turn influences the maximum undistorted output, these stages should normally be left uncontrolled. Thus AGC control should be restricted to the middle stages of the IF amplifier.

AGC is based on negative feedback controlling the gain of one or more IF amplifier stages. Signals appearing at the output of a given IF stage are first rectified and then averaged to produce the negative signal for control of the previous stage. The time constant of the AGC loop is usually a few pulsewidths. A target passes with little attenuation, but longer pulses such as those that may be generated from extended chaff targets are attenuated. Thus, in addition to controlling the IF amplifier gain, the AGC tends to attenuate extended clutter targets as well. Since instability may result if the AGC feedback is across many stages, only one or two stages may be controlled by a single AGC loop. Since the final IF stage of the receiver is usually the first to saturate, it or the next to last stage are the ones most likely to contain an AGC loop.

5.3.2.6. *Logarithmic Receivers*

References 29–34

Another means of preventing overloading or saturation of the IF amplifier is by designing the amplifier output to be proportional not to the receiver input but rather to the logarithm of the input. This requires much greater variation in the receiver input to effectively saturate the last amplifier stage.

Logarithmic receivers are also useful for reducing clutter or chaff to a certain extent. One difference between clutter or chaff on the one hand and normal receiver noise on the other is that the former varies with range. The logarithmic receiver eliminates or reduces this range variation and provides at the receiver output a clutter or chaff signal amplitude which is independent of range, thus permitting the easier display or automatic processing of legitimate targets. It should also be noted that the elimination of range variation also permits the convenient detection of targets with a single threshold.

5.3.2.7. Forward Biasing

A circuit which automatically controls the receiver gain in a manner of AGC is known as forward biasing. It differs from feedback AGC in several important respects; first, it does not employ feedback and, second, it controls the diode detector rather than an amplifier IF stage. The main utility of forward biasing is the lack of a finite time delay for AGC control. In all other respects forward biasing performs the same function as does feedback AGC.

5.3.2.8. Fast Time Constant (FTC)

The use of a blocking capacitor between the second detector and video amplifier to remove dc bias caused by CW interference represents good design practice. If the CW interference is accompanied by amplitude and frequency modulation, the modulation will appear at the output of the second detector along with the dc bias from the carrier. To eliminate the modulation, the simple blocking capacitor must be replaced by a high pass filter inserted between the second detector and the first video stage. The high pass filter when used for this purpose is called a fast time constant (FTC) circuit. The filter removes the CW carrier and interfering modulation, provided the interference is at a slightly lower frequency than is the radar carrier frequency. Since the filter may also remove a portion of the desired signal energy, as well as the interfering modulation, radar pulses are passed by the FTC filter with a certain amount of attenuation. Therefore, the FTC should be switched into the receiver only when needed.

Chaff returns manifest themselves as long video pulses caused by the overlap of many signal returns. With this type jamming, the FTC circuitry will reduce or remove the brightly displayed areas that would otherwise be present on the displays, though it will not improve subclutter or subchaff visibility. On the other hand, there is a possible loss of detection capability with the use of FTC, due to target shrinkage and small targets riding immediately behind preceding differentiated targets. Therefore, FTC should be used only at the discretion of a trained operator.

5.3.2.9. Sensitivity Time Control (STC)

Receiver saturation can be reduced by programming the receiver gain as a function of time so that it is low for targets at close range and high for targets at long range. This technique is called sensitivity time control (STC). Large target echoes and large clutter echoes result at close range because of the inverse fourth power range relation between the received power and range. At long ranges, the effects of clutter echoes are less important and the gain can be raised to detect the weaker targets. Since all anti-interference circuits attenuate the desired signal along with the interference, switches are included to remove them when they are not needed. Thus, STC is used only where needed and should be switched off when not in use.

5.3.2.10. Interference Suppression

Considerable interference might appear at the radar receiver even though the jamming frequencies are far removed from the radar frequency. For example, a rectangular jamming pulse produces a large amount of interfering energy at many adjacent frequencies to the jamming carrier. The shorter the jamming pulse the wider is the band of adjacent frequencies containing the interference. In general, the radar receiver which operates in a heavy interference environment should be well shielded to reduce stray pickup. Where necessary, filters should also be employed to further attenuate interfering radiation from jammers operating at adjacent frequencies. In addition to good engineering design other techniques for reducing interference are based on detecting differences in frequency, time of arrival, angle of arrival, or waveshape characteristics between the desired and interfering signals.

5.3.2.11. Synchronization and Blanking
Reference 35

Interference between a radar and a jammer operating in its vicinity can be reduced if the radar pulse repetition frequency can be somehow synchronized to the jammer PRF. A form of synchronization is accomplished by the technique of PRF discrimination.

When the PRFs of the radar and jammer cannot be synchronized, interfering pulses may be removed by turning off or blanking the receiver on arrival of an offending pulse. In practice, a separate auxiliary radar receiver at the victim radar can determine when the level of interference exceeds a threshold value. When the threshold is exceeded, the normal radar receiver is blanked. The auxiliary receiver might have a frequency response function which is a minimum at the radar carrier frequency but which peaks either side of the carrier. The auxiliary receiver will respond to broadband interfering signals and cause the radar receiver to be blanked but will not respond to the radar's own signals since its response is minimum at the radar frequency.

5.3.3. Signal Enhancement

The enhancement of a signal in a radar receiver increases its probability of detection. The signal can be enhanced in one of several ways; it can be increased at the transmitter, integrated in the receiver, or it can be received in a narrowband receiver. Signal enhancement encompasses those techniques which increase the signal-to-noise ratio apart from increases that can be obtained by optimizing radar parameters and is usually centered about the problem of selecting the radar waveform.

5.3.3.1. Pulse Compression
References 36–39

Pulse compression is a means of operating a radar with long pulses to obtain the resolution and accuracy of a short pulse having increased energy, and therefore detection capability, of a long pulse. This is accomplished by modulating the transmitted pulse with a linear or quadratic frequency modulation and designing the receiver to compress the received pulse into a much shorter one. In practice, pulse compression effectively increases the transmitted energy, especially when the transmitter is peak-power limited. Antenna feeds are normally limited to peak powers less than about 100 megawatts at S-band with $1/f^2$ scaling for other frequencies. Thus, pulse compression can be designed to operate at the single feed limit using long pulses to obtain increased range performance.

For a given repetition rate, peak-power output, and range resolution, pulse compression permits greater burn-through range in a jamming environment with but modest peak-power outputs. As an example, consider a 10 microsecond 1 megawatt transmitted pulse during acquisition. By utilizing a pulse compression ratio of 10:1, the same range resolution and maximum range, a burn-through range can be obtained identical to that of an uncompressed 1 micro-second 10 megawatt transmitted pulse of the same repetition rate. Therefore, the use of a pulse compression ratio of 10:1 in a radar imposes a 10 dB penalty on a noise jammer, extending the burn-through range accordingly by a factor of $10^{1/2}$ or about 3.3. As a result, the pulse compression radar can effectively outshout the jammer.

5.3.3.2. *Frequency Modulated (FM) Pulse*

The utilization of a frequency modulated transmitted pulse in a radar requires a prohibitive amount of sophistication on the part of the jammer; to jam effectively, the jammer would have to possess a wide-band, high-fidelity response. Of course, the use of FM would also require the use of a more sophisticated receiver. As example, consider a transmitted pulse whose carrier is modulated in a band which equals 5 percent of the radar center frequency. To be effective, the jammer must now cover the same band and is thus forced to spread its available power. The use of an FM pulse is one form of frequency agility.

5.3.3.3. *Predetection Integration*
Reference 40

Integration of radar pulses may be accomplished in the radar receiver either before the second detector in the IF or after the second detector in the video. Integration before the second detector is called predetection or coherent integration and is accomplished by preserving the radio frequency phase of the individual pulses. If n pulses, all of the same signal-to-noise ratio, are integrated by an ideal predetection integrator, the resulting or integrated signal-to-noise power ratio will be exactly n times that of a single pulse. Thus, if 10 pulses are integrated in a predetection type integrator the radar will impose a 10 dB penalty on the noise jammer, extending the burn-through range by a factor $10^{1/2}$ or about 3.3. As a result, the radar which uses predetection or coherent integration of radar pulses is less sensitive to the effects of noise jamming, when compared to the radar which does not integrate radar pulses.

5.3.3.4. Postdetection Integration

Reference 40

Integration of radar pulses after the second detector is called postdetection or video integration and is accomplished by integrating video pulses following their detection by the second detector. By utilizing video integration techniques, the signal-to-noise ratio at the output of the radar receiver can be ideally improved by a factor approximately equal to the square root of the number of pulses being integrated. If n pulses, all of the same signal-to-noise ratio, are integrated by an ideal postdetection integrator, the resulting or integrated signal-to-noise power ratio will be exactly $n^{1/2}$ times that of a single pulse. Thus, if ten pulses are integrated in a postdetection-type integrator then the radar will impose a $10^{1/2}$ or 5.2 dB penalty on the noise jammer, extending the burn-through range by a factor of $10^{1/4}$ or about 1.8 dB. As a result, the radar which uses postdetection or video integration of radar pulses is less sensitive to the effects of noise jamming when compared to the radar which employs no integration whatsoever. However, the radar which uses postdetection integration of radar pulses is less sensitive, by a factor of $n^{1/2}$, when compared to the radar which uses predetection integration of pulses.

From the standpoint of a radar operator, a direct-view storage tube affords the best opportunity to observe pulse integration, whether coherent or video, and to use his judgement to discriminate against the interfering pulses or noise which does not integrate simply by enabling him to study the entire display at full brightness.

5.3.3.5. Receiver Range and Angle-Tracking Bandwidths

To minimize the effects of noise jamming, the bandwidths of the range and angle-tracking circuits should be designed to be as narrow as possible, but should be kept consistent with the information rate to be handled.

Both range and angle-tracking circuits depend on the generation of difference or error signals needed to position range and angle circuits on the target. The width of range-gates or bandwidths employed in the tracking loops must be sufficiently narrow to minimize extraneous noise. On the other hand, they must not be so narrow that an appreciable fraction of the signal or modulation energy is excluded. A reasonable compromise is to make the gate width of the order of the pulse width being processed by the tracking circuit.

5.3.4. Signal Discrimination

The discrimination between signals and noise can be made on the basis of such signal characteristics as frequency, pulse-width, pulse-repetition frequency (PRF), polarization, and other waveform characteristics.

5.3.4.1. Doppler

Radars can be designed to track the doppler frequency shift of targets. These radars usually track the doppler frequency shift with a narrowband doppler filter which is wide enough to encompass the frequency spectrum of the signal. Doppler radars are also called coherent radars and have at least two major advantages over their equivalent incoherent counterparts; first, the signal-to-noise ratio is greater since the coherent radar bandwidth is usually much smaller than the incoherent radar bandwidth and, second, target discrimination or resolution is made on the basis of radial target speed.

Target detection in the presence of large dense chaff clouds, for instance, is a difficult problem for most monopulse radars.

By means of CW or pulse-doppler radars, however, velocity discrimination can be obtained between targets and the relatively slower moving chaff.

Prior to reentry a swarm of ECM devices including chaff and decoys may accompany the target, with all objects in the swarm moving at the same speed. At reentry, however, slight velocity differences occur between the target and the generally lighter ECM devices. It is precisely these small differences in velocity that can be measured by the doppler radar and thereby discriminate the target from the ECM devices.

5.3.4.2. Pulsewidth

References 17, 35, 41

If the pulsewidth of the interference differs from that of the radar, it may be used as the basis for discrimination. This can be accomplished by differentiating the pulse to determine the location of its leading and trailing edges and then using these edges to determine the pulsewidth. The pulse is rejected if its width differs from that of the radar. The ideal pulsewidth discriminator circuit rejects all pulses whose widths do not exactly match the width of the radar pulse.

The pulsewidth discriminating technique offers some improvement in interference rejection but the rejection is not always ideal since practical pulses are not usually rectangular. Repeater-type jammers with pulses differing from the radar pulse are particularly vulnerable to pulsewidth discrimination.

5.3.4.3. Pulse-Repetition Frequency (PRF)

If the pulse-repetition frequency (PRF) differs from that of the radar, it may be used as the basis for discrimination. This can be accomplished by delaying received pulses by the exact delay time between radar pulses. The interfering pulses will be rejected if their pulse repetition frequency differs from that of the radar. The ideal PRF discriminator circuit rejects all pulses whose PRF does not exactly match the PRF of the radar pulses. Repeater-type jammers and swept-frequency-type jammers with pulses whose PRFs differ from that for the radar pulses are particularly vulnerable to PRF discrimination.

5.3.4.4. Polarization

The radar-scattering properties of a target may be described as transmitting each of two orthogonal polarizations and receiving with the same polarization or the orthogonal polarization. In many cases, the target dimensions are such that maximum radar scattering occurs for a certain transmitting and receiving polarization of the radar antennas. This effect may be used as the basis for discrimination.

Linear polarization is most often used in conventional radar antennas since it is the easiest to achieve. The choice between horizontal and vertical polarization is a matter of knowing at which polarization maximum target-scattering occurs. An example of polarization discrimination is the simple dipole which is exactly tuned to the radar frequency. The dipole reflects the maximum signal when its orientation coincides with that of the linear antenna polarization. If the dipole is orthogonal to the antenna polarization, essentially no signal is reflected back to the radar. If the dipole is rotating and the linear antenna polarization is fixed, the reflected signal which appears at the radar resembles a sine wave with its maxima occuring at times of coincidence of the dipole and linear antenna polarization. For the rotating dipole, the reflected signal can be improved by using circular polarization, since now the dipole can be viewed in both orthogonal positions during its rotation. Circular polarization is often desirable in radars that must see through clutter or chaff-type jamming.

5.3.4.5. Balanced Mixers

A possible jamming threat utilizes two CW jammers which are separated in frequency by the radar IF amplifier frequency. When these two frequencies beat together in the radar mixer, they produce narrowband signals centered at their difference frequencies. The harmful effects of this type of jamming can be reduced by the use of a balanced mixer. A discrimination of several dB, which can be achieved in the balanced circuit, imposes an almost prohibitive penalty on this form of jammer. The radar should have the sum and difference channels of the balanced mixer electrical path lengths essentially equal across the radar band. This will result in the radar tracking the image as readily as it will track the real frequency.

In a sense, the balanced mixer discriminates on the basis of frequency, since RF signals of slightly different frequency will be rejected by the mixer whereas target signals will be preserved. A chief advantage of the balanced mixer is that it can be made relatively broadband.

5.3.5. Radar Parameters Optimization

Reference 42

If the radar is to maintain its performance in the presence of jamming, the signal-to-jamming ratio must always remain high enough to achieve the specified false-alarm rate. If the jamming noise is increased, the signal must be increased correspondingly. The signal-to-jamming ratio is given by equations (5.2), (5.3), and (5.5), and can therefore be increased by optimizing the various parameters in these equations. Thus, the signal-to-jamming ratio can be increased by using a higher power transmitter, by using higher gain antennas, by increasing the time the radar illuminates the target, by suppressing antenna sidelobes, by reducing ohmic and propagation losses, by siting the radar, by reducing the receiver noise figure, and by using a matched filter. The optimization of radar parameters encompasses those techniques which increase the signal-to-noise ratio apart from increases that can be obtained by optimizing the radar waveform itself and is usually centered about the problem of maximizing the parameters given in the appropriate radar equation being used. These techniques will now be briefly discussed.

5.3.5.1. Transmitter Power

Reference 43

Examination of the radar equation in the form most appropriate for operation under heavy active jamming generally leads to one conclusion: the most universally useful defenses against jamming are the transmitted power, transmitting antenna gain, and the receiver's usable bandwidth. Regarding the transmitter power, it turns out that singly-fed reflector-type antennas are limited to the transmission of peak powers of about 100 megawatts at S-band with $1/f^2$ scaling for other frequencies. As mentioned earlier, one method of increasing the radar energy is to design the radar as a pulse compression system. Another method for increasing the transmitted energy is to use an array antenna. An outstanding ability of the array radar is its capacity for handling vast amounts of power. Since the array elements are in parallel for purposes of power transmission, each element in principle could have its own transmitter. In a fairly large array, this would amount to several hundred thousand transmitters, so that even if each transmitter were to transmit 1 kw average power, the result would greatly exceed the power handled by singly-fed mechanically-scanned radars. In practice, fewer transmitters of higher average power are used, tying a number of transmitting elements to each transmitter. It is apparent that the tremendous power capability associated with an active-element transmitting array represents a real advantage in an active jamming environment.

5.3.5.2. *Antenna Gain*

Reference 43

Increasing the gain of the radar transmitting and receiving antennas is another useful technique for maximizing the target signal in an active jamming and chaff environment. This can be seen through examination of equations (5.2) and (5.3). Since the gain of an antenna can be increased by increasing the size of the antenna aperture and increasing its frequency of operation, both methods may be used if the target cross-section itself is a constant. For the case of reentry vehicles, the cross-section is proportional to the wavelength squared when the vehicle is viewed from near nose-on aspects and, as a consequence, cancels the effect of using a higher frequency for chaff operations and diminishes the effect of higher frequency assignments in active jamming operations. In any case, the method of increasing the antenna gain by increasing its size is the best of the two methods for operation in jamming environments. It does, however, impose the disadvantage of building and operating larger structures for the antenna.

To the extent that an array antenna has a lower achievable gain than a conventional antenna radar, it appears that it is more vulnerable in this respect to operation in jamming environments. This disadvantage, however, is overcome by the tremendous power capacity of the array and its flexibility in scanning rapidly, thus enabling the surveillance and tracking of many targets.

5.3.5.3. Duration of Observation Time

As used in equation (5.2), the duration of target observation may be interpreted as the time duration of a single radar pulse or as a time duration which contains many radar pulses. The latter interpretation implies either predetection or postdetection integration by the radar receiver and has been already discussed. Increasing the radar pulsewidth increases the time the radar illuminates the target and can be an effective means of distinguishing the target in an active jamming environment, although at the expense of decreasing the radar resolution and measurement accuracy for range. This technique, however, will not work in a chaff environment, as can be seen through inspection of equation (5.3). Radar long-pulse operation should be switched on only when needed.

5.3.5.4. Antenna Sidelobe Suppression
References 13, 41, 43

Barrage jammers and large chaff clouds can be detected through the antenna sidelobes if they are of sufficient strength. There will always be undesired sidelobe radiation from the radar in directions other than the main beam. Although the sidelobe radiation may be small compared with that from the main beam, the relatively large radiating powers of barrage jammers and relatively large cross-sections of chaff clouds, especially if close to the radar, will cause the radar to see the noise even though the radar beam is pointed in a direction which is free from jamming altogether. The suppression of antenna sidelobes is therefore desired when operating in a jamming environment. This may be accomplished in one of two ways; tapering of the antenna aperture illumination and cancellation.

If the antenna illumination is uniform across its aperture then a sin x/x pattern results with rather high sidelobes. In fact, for this type illumination, the first sidelobe removed from the main beam is down only 13 dB. Much better sidelobe performance can be obtained using other illuminations of the aperture. Thus, improvements on the order of 10 dB can be obtained using parabolic, triangular, or circular illuminations whereas even 20 and 30 dB improvements can be obtained with higher power sine illuminations and other taperings of the illumination as well.

Sidelobe cancellation depends on the difference in the angle of arrival of signals in space. If a separate receiver is used with an omnidirectional antenna, the interfering signal will appear in both the omnidirectional antenna receiver and in the radar receiver. Interference signals in the auxiliary receiver may be used to blank the radar receiver or can be used to cancel the interference in the normal radar receiver.

To the extent that an array antenna has higher sidelobe levels than a conventional antenna radar, it appears that it is more vulnerable in this respect to operation in jamming environments. This disadvantage, however, may be considerably alleviated through sidelobe suppression and through the fact that the array has other desirable characteristics more effective perhaps for its operation in a jamming environment.

5.3.5.5. Ohmic Losses

The radiation efficiency of an antenna accounts for ohmic losses, RF losses, and mismatch losses. The presence of losses reduce the signal-to-jamming ratio for the case of active jamming but have no effect whatsoever in a chaff environment. Losses generally range from the order of 1 dB at L-band frequencies to about the order of 10 dB at X-band frequencies and are quite dependent upon the length of transmission lines, poor connections both mechanical and electrical, duplexer losses, etc.

At the lower radar frequencies the transmission line introduces little loss, unless its length is exceptionally long, as for the case of array antennas. At the higher radar frequencies, attenuation may not always be small. In addition to the losses in the transmission line itself, losses can occur at each connection or bend in the line. If the same transmission line is used for both receiving and transmitting, the loss to be inserted in the radar equation or equation for active jamming is twice the two-way loss. Basically, the losses can be reduced to their minimum values through good design and engineering practice.

5.3.5.6. Propagation Losses

See Appendices

Radio frequency waves are attenuated through absorption by the atmosphere, precipitation, and plasmas. Absorption and precipitation occur naturally while plasma is encountered both naturally in the ionosphere and in a nuclear environment. The lower radar frequencies are less attenuated by the atmosphere and precipitation but more attenuated by plasmas. Propagation losses are therefore strongly determined by the selection of the radar frequency of operation.

From a jamming point of view, the radar frequency should be made as low as possible to insure high signal-to-jamming ratios in an active jamming environment or should be made as high as possible to insure the visibility of targets in a nuclear environment. Obviously, a compromise frequency must be selected if the radar is to operate in both environments. Optimization of propagation losses will not help in a chaff environment.

5.3.5.7. Siting

Reference 44

Examination of equations (5.2) and (5.3) indicates that siting for operation in a jamming environment requires placing the radar as near as possible to the target trajectory. Examination therefore of the possible target trajectories and impact points is required preliminary to the selection of the radar site as well as the more conventional siting requirements for field-of-view, ground reflections, etc. In unjammed environments, radar siting requirements depend on such conditions as the line-of-sight visibility of targets and ground returns. The most exacting site requirements are presented by long-range surveillance radars which, to a large extent, depend on ground reflections for successful operation. The siting problem then requires the consideration of terrain effects such as limited reflection areas, cliff edges, obstacles, etc., which involve diffraction problems of great complexity. Clearly then, the siting of radars involves a tradeoff between clear visibility requirements for volume search operations and nearness of radars to target trajectories for optimized operation in jamming environments.

5.3.5.8. *Receiver Noise Figure*
References 45–47

The noise figure of a linear receiver is defined by:

$$F = \frac{N_o}{N_i G} \tag{5.7}$$

where

N_o = noise at the receiver output,
N_i = noise at the receiver input, and
G = gain of the receiver.

In the absence of jamming, the input noise is the so-called receiver thermal noise, namely $kT = 4 \times 10^{-21}$ watts/Hz of bandwidth. Equation (5.7) can also be written as:

$$F = 1 + \frac{\Delta}{N_i G} \tag{5.8}$$

where Δ is the additional noise introduced by the receiver itself. Equations (5.7) and (5.8) assert the fact that noise is generated in a practical receiver.

The noise figure F of a receiver in the absence of jamming is between 1–10 dB indicating that the receiver is a major contributor of noise. In a heavy jamming environment, however, the major noise source is the jamming noise, so that for this case the receiver noise figure approaches unity. In a light jamming environment, therefore, it can be expected that the receiver noise figure will assume some value between 1–10 dB, indicating that both the jammer and receiver noise levels are important. For this intermediate jamming situation, the input noise to the receiver can be stated as the sum of the two noise components; that from the jammer and that from the receiver:

$$F = 1 + \frac{\Delta}{(kT + J_o)BG} \tag{5.9}$$

where

J_o = the jammer spectral density, and
B = receiver bandwidth.

For the particular case of a light jamming environment, optimization of the radar receiver might provide a 1–10 dB improvement in the signal-to-jamming ratio and this might be sufficient to reduce the jamming to a more tolerable level.

In summary, although it appears that a sensitive receiver will not help in a heavy jamming environment, it should produce a useful improvement if the jamming is light.

5.3.5.9. Matched Filter
References 11–13, 48–51

The signal input to a radar receiver is not only amplified but is also, in a sense, processed by the receiver. Perfect processing in the receiver means that the signal appearing at the receiver output is just the input signal amplified. In general, the receiver degradation is given by:

$$D = \frac{S_o}{S_i G} \tag{5.10}$$

where

S_i = signal at receiver input,
S_o = signal at receiver output, and
G = receiver gain.

If the receiver is matched to the received signal, then $D = 1$. For mismatch, the degradation factor D is a number greater than unity. Therefore, matching the radar receiver insures the highest possible signal-to-noise ratio in both jammed and unjammed environments.

5.3.6. Computer Parameters Optimization
Reference 13

If the radar is to maintain its performance in the presence of deception ECM, the altitude at which the target can be discriminated must be high enough so that the ABM system can react properly once the target has been identified as a true threat. A target in ballistic flight can be measured by the radar in the sense that a series of M measurements can be obtained of the target over a span of its trajectory. Having obtained these measurements, the radar can discriminate the target on the basis of trajectory prediction, signature analysis, and atmospheric filtering. This is accomplished by the radar computer by solving sets of linear equations either directly or using algorithms. The application of digital smoothing techniques can be used to reduce the noise in individual measurements. The smoothing process consists of fitting a polynomial of sufficiently high order to the measured data on the basis of least squares deviation.

The trajectory, signature, and slowdown characteristics of an object are essentially deterministic. If the constants of motion are known, the object's trajectory, signature, and slowdown can be computed. In practice only the discrete measurements are known, not the constants of motion. The problem is how to process discrete measurements to give a best prediction of the object's trajectory, signature, and slowdown so that any one of these can be used as a basis of discrimination.

The altitude at which a target can be discriminated can be increased by using more computer capacity, higher computer speeds, and digital smoothing to handle the vast numbers of computations which must be accomplished. Thus, the optimization of computer parameters encompasses those techniques which increase the target discrimination altitude and are usually centered about the problems of increasing the computer hardware for capacity and speed and choosing the appropriate smoothing algorithm to perform the computations.

5.3.6.1. Memory Extension (or Time Compression)

The techniques of memory extension or time compression essentially store data for high-speed computations. They enable the radar to correlate events that seem to be independent of each other in short time intervals, but which obey deterministic laws if observed over longer intervals of time. The chief concern in memory-extension or time-compression systems is data storage, since even modest systems require very large memories.

If the deterministic portion of an object's trajectory, signature, or slowdown characteristic in a given time span of observation T can be sufficiently approximated by a polynomial of degree $\nu - 1$, then M discrete measurements in T

will result in a set of M linear equations with ν unknowns, the so-called constants of motion. The purpose of the computation process is to find a set of the constants of motion that may be used in turn to describe the object's trajectory, signature, or slowdown. For each input data point d_i, the error in the measurement e_i should be made as small as possible.

Consider the finite set of linear equations given by:

$$d_i = \sum_{j=0}^{\nu-1} c_j g_{ij} \tag{5.10'}$$

where

d_i = data measurements; $i = 1, 2, \ldots, M$,
c_j = constants of motion; $j = 0, 1, \ldots, (\nu - 1)$,
g_{ij} = known functions of time; ballistic and reentry trajectories usually require that $g_{ij} = t_i^j$ in which case equation (5.10′) is an ascending polynomial of the time t taken with respect to some convenient point in the observation span T; long-term target signature usually requires that $g_{ij} = \sin \omega_j t_i$ or even some other repeating function of time in which case equation (5.10′) becomes a repeating function of time.
M = total number of measurements, and
$\nu - 1$ = order of the polynomial.

Equation (5.10′) states that there are M equations corresponding to M instants of time at which the constants of motion best represent the data. The set of equations given by equation (5.10′) will usually be overspecified and inconsistent, since in typical measurements, with an ample supply of input data, it is usual that the number of measurements greatly exceeds the order of the equation, i.e., $M \gg \nu$. These equations do have a simple solution when $M = \nu$, since for this case the number of data points equals the number of unknown constants of motion, i.e., the system is not overspecified.

When the number of measurements is very large compared to the number of unknown constants of motion, the problem becomes that of obtaining the best solution or fit which minimizes the sum of squares of errors present in each measurement, i.e., a set of parameters C is found to minimize the sum:

$$\sum_{i=1}^{M} e_i^2 \tag{5.11}$$

When the choice of parameters has been optimized it can be shown, in matrix notation, that:

$$H(g, g)C = H(g, d) \tag{5.12}$$

where $H(g, g)$ is a $\nu \times \nu$ matrix of the sums of cross-correlations and auto-correlations of the known functions of time $g_{ki}g_{ji}$ and $H(g, d)$ is a $1 \times M$ matrix of the sums of cross-correlations of the known functions of time with the data $g_{ki}d_i$. The matrix C is a $1 \times \nu$ matrix of the constants of motion c_j. Equation (5.12) provides the solution for the constants of motion which minimize equation (5.11). The solution is found by solving equation (5.12) for the constants of motion:

$$C = H^{-1}(g, g)H(g, d) \tag{5.13}$$

One way of finding the constants of motion is to solve the set of equations given in equation (5.13). The solution is generally straightforward but presents serious computational problems when the number of unknown constants ν is large and when the data rates are high, i.e., when the order of polynomials $\nu - 1$ and the number of measurements M are large. In addition to the necessity of inverting the $\nu \times \nu$ $H(g, g)$ matrix, this method may require as many as $\nu(\nu + 1)/2$ auto-correlation and cross-correlation computations to obtain the elements of this matrix and $\nu(M + 1)$ computations to obtain the elements of the $1 \times MH(g, d)$ matrix. Furthermore, this process generally needs to be repeated in most practical situations where the input signal statistics change slowly. No perfect solution of equation (5.13) is possible in practice because of the fact that an infinite statistical sample would be required to estimate perfectly the elements of the correlation matrices. Specifically, forming the matrix $H(g, d)$ requires $\nu(M + 1)$ storage locations in the computer memory while forming the matrix $H(g, g)$ requires $\nu(\nu + 1)/2$ locations. In addition, inversion of the matrix $H(g, g)$ requires ν^2 locations for the computation of minors, neglecting arithmetic operations. It is therefore seen that the capacity which is needed to compute the constants of motion for a single measurement is given approximately by:

$$C_p = N\left[\nu M + \frac{3}{2}\nu(\nu + 1)\right]b \quad \text{bits} \tag{5.14}$$

where the quantity within the brackets gives the number of memory locations per measurement per target, and:

C_p = computer capacity,
N = number of targets, and
b = number of bits in the computer word/measurement.

The capacity which is needed in a moving arc filter with M positions is obtained by multiplying equation (5.14) by the number of bits for all M measurements. Since $M \gg \nu > 1$:

$$C_p = N\nu(Mb)^2 \tag{5.15}$$

When considering the additional capacity which is needed for arithmetic operations as well as data storage, the capacity needed is better represented more nearly by:

$$C_p = N(\nu Mb)^2 \tag{5.16}$$

Equation (5.16) reports the fact that the computer capacity needed by a moving arc filter is a linear function of the number of objects being simultaneously measured by the radar, and a square function of the filter order parameter ν, the number of measurements M, and the number of bits per measurement b.

The capacity which is needed in a simple reiterative filter with M positions is actually given by equation (5.14) which is simplified as:

$$C_p = N(\nu Mb) \tag{5.17}$$

Equation (5.17) indicates that for a reiterative filter the computer capacity needed is a linear function of the number of objects being simultaneously measured by the radar, the filter order parameter ν, the number of measurements M, and the number of bits per measurement b.

Equations (5.15) and (5.16) can be used to estimate the size of the memory which is required by the computer for the implementation of target discrimination techniques of trajectory prediction, signature analysis, and slowdown prediction. Although indicative of a lesser requirement for capacity, the use of a reiterative filter can be accomplished only at the expense of computation delays which may become intolerable if the time for such computations is not available. Also, not discussed are the various computational algorithms for reducing the number of actual computations. In any case, the defense planner must specify the parameters N, ν, M, and b. The parameters N and b involve time-on-target tradeoffs. The parameter ν is specified by knowing or estimating the target motion and assigning the degree of polynomial or equation which can describe this motion with a tolerable error. The number of measurements M are determined by the target visibility, expected measurement errors including those expected by the jamming, and the accuracy improvement desired by smoothing the data.

Although the computer speed has not been treated in the foregoing it is equally important. It is intuitive to expect that the computer should be capable of processing data at the rate:

$$R_p = \frac{C_p M}{T} \quad \text{bits/s} \tag{5.18}$$

where

R_p = computer speed,
C_p = computer data, and
T/M = time available between data measurements to process data.

Equation (5.18) indicates that the computer speed requirement is controlled by the same parameters which determine the computer capacity C_p and, in addition, the number of measurements M, and the time span T over which observations are made. In the latter respect, the availability of observation time reduces the requirement for computer speed.

5.3.6.2. Data Smoothing

Data smoothing is the computer technique which is used to reduce the effects of noise following a series of noisy measurements. This technique can be used both for improving the radar measurement accuracy in noise type (thermal, jamming, etc.) environments and in implementing target discrimination techniques for deception ECM. In both applications, smoothing or filtering the data provides better estimates for the target trajectory, signature, and slowdown.

Minimizing equation (5.11) with respect to the constants of

motion requires that:

$$\frac{\delta}{\delta c_k} \sum_{i=1}^{M} \left(d_i - \sum_{j=0}^{\nu-1} c_j g_{ij} \right)^2 = 0 \tag{5.19}$$

where

d_i = data measurements; $i = 1, 2, \ldots, M$,
c_j = constants of motion; $k, j = 0, 1, 2, \ldots, (\nu - 1)$, and
g_{ij} = known functions of time.

Evaluation of the partial derivatives in equation (5.19) yields:

$$\sum_{i=1}^{M} d_i \cdot g_{ik} = \sum_{j=0}^{\nu-1} c_j \sum_{i=1}^{M} g_{ij} \cdot g_{ik} \tag{5.20}$$

This last equation is a system of k linear equations ($k = 0, 1, \ldots, (\nu - 1)$) with unknowns c_j ($j = 0, 1, \ldots, (\nu - 1)$). It is therefore possible to solve for the constants of motion in terms of the data measurements, i.e.:

$$c_j = \sum_{i=1}^{M} w_{ij} d_i \tag{5.21}$$

where the weights w_{ij} are actually obtained by solving the system of equations (5.20) for the c's.

Having obtained the constants of the motion, the motion itself can be described by the next equation, in which discrete time is replaced by the continuous time t:

$$x(t) = \sum_{j=0}^{\nu-1} c_j g_j(t) \tag{5.22}$$

where $x(t)$ is a function which describes position for trajectory and slowdown prediction and signal amplitude for signal analysis. Substitution of the values for the constants of motion provided by equation (5.21) into equation (5.22) results in the well-known general filter equation for position or amplitude:

$$x(t) = \sum_{i=1}^{M} \sum_{j=0}^{\nu-1} w_{ij}(t) \cdot g_j(t)\, d_i \tag{5.23}$$

Similar filter equations can be developed for the velocity or amplitude rate, acceleration, etc. This may be accomplished by taking successive derivatives of equation (5.23) with respect to time. Thus:

$$\overset{(n)}{x}(t) = \sum_{i=1}^{M} \sum_{j=0}^{\nu-1} w_{ij} \cdot \frac{\delta g_j(t)}{\delta t^n} \cdot d_i \tag{5.24}$$

where $\overset{(n)}{x}(t)$ is the nth derivative of $x(t)$ with respect to time.

Equation (5.24) indicates that the nth derivative of the target parameter $x(t)$ is a linear function of the M data points d_i.

The noise reduction, also called the smoothing or filter factor, can be obtained, as the variance of equation (5.24), by:

$$\left(\frac{\overset{(n)}{\sigma}(t)}{\sigma}\right)^2 = \sum_{i=1}^{M} \sum_{j=0}^{\nu-1} w_{ij} \cdot \frac{\delta g_j(t)}{\delta t^n} \tag{5.25}$$

where

$\overset{(n)}{\sigma}(t)/\sigma$ = ratio of output noise for the nth derivative to the input noise in the data measurements; the input noise has been assumed equal for all measurements.

The noise reduction factor given by equation (5.25) is a function of time as reckoned from some convenient point in the time span T of observations.

If the known functions of time $g_j(t)$ are powers of time, as is the case for ballistic and reentry trajectory prediction, then the ratio of variances of the output and input noise given by equation (5.25) become increasing functions of time and when the time t becomes very large can be approximated by:

$$\left(\frac{\overset{(n)}{\sigma}(t)}{\sigma}\right)^2 = \sum_{i=1}^{M} \left[\sum_{j=0}^{\nu-1} \frac{j!}{(j-n)!} w_{ij} t^{j-n}\right]^2 \tag{5.26}$$

If, on the other hand, the known functions of time are repetitive ones, specifically if $g_j(t) = \exp(\sqrt{-1}\, w_j t)$, as may be the case for simple amplitude prediction, then the ratio of variances of the output and input noise, given by equation (5.25), themselves become repetitive functions of time.

The problem of smoothing is first to find the best estimate of the object's trajectory, signature, or slowdown, and then to assess how good the estimate is. Equations (5.24) and (5.25) may be used to attain these objectives. Furthermore, the solution of the smoothing problem will provide the parameters ν and M needed for the optimization of the computer's capacity and speed.

The general procedure for the optimization of computer parameters consists of, first, to determine the polynomial or equation which can be used to estimate the object's trajectory, signature, or slowdown characteristics, and second, to determine the goodness or accuracy of the estimate. This can be accomplished for measurements which can be taken at sufficiently high altitude to insure the discrimination of targets. The computer capacity and speed can then be determined to support the needed computations.

5.3.7. Jamming Parameters Suppression

Although the design parameters of the opponent's ECM equipment is not under the immediate control of the ABM system planner, it is to a great extent influenced by the radar counter-countermeasures environment. Thus, in his knowledge that the defense has a certain ECCM capability, the offense will design its jamming equipment for performance against all possible eventualities. This is particularly true in strategic planning since there will be no chance to discover an opponent's weakness. In this manner, the defense ECCM capability influences the offense ECM planning and *vice versa.*

5.3.7.1. Jammer Power

References 43, 52

The problem of designing radar to combat the effects of jamming has been considered using the concepts of the mathematical theory of games. If the assumption is made that both the radar and the jammer consider the radar post-detection signal-to-noise ratio as the measure of radar performance, then the optimum strategy for both the radar and the jammer is to spread their power evenly over the entire radar band and for the radar to employ a matched filter receiver. Any deviation from the optimum strategy by either the radar or the jammer can only produce a degraded performance. As a consequence, the stress on the jammer when called to jam a wide frequency band may become intolerable since, for increased jamming power requirements, the weight of the jamming vehicle itself may become unjustified.

5.3.7.2. Jammer Antenna Gain

In a manner quite analogous to that for allocating the jamming power it can be shown that the optimum strategy for the jammer is to widen its antenna beam to as great an angle as possible to cover as many possible radars on the ground. Widening the jammer antenna beam, of course, requires reducing its gain, with consequence that less jamming power is made available to each radar being jammed. In most applications, jammers employ antennas with omnidirectional gain.

A special case of interest is the jamming antenna which must survive the effects of the atmosphere during reentry. By the nature of reentry, the antenna itself will be constrained to some portion of the reentry vehicle carrier which may not be optimal from the standpoint of gain. Furthermore, the jammer antenna structure must be reinforced to survive the reentry environment. The communications blackout problem must also be considered.

5.3.7.3. Jammer Range

The range at which the jammer operates from the radar affects the radar in two different ways, depending on whether the jammer is an off-board or on-board jammer to the target itself. If the jammer is physically located on some other vehicle than the target itself then it becomes quite effective as it is located nearer the radar. A very special case of interest is the case for a small lightweight jammer placed exactly at the radar site. If, on the other hand, the jammer is physically located on the target then it becomes less effective as the target approaches the radar. The optimum strategy for the radar against an on-board jammer is to increase its burn-through range and this can be accomplished by increasing its transmitted power and transmitting antenna gain. These measures are not as effective, however, against the threat of off-board jammers and especially to such off-board jammers that can be delivered or placed in the vicinity of the radar. Apart from those measures which can be implemented to maximize the signal-to-noise ratio, it may well be advised to consider the problem of off-board jammers from the standpoint of delivery and emplacement of this type of jammer.

5.3.7.4. Chaff Cloud Volume
References 13, 53, 54

The dispensing and deployment of very large chaff clouds in strategic applications imposes a severe penalty on the chaff delivery and carrier systems. It is much simpler in these applications to dispense and deploy a number of small but widely separated chaff clouds, the separation between chaff clouds being determined by the requirement to prevent the destruction of adjacent clouds by a single defensive interceptor. As a consequence, small chaff clouds which move along in a swarm must be considered to be deception-type ECM devices and can therefore be discriminated on the basis of their trajectories, signatures, or slowdown rates. Trajectory prediction and signature analysis are particularly effective techniques for discriminating chaff clouds prior to their reentry because each cloud is on a slightly different trajectory from other clouds which, in view of dispensing and deployment constraints, may not impact at the same location the reentry vehicle does and, furthermore, because the concealment of a reentry vehicle or booster in a given cloud may result in a distinguishable signature for the concealing cloud.

5.3.7.5. Chaff Dipole Cross-Section
References 13, 54, 55

The radar cross-section of a chaff dipole is a function of the frequency. A randomly-oriented dipole in a chaff cloud will contribute cross-section to the total cross-section of the chaff cloud approximately as:

$$\begin{aligned} \sigma_d &= 0.15\,\lambda_R^2 && \text{at Resonance} \\ &= 0.15\,(f/f_R)^4\,\lambda_R^2 && \text{Raleigh Region} \\ &= 0.15\,(f_R/f)^2\,\lambda_R^2 && \text{Optics Region} \end{aligned} \tag{5.27}$$

Equations (5.27) express the fact that the radar cross-section of a dipole is a maximum at its resonant wavelength λ_R, and corresponding frequency f_R, and decreases for frequencies either side of the resonant frequency, as the 4th power of frequency in the Raleigh Region for which $f < f_R$ and as the 2nd power of frequency in the Optics Region for which $f > f_R$.

The resonant frequency of a dipole is determined by its length. For resonance, the dipole length l equals approximately the resonant half-wavelength, i.e., $l = \lambda_R/2$. Basically, in the design of a chaff cloud, the length of the dipole is selected to provide dipole resonance at the radar wavelength, and corresponding frequency, and for possible radar operation at adjacent frequencies in the radar band. Each dipole in the chaff cloud provides the broadband frequency coverage indicated by equations (5.27). If this coverage is not sufficient, different lengths of dipoles can be used. It should be noted that quite long lengths of dipoles are required if the radar frequency to be jammed is itself low. This fact imposes a severe constraint for the packaging, delivery, and dispensing of dipoles.

The dipole diameter d is determined primarily by the requirement that it can withstand the nuclear and reentry environments and the requirement that it remain rigid at all times following its dispensing, that dipoles maintain separations of at least half wavelength following their dispensing, and that dipoles not be too thin lest they have excessive ohmic losses. Equations (5.27) do not apply for non-rigid dipoles or rope. Long thin dipoles tend to collapse while thicker dipoles tend to maintain their rigidity. Thus, longer dipoles require larger diameters to stay rigid. If the density of the dipole metal is ρ, then its weight is given by.

$$w_d = \pi \rho l d \tag{5.28}$$

In view of the fact that both the length and diameter of the dipole increase with decreasing radar frequency, a corresponding increase in the weight of the dipole occurs as indicated by equation (5.28). Thus, extreme length and weight penalties are imposed on the chaff ECM device if its operation is required at very low frequencies. In addition, its capability to package, deliver, and dispense low frequency dipoles is stressed.

5.3.7.6. *Total Number of Chaff Dipoles*
References 13, 54

A means of increasing the total radar cross-section of a chaff cloud is to increase the total number of chaff dipoles in the cloud. This however cannot be accomplished without constraints being imposed on the chaff delivery system. The net weight of chaff in a chaff cloud with N dipoles is obtained by multiplying the weight of a single dipole, given by equation (5.28) by the total number of dipoles N. Thus:

$$W = \pi \rho l d N \tag{5.29}$$

In addition to the chaff net weight which is given by equation (5.29), more weight and volume are required to package, deliver, and dispense the chaff cloud. This total weight may impose a severe burden on the chaff system design, especially if the chaff cloud is to be effective against low-frequency radars. To be effective, the net chaff payload should be no less than 50 percent of the chaff system weight and volume.

A common problem which is encountered in the dispensing of large numbers of chaff dipoles is the so-called birdnesting problem. Birdnesting results from the entanglement and intermeshing of long thin dipole wires. Once mechanically meshed, the dipoles lose their individual electrical performance and, as a consequence, do not contribute radar cross-section to the chaff cloud. Equations (5.27) are valid only when the individual dipoles in a chaff cloud are separated by distances of at least half the wavelength of the radar observing frequency and remain rigid. Birdnesting of dipoles can be decreased by maintaining the rigidity of wires in the chaff cloud.

5.3.8. Non-Catastrophic Operation

Certain radar counter-countermeasures techniques have been developed and are based on the fact that the radar operation can indeed be disrupted by the presence of jamming. These techniques address the problem of using existing radars in such an environment and thus benefit by their partial or limited availability, either singly or in combinations with other radars in the ABM system, to prevent the altogether catastrophic loss of performance.

5.3.8.1. Wide Dynamic Range

An important measure for reducing the effects of ECM is to avoid saturating or overloading the radar receiver with large jamming or interfering signals. The receiver, therefore, should have a large dynamic range. The logarithmic receiver has already been discussed as a means of preventing overloading or saturation of the IF amplifiers. The logarithmic receiver is generally preferred over the linear and square law receivers for achieving large dynamic range. However, from the standpoint of detecting weak signals in noise, the square law receiver is preferred when used in conjunction with the technique of postdetection integration of pulses, discussed earlier.

5.3.8.2. Rapid Reacquisition

An important feature of a tracking radar is its capability to reacquire the target as soon as the jamming ceases. If the jamming is short-term, then the use of monopulse tracking is preferred over conical scanning since the former scanning technique is not degraded by amplitude fluctuations and is therefore less susceptible to electronic countermeasures than is conical scanning. With the monopulse tracker, it is possible in principle to obtain the tracking error on the basis of a single pulse whereas a minimum of four pulses are necessary with the conical scan radar. Thus, the monopulse tracker is theoretically capable of obtaining a radar measurement in microseconds as compared with tens of milliseconds for the conical scan radar. If the jamming is long-term then the radar must incorporate a memory extension which can be used to point the antenna to the predicted location of the target as soon as the jamming ends.

5.3.8.3. Passive Angle Tracking

During passive angle tracking a tracking radar tracks in angle and coasts in range. There are three ways or methods of placing a radar in the passive track mode of operation: (1) the operator may position a switch on the radar console which activates this mode of operation, (2) if the radar is in the acquisition mode and sufficient jamming is present and, if after one range strobe or search the radar fails to lock onto the target, then the radar can be switched automatically into the passive angle track mode of operation, and (3) the AGC circuits can be designed specifically to prevent a break in tracking by a barrage jammer when the radar is in the track mode. A noise AGC voltage can be developed from samples well removed from the returned signal, and a build-up in interference can be detected. Should this build-up continue, the noise AGC will exceed a certain preset level. When this occurs, the presence of a noise jammer is assumed. The system will then be switched automatically to passive angle track until the interference subsides.

5.3.8.4. Triangulation

The radiation from a jammer denies the radar range information but can be used nevertheless to determine the range to the jammer by means of triangulation, in which two radars separated by a known distance record the direction of the jamming source as seen by each radar. The range to the jammer can be obtained by the law of sines. Unfortunately, when more than one jammer is present, this technique results in the appearance of many false jammers or ghosts. These ghosts increase in number as the square of the number of jammers.

5.3.9. Trained Operators

The flexibility of a radar to maintain its performance in a variety of countermeasures environments can be enhanced in several ways; first, by incorporating automatic features and, second, by providing suitable monitoring and override displays and circuits which enable trained operators to make the proper decisions when the automatic features fail.

A trained operator fulfills a dual role in a countermeasures environment; first, as a mechanism for recognizing and integrating radar signals in jamming noise and, second, as a mechanism for recognizing automatic circuit failures. In the first case the operator uses his senses of sight and sound to detect the presence of signals. The most usual type of radar presentation is based on the sense of sight and is some type of the common cathode ray tube. In the second case, the operator uses his *a priori* training to detect the failure of automatic features to perform their designated functions.

5.3.9.1. *Operator-Aided Detection*

References 56–65

The ability of the operator to detect signals in the presence of noise is less than the corresponding ability of an electronic threshold detector. This is caused mainly by fatigue, guessing, or the overloading of the limited bandwidth or capacity of a human. Thus, an operator's capacity for searching a CRT display and recognizing the presence of weak targets is limited. The information bandwidth or capacity of a human operator is of the order of 10 Hz. This means that the operator can handle up to twenty weak targets each second on the CRT display. This rate could be easily exceeded even in unjammed operations. The presence of jamming would degrade the operator even further. This mismatch in bandwidth between the operator and the high target traffic, especially when jamming is present, can result in a loss in operator performance. Another factor contributing to operator loss occurs when he resorts to guessing, as when be becomes overloaded or panicky. In a sense then, the implementation of automatic features is intended to relieve the operator from the burdens of detecting targets in the presence of noise. The use of an operator to detect targets can therefore be justified only as a monitor or assist function to the normal operation of automatic feature circuits. By normal operation of automatic feature circuits is meant the operation of the circuit against its intended environment. The circuit, of course, can fail either because it becomes overwhelmed by its environment or because of the presence of some new environment. It is with respect to this last situation that the operator fulfills an important role.

5.3.9.2. Operator Over-ride

Automatic processing can be designed to operate only against those interfering or jamming signals which are known *a priori*. Anti-jam capability therefore is obtained by programming the equipment beforehand. New jamming situations which have not been designed into the equipment as anti-jam features may not be readily handled by the radar. For these types of situations, the trained operator is uniquely needed for his ability to adapt to new and varied situations; he is more capable of recognizing and coping with an altogether new form of interference than is the machine. Therefore, a trained operator is himself an important counter-countermeasure for maintaining radar operation and performance in the presence of unforseen jamming situations.

The flexibility of radar operation can therefore be accomplished by incorporating automatic features and by providing suitable monitoring displays and over-ride circuits which enable trained operators to make proper decisions in particular counter-measures environments. The operator should have the complete and positive control over the radar for its operation in either the automatic or manual modes.

References

1. M. Skolnik, *Introduction to Radar Systems*, McGraw-Hill, 1962. ECM and ECCM are discussed in Section 12.10.
2. Eds., "The silent war, an EW primer", *Countermeasures*, March, 1975. Few basic articles exist on Electronic Warfare (EW). The general tendency for the EW community is to mask its efforts in secrecy even to the point of not discussing information requiring no security classification. Consequently, individuals entering industry, workers in peripheral areas, and non-technical managers are relatively uninformed about EW even though they serve as program and hardware development managers. This article provides a general understanding and evaluation of the EW role, along with a glossary of associated terms.
3. J. Kamstra, "The History of ECM", *Countermeasures*, May/July, 1975.
4. J. Klass, "Avionic War Aims at Deceit, Confusion", *Aviation Week*, November 25, 1957.
5. J. Holahan, "Tools and Techniques of Electronics Warfare", *Space/Aeronautics*, April, 1960.
6. R. J. Schlesinger et al., *Principles of Electronic Warfare*, Peninsula

Publishing, P.O. Box 867, Los Altos, CA 94022. A timely reprint of a classic 1961 text. See Reference 4.14.
7. News item, "U.S. Weighs Israeli Request for Latest ECM Equipment", *Electronic Design*, November 8, 1973. Describes the performance of deceptive versus noise jammers.
8. H. David, "Mideast War is Forcing Changes in American Military Electronics", *Electronics Design*, December 6, 1973. Describes the performance of deceptive versus noise jammers.
9. Ed., *International Countermeasures Handbook*, EW Communications, Inc., 3921 E. Bayshore Rd., Palo Alto, California, 94303, May, 1975.
10. Ed., *Countermeasures Sources*, Hamilton Burr Publishing Co., Box X, Cupertino, California, 95014, October, 1975.
11. W. M. Hall, "General Radar Equation", *Space/Aeronautics R&D Handbook*, 1962–1963. This is an excellent engineering reference for the derivation and applications for the General Radar Equation. This equation is given in several forms usually encountered by radars. Particular attention is directed to the discussion of the jamming environment.
12. M. Simpson, "Counter-Countermeasures: Good Design Can Beat the ECM Threat", *Space/Aeronautics*, Vol. 33, pp. 138–142, April 1960. Describes ECM and ECCM design techniques.
13. J. N. Constant, *Introduction to Defense Radar Systems Engineering*, Spartan, 1971. Outlines the functions of a defense radar system to detect, track, discriminate or select, designate, and intercept unfriendly missiles. Particular attention is directed to the chapters on target characteristics, ballistic reentry trajectories, trajectory prediction, and radar data processing for discussions of techniques for trajectory prediction, signature analysis, and atmospheric filtering.
14. S. Adelman and S. M. Shimers, "Military Weapons Systems in a Countermeasures Environment", Paper presented to the National Winter Convention on Military Electronics, Los Angeles, California Feb 7–9, 1962. Presents a brief summary of radar countermeasures and radar counter-countermeasures techniques.
15. S. Letzter and N. Webster, "Noise in Amplifiers", *IEEE Spectrum*, August 1970. Describes factors which contribute to the noise problem and which can drastically limit the effectiveness of amplifiers and other instrumentation, especially at low signal levels.
16. H. Kaden, "The compensation of Interference in Carrier-frequency Receivers by Opposition Cancellation of a Parallel Receiver", *Frequenz*, Vol. 10, pp. 76–87, March 1956. Describes a technique for the cancellation of broadband interfering signals.
17. D. B. Brick and J. Galejs, "Radar Interference and its Reduction", *Sylvania Technologist*, Vol. 11, No. 3, pp. 96–108, July, 1958. Describes a technique for reducing the effects of strong impulsive and pulse-like interference.

18. J. J. Lamb, "A Noise Silencing IF Circuit for Superheterodyne Receivers", *QST*, Vol. 20, p. 11, Feb. 1936. Describes the Lamb Noise Silencing Circuit for reducing strong impulsive type noise.
19. I. M. Tepliakov, "The Improved Noise Immunity of Receivers with a Wideband Limiting Circuit", *Radio Engineering*, Vol. 16, No. 4, pp. 111–115 (1962?) describes the noise immumity of receivers using hard limiting. Translated from the Russian *Radiotechnica*, Vol. 16, No. 4, pp. 72–74, 1961.
20. I. Schmideg, "The Effects of Limiting on Angle- and Amplitude-Modulated Signals", *Proc. IEEE*, Letters, p. 1302, July 1969. Describes the AM compression, AM-PM and PM-AM conversion of a limiter. Additional recent references are given on the subject of the effect of limiters on modulated signals.
21. L. Wilson, "Asymptotes and Bandwidths for Spectra at the Output of a Hard Limiter", *Proc. IEEE*, Letters, p. 1676, Sept. 1960. Derives mathematical expressions which estimate the bandwidth increase at the output of a hard limiter with a Gaussian input.
22. H. Ward, "The Effect of Bandpass Limiting on Noise with a Gaussian Spectrum", *Proc. IEEE*, Letters, p. 2089, Nov. 1969. Computes the effect of limiting on the input spectrum of Gaussian noise. The output spectral density is plotted for various noise-to-limit level ratios.
23. J. Van Vleck and D. Middleton, "The Spectrum of Clipped Noise", *Proc. IEEE*, Vol. 54, No. 1, Jan. 1966. Calculates in great depth the spectrum which results after noise has been limited.
24. N. Blachman and S. Mayerhofer, "An Astonishing Reduction in the Bandwidth of Noise", *Proc. IEEE*, Letters, July, 1975. When a band of Gaussian noise is hard-limited and its frequency is divided by N, its bandwidth is divided by about N^2. This experimental observation is verified by theoretical analysis and is explained heuristically.
25. W. M. Siebert, "Some Applications of Detection Theory to Radar", IRE Natl. Conv. Record, Vol. 6, pt. 4, pp. 5–14, 1958. Describes the constant false alarm receiver (CFAR).
26. B. Oliver, "Automatic Volume Control as a Feedback Problem", *Proc. IRE*, Vol. 36, pp. 466–473, April 1948.
27. J. Field, "The Design of Automatic Gain Control Systems for Auto Tracking Radar Receivers", *Proc. IEE*, pt. C, Vol. 105, pp. 93–108, March 1958.
28. A. Locke, *Guidance*, D. Van Nostrand Co., Princeton, N. J., 1955. Particular attention is directed to the discussion of Automatic Gain Control given on pp. 402–408.
29. J. Croney, "The Reduction of Sea and Rain Clutter in Marine Radars", *J. Inst. Navigation* (London), Vol. 7, pp. 175–180, 190–192, 1954. Shows that the logarithmic characteristic must be maintained to about 20 db below the RMS noise level for effective reduction in the intensity of the clutter.

30. J. Croney, "Clutter on Radar Displays", *Wireless Engr.*, Vol. 33, pp. 83–96, April 1956. Shows that the output of a logarithmic receiver is independent of the input amplitude.
31. S. N. Van Voorhis, ed., *Microwave Receivers*, M.I.T. Radiation Laboratory Series, Vol. 23, pp. 583–606, McGraw-Hill, New York, 1948. Describes the design of receivers with logarithmic characteristics.
32. R. V. Alfred and A. Reiss, "An Anti-clutter Radar Receiver", *J. IEE*, Vol. 93, pt. 3, p. 459, Nov. 1948. Describes the design of receivers with logarithmic characteristics.
33. J. Croney, "A Simple Logarithmic Receiver", *Proc. IRE*, Vol. 39, pp. 807–813, July 1951.
34. S. Rosenstein, "Design of a Logarithmic Receiver", *Proc. IEE*, Vol. 102, pt. B, pp. 69–74, January 1955.
35. K. H. Chase and J. L. Pierzga, "Reducing Mutual Radar Interference", *Electronics*, Vol. 32, No. 28, pp. 39–43, July 10, 1959. Describes the technique of synchronization and blanking.
36. W. Howells and S. Applebaum, "Waveform Design for Tomorrows Radars", *Space/Aeronautics*, Vol. 32, No. 4, pp. 186–190, Oct. 1959.
37. C. E. Cook, "Pulse Compression: Key to a More Efficient Radar Transmission", *Proc. IRE*, Vol. 48, pp. 310–316, March 1960.
38. J. R. Klauder et al., "The Theory and Design of Chirp Radars", *Bell System Tech. Journal*, Vol. 39, pp. 745–808, July 1960.
39. C. E. Cook and J. E. Chin, "Linear FM Pulse Compression", *Space/Aeronautics*, Vol. 34, pp. 124–129, 148, Sept. 1960.
40. J. I. Marcum, "A Statistical Theory of Target Detection by Pulsed Radar", Math Appendix, *IRE Trans.*, Vol. IT-6, pp. 145–267, April 1960. Provides the basis for the theory of target detection. Particular attention is directed to the comparison of the efficiencies for both predetection and postdetection integration.
41. J. L. Lawson and G. E. Uhlenbeck, *Threshold Signals*, M.I.T. Rad. Lab Series Vol. 24, Chapter 12, McGraw-Hill, New York, 1950.
42. P. R. Dax, "ECM vs ECCM in Search Radar", *Space/Aeronautics*, Vol. 33, pp. 134–137, April 1960. Describes various ECM and ECCM techniques.
43. J. L. Allen, "Array Radars, a Survey of their Potential and their Limitations", *The Microwave Journal*, May 1962. Explores in a qualitative way the potentials and limitations of array radars, with emphasis on ground-based, long-range radar applications. Particular attention is directed to the discussion of array operation in adverse environments.
44. C. Burrows and S. Attwood, *Radio Wave Propagation*, Academic Press, 1949. Describes siting of radars in Vol. II, Chapter 11 and Vol. III, Chapter 10.
45. H. T. Friis, "Noise Figures of Radio Receivers", *Proc. IRE*, Vol. 32, pp. 419–422, 1944.

46. *IRE Dictionary of Electronics Terms and Symbols*, IRE, New York, 1961.
47. H. Goldberg, "Some Notes on Noise Figures", *Proc. IRE*, Vol. 36, pp. 1205–1214, Oct. 1948.
48. T. S. George, "Fluctuations of Ground Clutter Return in Airborne Radar Equipment", *Proc. IEE*, Vol. 99, pt. IV., No. 2, pp. 92–98, April 1952.
49. H. Urkowitz, "Filters for the Detection of Small Radar Signals in Noise", *J. Applied Phys.*, Vol. 24, pp. 1024–1031, Aug. 1953.
50. E. C. Westerfield et al., "Processing Gains against Reverberation (Clutter) Using Matched Filters", *IRE Trans.*, Vol. IT-6, pp. 342–348, June 1960.
51. J. L. Stewart and E. C. Westerfield, "A Theory of Active Sonar Detections", *Proc. IRE*, Vol. 47, pp. 872–881, May 1959.
52. N. J. Nilsson, "An Application of the Theory of Games to Radar Reception Problems", IRE Natl. Conv. Record, Vol. 7, pt. 4, pp. 130–140, 1959. Employs concepts of the mathematical theory of games to investigate optimum strategy for both the radar and jammer design optimization.
53. G. Schivley, "History of Chaff Development", WADC Tech. Note 59-6, ASTIA Doc. No. AD 208853, January, 1959.
54. Proc. IEEE, Vol. 52, No. 5, May 1964. A summary of the state of art for using chaff as an orbital scatterer. The articles by Mack and Reiffen, Lebow et al., Waldron et al., may be found particularly useful in describing RF characteristics of dipoles, their visibility by radar, and their dispensing.
55. D. Mott, "On the Radar Cross Section of a Dipole", *Proc. IEEE*, Letters, p. 793, May 1970. Considers the radar cross-section of a randomly-oriented dipole.
56. J. L. Lawson and G. E. Uhlenbeck, eds., *Threshold Signals*, M.I.T. Rad. Lab. Series, Vol. 24, Chapter 8, McGraw-Hill, New York, 1950.
57. R. Payne-Scott, "The Visibility of Small Echoes on Radar PPI Displays", *Proc. IRE*, Vol. 36, pp. 180–196, 1948.
58. D. G. Tucker, "Detection of Pulse Signals in Noise: Trace-to-Trace Correlation in Visual Displays", *J. Brit. IRE*, Vol. 17, pp. 319–329, June 1957.
59. P. M. McGregor, "A Note on Trace-to-Trace Correlation in Visual Displays: Elementary Pattern Recognition", *J. Brit. IRE*, Vol. 15, pp. 329–331, June 1955.
60. M. I. Skolnik and D. G. Tucker, "Discussion on Detection of Pulse Signals in Trace-to-Trace Correlation in Visual Displays", *J. Brit. IRE*, Vol. 17, pp. 705–706, Dec. 1957.
61. J. H. Van Vleck and D. Middleton, "A Theoretical Comparison of the Visual, Aural, and Meter Reception of Pulsed Signals in the Presence of Noise", *J. App. Phys.*, Vol. 17, pp. 940–971, Nov. 1946.

62. J. C. R. Licklider, "Basic Correlates of the Auditory Stimulus", in S. S. Stevens, ed., *Handbook of Experimental Psychology,* p. 1009, John Wiley and Sons, 1951.
63. W. P. Tanner, Jr., and J. A. Swets, "The Human Use of Information. I. Signal Detection for the Case of Signal Known Exactly", *IRE Trans.,* No. PGIT-4, pp. 213–221, Sept. 1954.
64. W. P. Tanner, Jr. and R. Z. Norman, "The Human Use of Information. II. Signal Detection the Case of an Unknown Signal Parameter", *IRE Trans.,* No. PGIT-4, pp. 222–227, Sept. 1954.
65. J. A. Swets and T. G. Birdsall, "The Human Use of Information III. Decision Making in Signal Detection and Recognition Situations Involving Multiple Alternatives", *IRE Trans.,* Vol. IT-2, pp. 138–165, Sept. 1956.
66. S. Johnston, "Simulation of the Effects of ECM on Radar Performance", 14th Annual Joint ECN Planning Conference Proceedings, USAE COM Report, pp. 2–29 through 2–39.
67. S. Johnston, "A Survey of Radar/ECM Simulation", 10th Annual NORAD Electromagnetic Warfare Conference, pp. 227–236.
68. S. Johnston, "Burnthrough Range—Bah Humbug", *Electronic Warfare,* November/December 1974.
69. S. Johnston, "The Effects of Various Types of Jamming on Radar Detection and Indicated Target Position", Proceedings 11th International Electronic Warfare Symposium, ADC000551, Vol. I.
70. P. Dax, "Noise Jamming of Long Range Search Radars", *Microwaves,* Sept. 1975. Jammer noise usually exceeds the receiver front-end noise by many tens of dB. Of the various ECCM techniques to combat standoff noise jamming, very low sidelobes are the most effective.
71. "Expendable Jammers—A New Approach to Low Cost ECM", *Countermeasures,* October/November 1975. A report on the TEDs program—the Air Forces' efforts to develop low-cost expendable jammers.
72. Special Report, "Radar Countermeasures", *Microwave Systems News,* December/January 1976. "Electronic Warfare: A Soviet View"; H. Eustace, "ECM vs Radar: A Lethal Scenario"; D. Anderson, "Postulating an Airborne ECM Threat"; L. Robbins, "Dynamic Range Considerations in Electronic Countermeasures Receivers".
73. Vakin et al., "Principles of Jamming and Electronic Reconnaissance", NTIS AD 692–642. Soviet publication.
74. Leonov et al., "Monopulse Radar", NTIS AD 742 696. Soviet publication.
75. "Radar ECCM", Session 30, ELECTRO 76, Boston, May 11–14 1976.
76. *International Countermeasures Conference 1976,* EW Com-

munications Inc. 3975 E. Bayshore Rd, Palo Alto, California 94303.

77. S. Johnston, "Radar ECCM Techniques", Proceedings International Countermeasures Conference, London 1976.
78. S. Johnston, *Master EW Glossary*, 2nd International Countermeasures Handbook, EW Communications Inc., 3975 E. Bayshore Rd, Palo Alto, California 94303.
79. S. Johnston, "A Complete Bibliography—Soviet Literature on Radar and EW", *Electronic Warfare*, September/October 1976.
80. Special Report, "Radar Countermeasures", *Microwave Systems News*, October/November 1976. "The Focus is on EW and radar countermeasures"; L. Meadows "Electronic Countermeasures Technology—An Overview (Four types of ECM systems are reviewed—self-protection, escort, standoff and expendables)"; "How chaff protects aircraft".
81. "Airborne Jammers get smarter over the years", *Electronic Warfare*, January/February 1977. Describes how Raytheon's jamming systems have evolved from the early spot noise maggie to frequency agile master-oscillator power amplifiers.
82. A. Arnot, "Countering the IR Threat", *Countermeasures*, February 1977.
83. R. Davis, "Anatomy of a Jammer", *Microwave Systems News*, July 1977. Describes rundown of the key components in the AN/ALQ-137 deception jammer developed by Sanders for the U.S. Air Force. Principles of operation in both the cw and pulsed modes are described.
84. "Metal coated glass chaff improves shielding", *Microwave System News*, September 1977.
85. H. Rainwater, "Weather Affects MM-wave Missile Guidance Systems", *Microwaves*, September 1977.
86. "Swedish ship defense system can deploy chaff clouds", *Electronic Warfare*, April 1978.
87. T. Walsh, "EW: from RF to EO", *IEEE Spectrum*, February 1978. Electro-optics (EO) gains on RF and acoustic technologies in electronic deception, jamming and neutralization.
88. *The International Countermeasures Handbook*, 4th Ed., 1978–79, EW Communications, Inc., 3975 E. Bayshore Rd, Palo Alto, California 94303.
89. S. Johnston, "ECCM Improvement Factors", *Electronic Warfare*, May/June 1974.
90. J. Frank and P. Tong, "ECCM Measures of Effectiveness", *Countermeasures*, January 1977. A considerable amount of work has been done in analyzing ECCM techniques, but there is no consistent way of evaluating them. This problem arose because of the wide variety of ECCM techniques, along with the variety of jamming methods they are designed to counter. This article is an attempt to provide several methods for comparative evaluation of

ECCM techniques. The comparisons are useful in masking-jamming, and particularly for barrage noise jamming.

91. "Radio Frequency and ECM Characteristics", Sanders Associates, Federal Systems Group, 95 Canal St, Nashua, New Hampshire 03060.
92. "U.S. Frequency Allocation Chart", Varian Electronic Device Group, 611 Hansen Way, Palo Alto, California 94304.
93. R. Hartman, "The Navy's New Jamming System: Big Business", *Microwave System News,* July 1978. The Navy's Advanced Self-Protection Jammer (ASPJ) system is a candidate for 3,000 tactical aircraft in the Navy and Air Force.
94. "Mini-Jammer Suited for Drone", *Microwave System News,* July 1978.
95. "Beam Weapon Conceivable—Possibly", *Microwave System News,* July 1977.
96. R. Hartman, "Shipboard EW", *Countermeasures,* May 1978.
97. K. Perry, "Frequency Diversity in Radar Systems", *Microwave System News,* Feb/March 1975. Multiple-frequency operation against radar's long-range performance enhances its ECCM capability, and offers clear benefits in mapping applications.
98. D. Kasparek, "Laser Guided Weapons Countermeasures Testing", *Countermeasures,* May 1975. Describes ECM/ECCM testing of laser-guided weapons at White Sands Missile Range. This highly successful test program is aiding system developers and testing agencies to improve the effectiveness of laser-guided weapons against countermeasures.
99. G. Johnson, "Understanding Low Power Spread Spectrum Radars", *EW/DE,* November 1978. Low Power Spread Spectrum radars differ from conventional matched filter and pulse compression radars only by their extended spectral spreading.
100. J. Gilder, "Efficient Switch-Tube makes Ground Radars more Jam-Proof", *Microwaves,* September 1978.
101. E. Drogin, "ALQ-161 Time Sharing Jamming Power", *MSN,* October 1978. Semiconductor processors are married to microwave hardware to provide pulse-by-pulse jamming in this power managed system.
102. R. Davis, "Microwaves Play Key Role in ASPJ", *Microwave Systems News,* July 1978. The Advanced Self-Protection Jammer, the next-generation airborne ECM system, will contend with some fairly sophisticated and difficult to jam Soviet radars.
103. *Electronic Countermeasures,* Peninsula Publishing. P.O. Box 867, Los Altos, CA 94022. Re-release of a classic text originally prepared under Government auspices. Formerly Secret and now fully declassified, it presents more than 1,100 pages of fundamental ECM and ECCM material by 50 authors. Amply illustrated with photos, diagrams and equations, this book addresses its subject from three major standpoints: signal intercept, jamming and

deception, and components. In addition, it features historical perspective plus more than 600 bibliographic references.

104. L. Van Brunt, *Applied ECM,* EW Engineering Inc. P.O. Box 28, Dunn Loring, VA 22027. Presents terms, fundamentals, tactics and techniques of ECM.

105. "Pentagon budget stresses electronic air warfare", *Electronic Design,* 15 February, 1979.

106. S. Johnston, "Radar Electronic Counter-Countermeasures", *IEEE Transactions on Aerospace and Electronic Systems,* January 1978. Employment of electronic countermeasures (ECM) against radars make the use of electronic counter-countermeasures (ECCM) in the radar a necessity. ECCM techniques have evolved over the last several decades. Many ECCM techniques have been described in the open literature. Unfortunately, there have been no recent comprehensive survey papers on this topic. This paper defines ECCM and the various parts of electronic warfare (EW) and shows the interrelationships between these parts. Principal effects of the parts of EW are described as a prelude to discussion of ECCM. ECCM techniques are described under two categories: electronic and operational. Electronic ECCM's are presented as those in the transmitter, the antenna, the receiver-signal processor, and in the system. Typical ECCM techniques in each part of the radar are listed. An ECM-ECCM matrix which lists some typical ECCM's for representative ECM's is given. Assessment of the efficacy of ECCM is briefly treated. A bibliography of studies containing descriptions of many of the approximately 150 ECCM techniques listed is included.

107. S. Johnston, *Radar Electronic Counter-Countermeasures,* Artech House, 1979. Collection of 51 key papers from the international literature spans the post war period through the present as electronic warfare (EW) and its three principal divisions are introduced: electronic countermeasures (ECM), electronic counter-countermeasures (ECCM) and EW support measures and their interrelationships.

108. M. Maximov *et al., Radar Anti-Jamming Techniques,* Artech House, 1979. Originally published in Russian in 1976, the book provides an original exposition of ECCM techniques—from basic engineering principles through a detailed analysis of radar performance in man-made and natural interference, with and without ECCM provisions. Several techniques are described which reduce radar system susceptibility to various types of jamming and interference.

6

Bomber Defense

There is little evidence that Russia is developing a long-range bomber force. The present force of long-range Russian bombers, consisting of some 100 TU-95 Bears and 40 Mya-4 Bisons is small and has not expanded since the mid-1950s. While the Russians have developed and already deployed their new twin-engine, sweep-wing bomber, designated "Backfire" by NATO, reports indicate that this aircraft has a less than intercontinental range. Against this force, the USA maintains seven manned interceptor squadrons equipped with F-106 aircraft in the active Air Force and sixteen Air National Guard squadrons equipped with F-101s, F-102s, and F-106s. Assuming an average of twenty-five aircraft per squadron gives a total number of 675 interceptors in a 4:1 ratio of U.S. interceptors to Russian bombers. In addition, the USA maintains twenty-one Nike-Hercules missile batteries in the active Army and twenty-seven Nike-Hercules batteries in the National Guard. It should be noted however that whether Backfire is a tactical or strategic bomber remains a live issue in the present ongoing SALT negotiations.

By way of comparison, the Russians have some 3,000 interceptors to counter the 565 U.S. long-range bombers in a 5:1 ratio of Russian interceptors to U.S. bombers. Apart from the older SA-2 and SA-3 surface-to-air missiles (SAMs), the Russians have had excellent results with their new SA-6 low-altitude SAM system.

6.1 Bomber Defenses

References 1–3, 6–8, 30–37

The present continental air defense is the responsibility of the North American Air Defense Command (NORAD) organized in 1957 and including both U.S. and Canadian forces. It comprises four components; the radars, which detect and track enemy bombers; the command-and-control centers known as Semi-Automatic Ground Environment (SAGE) and Back-Up Interceptor Control (BUIC), which direct battle forces against invading aircraft; and two types of weapons, the manned interceptor and surface-to-air missiles (SAMs). The interceptor is a piloted aircraft equipped with air-to-air missiles, sometimes armed with a nuclear warhead. SAMs are usually guided from the ground although newer types have an on-board guidance system that takes over after the missile has neared the target area. To a large extent, the effectiveness of the present bomber defense depends on whether it could survive the missile strike that would almost certainly precede the bombers.

6.1.1. Early Warning Radar

References 40–42

The radars and command centers, which are immobile ground-based installations, are ideal targets for Russian ICBMs and SLBMs. Because they are not encased in concrete or otherwise protected from a near nuclear explosion, they are soft and even a relatively inaccurate missile could destroy them. There are fifty-six long-range radar stations feeding data into fourteen SAGE units and these are vulnerable to a very small force of ICBMs. The loss of the radars and the control centers would make it impossible to detect and plot the course of attacking bombers and, lacking this information and the command centers to utilize it, it is impossible to guide interceptors to their targets or to alert SAM batteries of the bombers' approach. Complicating matters is the fact that present interceptors and SAMs are themselves also vulnerable to missile attack. Apart from their vulnerability to missile attack, present radars cannot track aircraft flying close to the ground in the maneuver called "terrain avoidance". Present Russian bombers can use this technique and the Backfire is designed expressly for it.

Over-the-horizon forward scatter (OTH-F) radar already forms a part of the U.S. early-warning network. Signals generated by transmitters in Japan, Taiwan, and the Philippines are reflected by the ionosphere over the USSR and are detected by receivers in Western Europe. Activity in the intervening atmosphere, such as the launching of missiles or bombers, produces signals which, when interpreted, give a general picture of events in the Russian airspace.

Now being deployed is the over-the-horizon backscatter (OTH-B) radar in which signals return to the vicinity of receivers located near the transmitters. The received signals are affected in the same way by events over the USSR and would be interpreted to provide the same information provided by OTH-F radar. Two units are emplaced, one looking east and one looking west. OTH-B could give a thirty minute warning, which would allow U.S. missile forces to launch in time and U.S. air forces on ground alert to be launched and therefore to survive the missile attack.

6.1.2. Airborne Warning and Control System (AWACS)
References 9–13

The replacement for the present ground-based radar and control stations is the airborne warning and control system (AWACS). As shown in Fig. 6.1 the AWACS radar is installed in a modified version of the Boeing 707 commercial jetliner; a rotating radome, or radar pod, is mounted on two struts above the rear section of the fuselage. The radome is fiberglass, 9 m in diameter and 1.8 m thick at the center, and contains antennas for surveillance radar and for Identification Friend or Foe (IFF) transceiver, which discriminates between friendly and hostile aircraft. Outputs of these systems are displayed on video consoles. While the horizontal range of ground-based radar is limited by the curvature of the earth, the AWACS aircraft flying at 12,000 m can survey the sky from ground level to about 450 km and therefore has a horizontal range much greater than that of ground-based radars. In addition to extending the range of aerial surveillance it provides more reliable tracking of bombers flying in terrain avoidance. The AWACS aircraft could survive a missile attack simply by staying aloft. With an endurance of at least seven hours, it could probably remain airborne long enough to meet the incoming Russian bombers. If necessary it could be refueled in flight. Twelve AWACS aircraft have been delivered and now serve both the tactical and strategic warning and control functions.

6.1.3. Interceptor Aircraft
References 14–21

Interceptors which have been proposed to replace present interceptors are the F-111-X-7, the NR-349, the F-14, and the

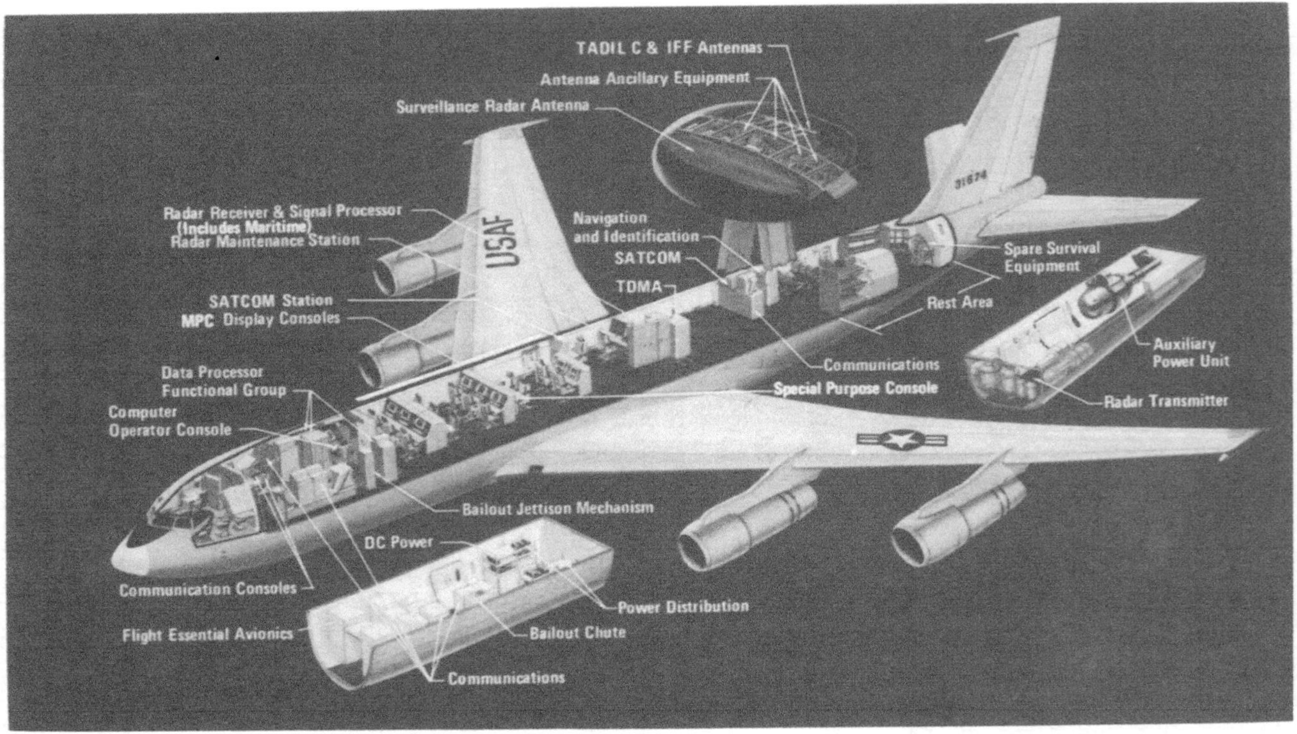

SOURCE: See reference 22.

F-15. The F-14 is a Navy plane that has a 160 km radar and mounts the long-range Phoenix air-to-air missile while the F-15 is an Air Force plane with a 80 km radar and additional fuel tanks for extending its range of operation. Estimates of the strength of the present interceptor force set the number of aircraft at 200. For the F-14 and F-15, 200 airplanes cost about $2–$3 billion.

Regardless of which airplane is chosen, its ability to survive a missile attack is doubtful. Interceptors are kept on the ground, and like ground-based radars they are probable targets for Russian ICBMs. Although several counter-measures are available to protect them, none seems entirely workable. To begin with, the interceptors could take off when the attack began and could wait out the missile attacks in the air. However, neither the F-14 nor the F-15 could stay aloft long enough (7–12 hours) to meet the arriving enemy bombers. Aerial tankers of course could be used to refuel them, but the present tanker force is committed to the bomber fleet of the Strategic Air Command (SAC) and new tankers are costly to procure. The interceptors could also fly to alternate airfields on receiving an attack warning. Any fields with the facilities to service them, however, would also be likely targets. Yet the interceptors could fly to secondary or clandestine fields, accompanied by their transport carrying the men and equipment needed to service and control the planes.

6.1.4. Surface-to-Air Missiles (SAMs)

References 16, 21, 26–29

The Army's developmental SAM-D system is the surface-to-air missile in the new bomber defense system. Four missiles, each in a protective canister that also serve as a launching tube, are carried on a large wheeled trailer. A similar trailer carries a phased-array radar, and a third trailer carries communications and command facilities. The Army estimates the cost of a planned SAM-D force at $3.9 billion, in addition to R&D costs of $1.2 billion, and operating costs of about $1 billion a year.

SAM-D has a range of about 160 km, which restricts it to areas near cities and to military sites being defended. Thus the SAM-D seems as vulnerable to missile attack as are interceptor aircraft; and as unlikely to survive to repel a bomber attack.

6.1.5. Projected Costs

References 5, 22

Table 6.1 gives some estimates of the projected size of strategic air defenses, warning, and control forces through 1980 and their costs. Major system acquisition costs are relatively small in comparison to other investment and operating costs. The rate of expenditures in the projected air defense program are estimated to be about $3–$5 billion a year through 1980. About two thirds of this amount—some $2–$3 billion a year—consists of direct and indirect costs. Direct costs include personnel pay, fuel, replacement parts, a share of base flight operations, and a share of intelligence and communications. Indirect costs include maintenance of central depots for aircraft overhaul, shared costs of base support such as hospitals, training of aircrews, launch-crews and maintenance personnel, and part of the costs of higher headquarters. Perhaps one third of the indirect costs therefore are fixed and would not vary over the near term with small changes in the size of the air defense force, but others, such as training costs, will vary with force size.

6.2. Russian Bomber Defense

References 22–25, 39

During the past two decades the USSR has made large investments in its air defense system, building an extensive network of ground radars and a force of about 3,000 interceptor aircraft. The most effective of the interceptors is the MIG-23, designated "Foxbat" in the North Atlantic Treaty Organization (NATO) code word system. However there is no indication that the Russians have achieved or attempted to achieve a "look-down, shoot-down" capability in their interceptors. Until they develop that technology, low-flying U.S. bombers and air-to-surface missiles will remain relatively invulnerable to Soviet air defenses.

Since 1964, the Russians have been deploying highly sophisticated defense systems. Their willingness to commit advanced defenses to their clients indicates the seriousness of their regard for such systems. Some 10,000 Sam launchers, including high-altitude SA-2s, medium-altitude SA-3s and low-altitude SA-6s have been set up to cope with any attack by the U.S.

Table 6.1. *Projected size and costs of strategic air defense, warning and control forces,* 1973–1980

CATEGORY	1973	1974	1975	1976	1977	1978	1979	1980
INTERCEPTOR SQUADRONS	7	7	7	7	7	7	7	7
ACTIVE AIR FORCE	7	7	7	7	7	7	7	7
AIR NATIONAL GUARD	20	20	20	20	20	20	20	20
SAM BATTERIES								
ACTIVE ARMY	21	21	21	21	21	21	21	21
NATIONAL GUARD	27	27	27	27	27	27	27	27
AIRBORNE WARNING AND CONTROL SQUADRONS								
EC-121	2	2	2	2	1	0	0	0
AWACS	0	0	0	0	1	2	4	4
COSTS (BILLIONS)								
MAJOR SYSTEM ACQUISITION	320	290	500	800	1,200	1,200	1,500	1,500
OTHER INVESTMENT COSTS	570	570	940	980	1,100	1,150	1,200	1,200
DIRECT OPERATING COSTS	1,130	1,140	1,100	1,100	1,100	1,100	1,100	1,100
INDIRECT OPERATING COSTS	1,220	1,240	1,200	1,200	1,200	1,200	1,200	1,200
TOTAL	3,240	3,240	3,740	4,080	4,600	4,650	5,000	5,000

SOURCE: See reference 35.

bomber force; the USA, by way of contrast, has only 1,136 SAM-D launchers. There is a controversy whether or not the SA-5, the newest Russian SAM-D system, can be used in an anti-ICBM role.

The new Russian SA-6 low-altitude SAM system includes several radars for tracking, guidance and command, operating in conjunction with an acquisition radar. The missile operates on command guidance and apparently has alternate guidance modes, such as infrared or active radar. The highly mobile SA-6 radar is mounted on two tracked vehicles, one for weapons, the other for radar. A tracking radar follows the target and after missile launch a second radar tracks the missile based on signals from a transponder in the rear of the missile. Commands can be sent to the SA-6 to make the target tracking and missile radars coincide and lock onto the target. There are four frequency bands in which the SA-6 operates. The low-altitude acquisition and tracking radar operates in a band around 5 GHz (giga-Hertz) and has a 24 kilometer slant range. The high-altitude detection and acquisition radar operates around 6 GHz and has a range of 40 to 80 kilometers, depending on target altitude. Once lock-on is obtained, the target is tracked in a band up around 8 GHz. The command guidance frequency is believed to be lower than these. Although these frequencies have been previously used in other Russian radars, few jammers stretch in frequency over all bands of SA-6 operation. During the 1973 Mid-East war, Israeli pilots were unable to suppress the Egyptian manned SA-6 batteries and as a consequence the latter were able to shoot down many Israeli airplanes.

6.3. Multipurpose Weapons

AWACS and SAM-D have also been proposed as multipurpose weapons, with value both as strategic components of the bomber defense and in tactical operations. Although forty-two AWACS aircraft ordered have been assigned to the Aerospace Defense Command, the Air Force still gives a high priority to AWACS as a tactical weapon. AWACS, together with remotely-piloted vehicles (RPVs) and other automatic weapons developed in the course of the Vietnam conflict, could facilitate the prosecution of limited non-nuclear wars. Areas for their possible tactical deployment range from Southeast Asia to

Western Europe, where ground radars would not be normally available for the control of interceptors. SAM-D, on the other hand, is primarily a battlefield unit for the defense of Army installations but the cost of its development is included in the continental air defense section of the defense budget.

References

1. R. English and D. Bolef, "Defense Against Bomber Attack", *Scientific American*, August, 1973. Describes the present bomber threat and bomber defenses.
2. H. Scoville, "The Limitation of Offensive Weapons", *Scientific American*, January, 1971. Enumerates strategic offensive forces.
3. R. Hudock, "Strategic Posture: Political Necessity, Budget Realities", *Astronautics and Aeronautics*, April, 1972. Reports testimony by Admiral T. Moorer, Chairman of the Joint Chiefs of Staff, before the Senate Armed Services Committee, and testimony by Defense Secretary M. Laird before Congress.
4. A. Quanbeck and B. Blechman, *Strategic Forces: Issues for the Mid-Seventies*, The Brookings Institution, 1775 Massachusetts Ave, N.W., Washington, D.C. 20036, 1973. The Institute publishes a series of studies as monograms. This is a staff study of strategic nuclear weapons, their uses and impacts.
5. B. Miller, "Rising Costs Stimulate the Search for Better Ground Surveillance Radar", *Electronic Warfare*, February 1978. In an effort to cut soaring maintenance expenses, defense planners everywhere are reevaluating their surveillance radar systems with an eye to refinement.
6. "Radar Requirements on the Upswing", *Microwave Systems News*, March 1978. Radars for test ranges, early-warning air defense, intelligence, and shipboard radars are being planned. PAVE PAWS, the new SLBM warning system, is to be erected at Otis Air Force Base, Mass.
7. "Air Force Alaskan Radars to Go Modern", *Countermeasures*, May 1978. The manned bomber watch was initiated more than twenty years ago. Strung along Alaska's western coast is a 7-site linkup. Other network radars are located more inland. This is the SEEK IGLOO system.
8. "Alaskan Radar to be Updated", *Microwave Systems News*, July 1978.
9. E. Ebersol, "Rush is on to complete first AWACS radars", *Microwaves*, September, 1973. Westinghouse is building three preproduction units for early flight testing.
10. News item, "AWACS flies through first trials", *Microwaves*, February, 1975.
11. R. Levine, "Boeing Co Wins Air Force Award of $247.6 Million",

Wall Street Journal, 29 April 1975. Six production models set in midst of hard sell for controversial radar jet.

12. J. Rhea, "Takeoff Clearance for AWACS", *Southern California Electronic News,* 19 May 1975.
13. B. Walsh, "An Eagle in the Sky", *Countermeasures,* July 1976. The E-3A Airborne Warning and Control System (AWACS) provides the "big picture vision and understanding for battle management through a unique all-altitude surveillance, command, control and communications capability". The AWACS consists of special avionics and a large surveillance radar. The U.S. Air Force is developing it to fulfill tactical and air defense forces' needs.
14. R. Davis, ed., "Despite record Performance, Phoenix Missile faces shakey future", *Microwaves,* September, 1973.
15. "Long Range Fire Control", *Countermeasures,* March 1975. The AN/AWG-9 is a long-range fire-control defensive and offensive Airborne Weapon Control System (AWCS) developed for the U.S. Navy's F-14 aircraft. It includes on-board target acquisition, long-range search and detection, positive look-down identification, multiscan and missile launch, IR detection and ECM immunity.
16. "Advances in Airborne Laser Systems", *Countermeasures,* March 1976. Recent advances in laser-augmented target acquisition and recognition systems offer greatly enhanced air-to-ground and air-to-air attack capability for single-seat aircraft.
17. B. Walsh, "Countdown on a New International Deterrent", *Countermeasures,* January 1977. A look at the F-16.
18. "Commonopoly: The Billion Dollar Game", *Electronic Warfare,* Jan./February 1977. DODs quest for F-16/F-18 EW suite.
19. "Avionics", *Microwave Systems News,* May 1977.
20. "JTIDS to Hand Off Airborne Missiles", *Electronic Warfare,* February 1978. Fighter aircraft of the future that have run out of missiles during combat may get help through a "forward pass" of weapons launched from other planes.
21. A. Ivanov, "Improved Radar Designs Outwit Complex Threats", *Microwaves,* April 1976. Three guidance techniques for air-defense guided missiles are described and the widely-used cw semi-active homing system is detailed.
22. News item, "U.S. Weighs Israeli Request for Latest ECM Equipment", *Electronic Design,* 8 November 1973. Describes the performance of U.S.-made ECM equipment against the Russian-made SA-6 SAM system.
23. H. David, "Mideast War is Forcing Changes in American Military Electronics", *Electronic Design,* 6 December 1973. Describes the performance of U.S.-made ECM equipment against the Russian-made SA-6 SAM system, also described.
24. News item, "Soviet missile gains spur ECM activity", *Microwaves,* December, 1973. Describes the performance of U.S.-made ECM

equipment against the Russian-made SA-6 SAM system, also described.
25. "SA-8 Russias Newest Microwave Technology", *Microwave Systems News*, April/May 1976.
26. C. Nolan, "Improved HAWK Simulator AN-TPQ-29," *Countermeasures*, October 1976.
27. B. Walsh, "ROLAND", *Countermeasures*, March 1977. The U.S. Army's New International Missile System provides all-weather defense against attacking aircraft in forward combat areas.
28. "U.S. ROLAND Test Fire Due", *Electronic Warfare*, February 1978.
29. R. Hartman, "SAM-D is a Patriot", *Countermeasures*, July 1977. Being developed to replace both Nike-Hercules and Hawk systems, the highly mobile, all-weather Patriot will be the field army's air-defense cornerstone against medium- to high-altitude targets in the sophisticated land warfare environment of the 1980s.
30. "PAVE PAWS bounded by another lawsuit", *Microwaves*, September 1979.
31. "USAF's SEEK IGLOO Ready, Soon", *Microwave System News*. November 1979.
32. G. B. Panero, "Survivability of Air Defense Systems", *Mitre Corp.*, November, 1959. See Reference 4.3.
33. "Airborne Navonics", *Countermeasures*, February, 1976. See Reference 1.30.
34. W. Shockley, "The E-4 Airborne Command Post", *Countermeasures*, July, 1976. See Reference 1.31.
35. B. Walsh, "An Eagle in the Sky", *Countermeasures*, July, 1976. See Reference 1.33.
36. P. Nahin, "Can Land-Based Strategic Bombers Survive an SLBM Attack?", *IEEE Transactions on Aerospace and Electronic Systems*, March, 1977. See Reference 1.51.
37. "PAVE PAWS Tests Scrutinized", *Microwave Systems News*, June, 1978. See Reference 1.40.
38. A. Arnot, "SM-2 Passes Naval Tests", *Countermeasures*, June, 1977. See Reference 1.72.
39. "Latest Soviet Threat Intelligence", *International Report EW/DE* (Electronic Warfare/Defense Electronics), May, 1978. See Reference 2.41.
40. Eds., "Backscatter Radar on 2 Coasts to Detect Planes over Horizon", *Electronic Design*, July 6, 1972.
41. J. Headrick and M. Skolnik, "Over-the-Horizon Radar in the HF Band", Proc. IEEE, June, 1974, pp. 664–673. See Reference 7.7.
42. Eds., "New OTH-B radar to be CW bistatic", *Electronic Design*, 12 April 1975.

7

Surveillance and Reconnaissance

By 1962, and probably even earlier, the USA had detailed information on the number and location of Russian strategic missiles. In 1967 President Johnson stated that satellite reconnaissance was worth ten times the money the USA had spent in space. He suggested that this knowledge had prevented the country from harboring fears that otherwise may have arisen. From the late 1960s to the present the Department of Defense has regularly published information on the level of Russian ICBM, SLBM and ABM deployment.

The major task of reconnaissance and surveillance systems is to monitor the quantitative and qualitative features of the opponent's offensive forces. This is a task for which observation satellites are well suited. From satellite photographs intelligence analysts are able to monitor silo construction and the transport of missiles to their deployment silos. With multispectral and infrared photography they can detect or penetrate camouflage and monitor nighttime activity. Submarine shipyards are observed on a routine basis to monitor new construction. Similarly, ABM launchers and radar deployments can be observed. There seems to be no doubt that such activities can be verified with a high degree of confidence.

The more important question, however, is whether the Russians can find ways clandestinely to circumvent or nullify these observations and thereby achieve a military or political advantage over the USA. The continuous monitoring of Russian transportation networks, power generation plants and manufacturing facilities by observation satellites would make it unlikely that the USA could not detect such clandestine activity in time to react in some way. The decrease in delay time provided by the most recent U.S. satellites between the first indication of suspicious activity in an area surveillance photograph and further photography with a high resolution camera will certainly reduce the uncertainties involved. Although photography cannot penetrate buildings, infrared and multispectral techniques can often reveal a great deal about activities inside, particularly since activities of special interest may be indicated by changes in standard operating procedures.

Cloud cover remains an impediment to observational activities, but with longer orbital lifetimes and the capability of changing orbit the newer satellites are less constrained by their predecessors. The clarification of ambiguities in the observations depends primarily on the use of the unilateral techniques of electronic and communications intelligence, although any bilateral apparatus resulting from the SALT treaties may also be useful. To summarize, the very size and complexity of construction and industrial activity required to build and deploy modern strategic weapons, combined with the breadth of coverage, resolution and multispectral aspects of observation satellites guarantee that much can be verified unilaterally with high confidence. It can be assumed therefore that the forthcoming SALT III agreements, like the SALT I and II agreements, will cover areas wherein both opponents already have obtained a high-confidence unilateral capability. Thus while the quantitative deployments of an opponent's offensive force can be readily observed, it is with respect to qualitative improvements that the unilateral capability of each side falls short. In any case, the cost of hardware and manpower for photo-interpretation will remain high.

If a future agreement in SALT were to impose qualitative restrictions on strategic systems, the USA would have to rely on other verification techniques to augment the observation satellites. The usefulness in this regard of systems that monitor missile tests is indicated. Observation satellites would also, however, have a role to play in monitoring such agreements. Any qualitative improvement in missile weaponry, for new guidance systems, multiple warheads, improved ABM radar, new SLBMs etc., must be manufactured and any new hardware must be delivered to an operational site installed and tested. These activities can be monitored by satellites.

7.1. Advanced Surveillance

References 1–4, 40–52

The present horizon-limited radar provides about ten to fifteen minutes warning of long-range missile attacks. Three other surveillance techniques are possible to increase this time even further: over-the-horizon (OTH) radar, satellite ground surveillance and satellite early warning. Such techniques are expected to increase the warning time to about thirty minutes, which is about the time it takes an ICBM to fly between continents. The increased warning time should better enable the ICBM and bomber forces to be launched earlier for a counterstrike and, of course, should enable the civilian population to take cover.

Reconnaissance and surveillance systems currently used by the USA are represented schematically in Fig. 7.1. U.S. photo-

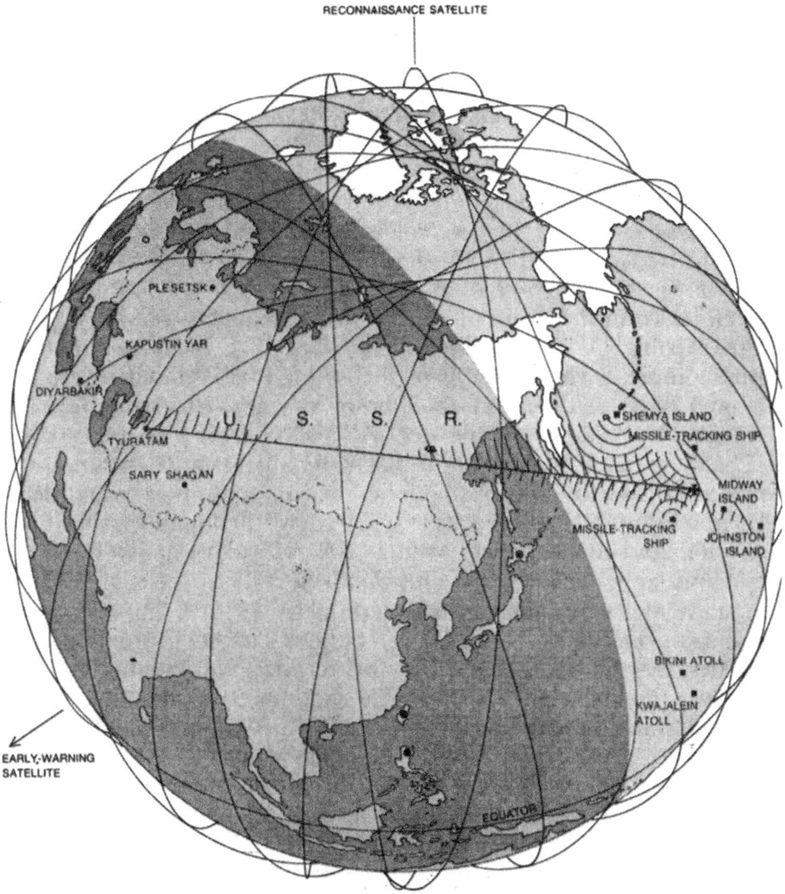

SOURCE: See reference 1.

Fig. 7.1. *U.S. reconnaissance and surveillance systems*

reconnaissance satellites are routinely launched from Vandenberg Air Force Base in California into a near polar elliptical orbit with an orbital period of approximately ninety minutes and a perigee (lowest point) on the order of 160 km. The latest, fourth-generation U.S. observation satellite, called Big Bird, combines the separate functions of area surveillance photography and close-look photography and hence is required to stay aloft for a much longer period than earlier close-look satellites; orbital times to date have averaged about seven weeks. The orbit of such a satellite remains essentially fixed in space while the earth rotates, with the result that to an earth-based observer the satellite appears to move westward on each successive orbit. Hence most of the earth's surface passes under the orbital path of the satellite. U.S. early-warning satellites, in contrast, are typically launched into near-equatorial near-synchronous "parking" orbits at altitudes of about 35,680 km. Two such satellites, launched into identical "figure eight" orbits at the same fixed longitude over the Indian Ocean but lagging each other by twelve hours, can provide continuos infrared coverage of most of the USSR and all of China, as shown by the lightly shaded area. The value of early-warning satellites from the point of view of arms control is that they are also capable of monitoring missile tests. The black circles in Russia indicate the locations of the major Russia missile-testing launch centers; the black line shows a typical trajectory for a Russian long-range missile test. Also shown are two types of U.S. radar used to monitor Russian missile tests; OTH transmitters shown by the black circles in Japan, Taiwan and the Philipines and their receivers shown by the open circles in Europe, and conventional, or line-of-sight, radars shown by the black squares in the Pacific and in Turkey.

Surveillance systems include certain land-based line-of-sight radars, over-the-horizon (OTH) radars, satellite systems and particularly shipboard sensors for terminal observations. As early as the summer of 1955 a U.S. radar at Samsun in Turkey was tracking missile tests from the Russian launch site at Kapustin Yar, northwest of the Caspian Sea. As a result of this monitoring, the launching of Sputnik I in October, 1957, and the Russian ICBM tests of the same year, came as no surprise to the U.S. intelligence community and Government officials with access to such data. They were well aware of the Russian capabilities in this area. By late 1963 or early 1964 a longer-range radar had been installed in Diyarbakir in Turkey, bringing into view missiles being launched from the newer Russian test center at Tyuratam, east of the Aral Sea.

Several fixed land-based radars are also available to observe Russian tests near the end of their flights. One of these has been operational on Shemya Island, far out in the Aleutian chain, since at least 1959. This radar can track and provide data on the Russian reentry vehicles (RVs) that impact in the test area on Kamchatka Peninsula or in the North Pacific area northwest of the Midway Islands. For the longer-range tests that end up in the Pacific southwest of Johnston Island, several radars that have been installed for other purposes at the Midway Islands, Bikini Atoll, Kwajelein Atoll and Johnston Island can be employed.

7.1.1. Over-the-Horizon (OTH) Radar

References 5–7

Unlike conventional radar, OTH radar is not restricted in its range by the curvature of the earth. By reflection from the ionosphere OTH radar can penetrate to great distances, making possible the detection of missiles soon after they are launched. The currently deployed "forward scatter" OTH radar detects the disturbances in the ionosphere caused by the ionized jet of gas emanating from a rocket's motor. Since each type of missile disturbs the ionosphere somewhat differently, a detected missile can be identified by its characteristic OTH signature. In the currently employed system three transmitters are deployed in Japan, Taiwan and the Philippines. These transmitters are matched with corresponding receivers in Italy, Germany and Turkey. Although the system was originally intended as an early-warning system for a massive missile attack, it has detected a high percentage of the known single events since 1968. All long-range missiles fired from test sites in the USSR are detectable.

The concept of OTH radar is based on the principle of long-range communication in which radars operate at high frequency (HF), about 2 to 20 MHz, and take advantage of ionospheric reflections. Very high transmitter powers are usually easier to achieve at such low radar frequencies. However, operation at HF results in poor angular accuracy and poor resolution, which result with antennas of reasonable size. The antenna half-power beamwidth is proportional to the radar's wavelength and inversely proportional to the size of its antenna aperture. Thus, at 20 MHz, an antenna diameter of more than 1 km would be necessary to achieve a 1° beamwidth. The antenna size would increase to 10 km if the frequency is reduced to 2 MHz.

7.1.1.1. Reflections from the Atmosphere

An electromagnetic wave incident on an ionized gas such as the ionosphere will be either absorbed, reflected, refracted or transmitted with negligible attenuation. A signal of frequency f_c will be totally reflected if:

$$f_c = 8970\, N_e^{1/2} \tag{7.1}$$

where

f_c = radar critical frequency; in Hz, and
N_e = number of free electrons in the ionized gas; number/cm^3.

The radar critical frequency given by equation (7.1) determines the highest frequency which is reflected by the ionosphere. Frequencies therefore which are less than the critical frequency are reflected back to earth; higher frequencies continue to propagate through the ionsphere. Since the ionosphere is composed of several different layers of ionization, the so-called D, E, F_1 and F_2 layers, there will be a different critical frequency associated with each.

Figure 7.2 illustrates the geometry of OTH operation. The

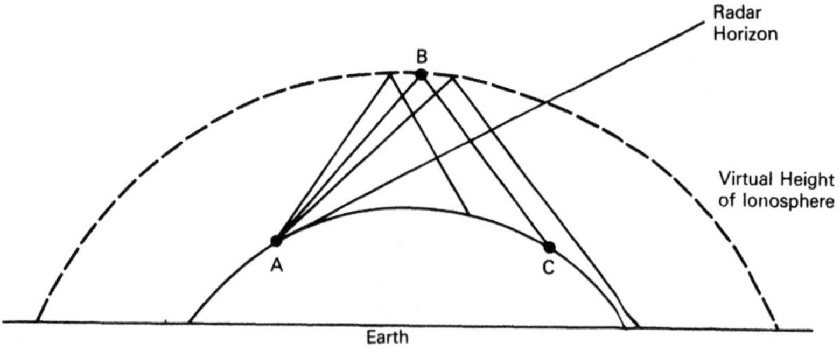

Fig. 7.2. *Geometry of OTH radar operation*

radar at point A points its beam at the ionosphere at B which in turn reflects the radar beam in the direction of a missile during launch at point C.

Shown in Fig. 7.2 is the virtual height of the ionosphere which is the height from which the radar beam appears to be reflected. A knowledge of the virtual height of the ionosphere along the propagation path from the radar at point A to the missile launcher at point C therefore permits the selection of the optimum point B for pointing the radar beam and also permits the optimum selection of the radar's frequency of operation.

The virtual height of the ionosphere increases as a function of frequency. A typical record of virtual height as a function of frequency, as might be obtained with a multifrequency recorder, is shown in Fig. 7.3. The critical frequencies are marked as the vertical dashed lines for each ionization layer and are indicated by a sudden jump in the virtual height. The horizontal dashed lines represent the approximate mean altitude of each ionization layer.

The ionosphere itself, as its name implies, is a series of layers composed of ions and free electrons lying at altitudes of between 100–500 km. These layers play an important role in the total reflection of electromagnetic waves at frequencies below

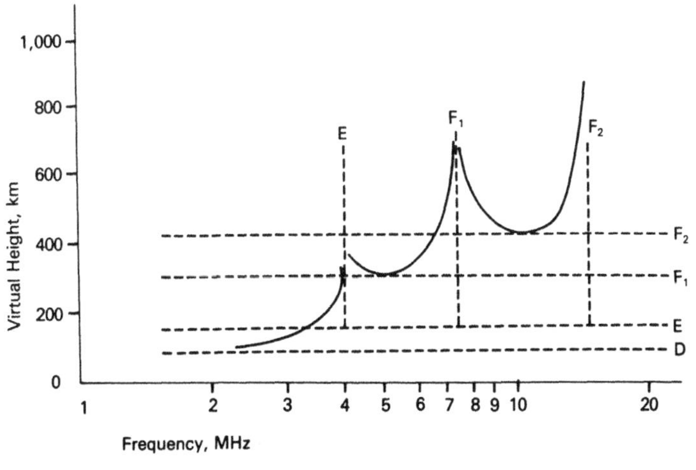

Fig. 7.3. *Virtual height of ionosphere as a function of frequency*

about 20-30 MHz and are responsible for long-range communications as well as OTH radar. The altitude of the maximum electron density concentration in a particular layer varies with the time of day and seasonally as a result of variations in the composition and temperature of the atmosphere and of solar radiation. The regions of ionization that are of interest to radio and radar transmission and the typically approximate altitudes of their maxima are: the D-layer, 70–80 km; the E-layer, 140–160 km; the F_1-layer, 270–290 km; and the F_2-layer, 380–460 km.

The path of a radio or radar wave enters and is gradually refracted in a given layer in the ionosphere. As a consequence, the apparent height of wave reflection, i.e., the virtual height, is much greater than the actual height of the refracting layer. Since the altitude and the thickness of a given layer in the ionosphere varies, the virtual height may also be expected to vary both seasonally and with the time of day.

Although OTH radar has been defined as being based on the ionosphere as the reflecting medium, the possibility exists for using satellites to reflect or relay radar beams over the horizon and back. This could be accomplished by using conventional microwave radar with extremely narrow beamwidths directed permanently at the satellite, which in turn either reflects or relays the radar beam through appropriate antennas in the direction of the missile launch. The reflection of a missile during launch and early flight will then retrace its path through the satellite and back to the radar.

7.1.2. Satellite Reconnaissance

References 8–11, 28, 29, 45, 52

Satellites employ optical and infrared (IR) sensors to view the ground in a variety of missions, for example, viewing the radiation from a missile's rocket motor. They can also carry a telescope television (TV) system and sensors to detect nuclear detonations in the atmosphere and to perform general military surveillance. Satellites can track a missile during the powered portion of its climb into space and in turn relay the missile's trajectory data to friendly command posts, thus increasing the warning time for attacks by ICBMs, SLBMs and FOBs.

Intelligence and warning satellite systems in operation or being studied include the Samos, ICBM Alarm, Vela, Integrated Multipurpose, Ocean Surveillance Discoverer family, Samos family, Big Bird and Early Warning satellites. Table 7.1 gives a summary of the missions, technical goals, status and milestones of these satellite reconnaisance systems.

Although the possibility of artificial earth satellites and their potential for military reconnaissance were recognized immediately after World War II, not until the mid-1950s was the future availability of rocket boosters assured and a satellite program actually initiated. On March 16, 1955 the USA issued a requirement for a strategic satellite system which culminated in the Agena vehicle family. This vehicle family was for many years the workhorse of the U.S. observation-satellite program.

The USA did not wait for the availability of satellites, however, before beginning a strategic-reconnaissance program. By 1956 U-2 reconnaissance aircraft were making their first flights over the USSR. Their photographic information supplemented the radar data on Russian missile tests, providing the basis for the general downgrading of estimates of the size of the Russian bomber force and convincing most informed observers that the Russian ICBM build-up was proceeding at a much slower pace than had been anticipated. In spite of the considerable usefulness of the U-2 flights, the aircraft had two weaknesses as a sensor platform. First, the limited range of the plane, the endurance of the pilot and the provocative nature of the mission imposed a severe limitation on the area of the USSR that could be photographed. As a result of this limitation, the lack of U-2 evidence for large numbers of deployed Russian ICBMs was not conclusive proof that only a few existed. Not until reconnaissance satellites had provided much greater coverage of the landmass of the USSR could the U.S. intelligence community be certain that missiles had not been deployed and remained undetected. The second weakness of the U-2 airplane was its vulnerability. Eventually the Russians developed an antiaircraft missile that could hit the U-2 at 21,000 meters. A U-2 was shot down in May, 1960, and U-2 overflights ended, except for certain minor incursions. Satellites, on the other hand, were known to be invulnerable, and would remain so for the then foreseeable future.

Among the many technical problems that had to be solved in order to develop an operational reconnaissance satellite were the stabilization and orientation of the spacecraft, the design and production of light cameras with a long focal length and a large aperture, and the recovery of the data. The possibility of

Table 7.1. Intelligence and warning reconnaisance systems

SYSTEM	MISSIONS AND TECHNICAL GOALS	STATUS AND MILESTONES
SAMOS	A complete reconnaisance system evolved from the Discoverer program for photographic and other electromagnetic surveillance of the earth from 160–480 km orbits. Photo data returned in recoverable capsules. Advanced Samos is equipped with IR and UV sensors, TV link readout. Weights: early SAMOS 1,800 kg; advanced SAMOS 3,600 kg. (USAF)	First launch 1960; advanced SAMOS 1966. Effective network requires nine satellites in orbit, twelve launches per year. High definition readout lacks real-time capability. Non-photographic capability may go to reconnaisance early warning satellites.
ICBM ALARM SATELLITE	Detects ICBM trails with IR and UV sensors as little as two minutes after launch. Random orbits. Piggyback launches on various boosters. (USAF)	Operational since 1963. Used as a sensor development testbed. Will be replaced eventually by the multipurpose satellite.
VELA	Carries radiation sensors for the detection of nuclear explosions in space. Later versions detect nuclear explosions in the atmoshphere; may develop the capability to detect underground explosions. 227 kg polyhedrons launched in pairs to fly 180 degrees apart in 80,000–96,000 km polar and equatorial orbits. Five pairs now flying. (USAF)	First pair launched in 1963 to near circular 88,000 km orbit, second pair in 1964. Third pair in 1965 was first to keep station, carry sensors for detecting atmospheric explosions. EM pulse, IR, optical, X-ray, gamma ray and particle detectors aboard. 1967 launch added advanced sensors for very long range detection of space explosions.

INTEGRATED MULTIPURPOSE WARNING RECONNAISANCE SATELLITE	Multipurpose operation in synchronous orbit combining ICBM alarm with functions of VELA, weather satellite, and portion of SAMOS in real time. Detects launch and possibly reentry points of ICBMs and SLBMs, nuclear explosions in space, atmosphere and underground, troop movements, and assesses post-attack damage. Real-time readout, perhaps adaptive data-processing (USAF).	System may employ a variety of orbits and spacecraft. ETR launch into elliptical orbit suggests Molniya-type surveillance being explored besides low circular orbits. Synchronous orbits with satellite-to-satellite relay in millimeter bands also may be studied. Integration with 496L and Safeguard under consideration.
OCEAN SURVEILLANCE SATELLITE	Worldwide surveillance of sea state for surface ship routing; monitoring of all surface ocean traffic to prevent sneak attack. Also may attempt detection of submarine trails (Navy).	Feasibility studies in progress; radar sensors and power supplies are limiting factors to date. Signal processing to distinguish ship and ocean noises requires improved sea-state measurements. Flight of some sensors by 1975; operational by late 1970s.
DISCOVERER FAMILY	Close-look satellite evolved from Discoverer program. Photo data returned in recoverable capsules.	Typical perigee 128 km. In orbit two weeks. Operational since 1963.
SAMOS FAMILY	Area-surveillance satellite. Radio transmission of data back to earth. The first successful satellite of this family, SAMOS II, was launched in 1961 and carried 150 kg of instruments.	Elliptic orbit. Perigee 100 km. In orbit three to four weeks. Operational in 1961.

Table 7.1. (*continued*)

SYSTEM	MISSIONS AND TECHNICAL GOALS	STATUS AND MILESTONES
BIG BIRD	Satellite for area-surveillance and close-look photography. Combines radio transmission of low-resolution data to earth with the return of high-resolution photo data in recoverable capsules. Weight 9,000 kg.	Launched into elliptical orbit. Perigee 160 km. Apogee 260 km. In orbit two months. Operational in 1971. Replaces the Discoverer and Samos families.
EARLY WARNING	Satellite placed in near-synchronous orbit over the Indian Ocean for early warning. Provides two minute response to the launching of ICBMs in Russia and China.	Two satellites required for effective coverage. Operational in 1971. Replaces the ICBM Alarm Satellite.

Reference 42.

using a TV camera was considered very early but was rejected because the desired resolution was not attainable with available technology. In 1957 the decision was made to pursue two parallel approaches: first, direct recovery of a film package, and second, on-board developing and scanning of film followed by radio transmission to ground stations.

The first successful reentry package was recovered from Discoverer 13 on August 11, 1960. Either this package or the one recovered from Discoverer 14 the following week probably yielded the first satellite photographs of the USSR. That was just three months after the last U-2 overflight of the USSR. The first successful radio-transmission observation satellite was Samos II, which was put in a polar orbit varying in altitude between 480 and 560 km on January 31, 1961; this spacecraft carried between 135 and 170 kilograms of instruments. As a result of these early successes, the USA was able to estimate the number of deployed Russian ICBMs by September, 1961, reportedly put at fourteen.

The Samos family of satellites have been comparatively small radio-transmission observation satellites whose sensors and orbital characteristics were chosen to maximize their degree of coverage. They stay in orbit for three to four weeks. Although their perigee (lowest orbital point) is about 160 km above the earth's surface, the ellipticity of their orbit greatly reduces their atmospheric drag, thereby increasing their lifetime. With an inclination of 80 to 92 degrees with respect to the Equator, they provide virtually full coverage of the USSR and complete coverage of China. These "area surveillance" satellites provide low-resolution coverage of wide regions, and the radio transmission of data makes possible the rapid recovery of intelligence information. Particular areas of interest identified on the basis of this information and higher-resolution pictures are later made by another satellite with a recoverable film capsule. By early 1962 the size and weight of the camera system had been reduced sufficiently to allow the area-surveillance satellites to be launched by the Thor/Agena booster rather than by the more powerful Atlas/Agena booster used earlier.

The use of the thrust-augmented-Thor/Agena D, beginning in May, 1963, indicates that a new generation of satellites was then introduced. The greater throw-weight of this booster rocket enabled the satellite to carry a larger camera and more of a consumable payload (including film), resulting in an increase in useful lifetime. These satellites were launched at roughly one-month intervals beginning in the middle of 1963. Until the end of 1965, however, two or more area-surveillance satellites were frequently in orbit at the same time, but occasional delays in the regular launch schedule were made. This suggests that problems of reliability necessitated the replacement of satellites that had failed before the completion of their mission and also that problems had sometimes occured at ground level. Since 1966, one of these satellites has been in orbit for almost six months, and there has been hardly any overlap. One can therefore probably fix the beginning of 1966 as the date of a fully operational status for the area-surveillance satellites.

The first test launch employing the even more powerful long-tank-thrust-augmented-Thor/Agena was in August, 1966, and that booster was introduced into regular service in May, 1967. The use of this booster also marks the arrival of a heavier, third-generation satellite. It was probably equipped with a camera of longer focal length, a larger film supply, an IR optical system and a new transmission system with an increased data rate.

The task of these satellites has been to survey wide areas with sensors of moderate resolution and to reveal targets that merit a closer look at higher resolution. In order to provide that closer look, a different type of satellite is used.

The Discoverer family of recoverable-capsule satellites are the close-look satellites. They are heavier than the area-surveillance satellites, reflecting the fact that they carry a camera with a longer focal length and a wider aperture. They are also in a lower orbit, with a perigee typically about 128 km, in order to maximize resolution. With this more powerful telescopic system, interesting targets identified by an earlier area-surveillance satellite can be rephotographed and examined more closely. To minimize information losses introduced by electronic data storage and transmission, the close-look satellites send their film packs back to the earth in a reentry capsule for midair recovery by aircraft.

The first launching of a close-look satellite appears to have been on April 26, 1962, when a Thor/Agena booster put into orbit a satellite carrying a recovery capsule designated E-6. Three days later the capsule and its film were recovered. Judging from the frequency of launchings this program seems to have achieved operational status by the middle of 1963. With their shorter lifetime in orbit, the close-look satellites had less stringent reliability requirements than the area-surveillance types and were therefore able to reach full operational status more quickly. By 1964 the Atlas/Agena had been introduced as the program's booster, ensuring that heavier satellites with improved capabilities could be placed in orbit. These second-generation high-resolution satellites were launched about once a month and remained in orbit for three to five days before sending their film package back to the earth.

The Titan 3B began to be used for test launches in July, 1966, and came into regular service in August, 1967. With the introduction of this still larger booster the lifetime of the close-look satellites began to increase until by 1968 they remained in orbit for a period averaging some two weeks. This clearly shows that a third-generation satellite had been introduced with a much greater film capacity and the ability to raise its orbit in order to avoid early burn-up. It has been suggested that the new satellite could alter its orbit to take advantage of breaks in cloud cover; moreover, it seems likely that several new types of sensor were included in the satellites. IR sensors and multispectral photography appear to be the most likely candidates because of their ability to discover and penetrate camouflage and in the case of the IR sensors to operate in the dark, in accurate mapping cameras for the purpose of pin-pointing the location of strategic targets in the USSR. It is launched by the powerful Titan 3D booster.

The extra weight and size of this system result from joining the separate functions of area-surveillance and close-look photography into one satellite. Big Bird is reported to carry an area-surveillance camera and an on-board film processor and scanner. The resulting data are reportedly transmitted by means of a new 6.1 meter unfurlable antenna, which could represent an increase in capacity by a factor 16 over the older 1.52 meter antennas. In the past, several months would go by before a close-look satellite could be launched to rephotograph an area of interest identified by a low-resolution photograph and its film pack could be recovered. Now, however, Big Bird can be directed to turn on its high-resolution camera during a subsequent pass. Film from this camera, said to have a resolution of less than 0.1 meters from an altitude of 160 km, is returned in one of several recovery capsules. The delay time should therefore be cut to several weeks.

This dual capability requires that Big Bird remain aloft for a much longer period than earlier close-look satellites. In order to accomplish this result the satellite is placed in a higher and more elliptical orbit. The orbital characteristics of the first Big Bird, launched on June 15, 1971, were a perigee of 177.6 km and an apogee of 288 km. To compensate for this higher altitude and to improve resolution, both the focal length and the aperture of the high-power camera had to be increased over those of earlier types. To further increase the satellite's lifetime it has been equipped with an on-board rocket to raise its orbit and prevent early burn-up. The times in orbit for the first three satellites were respectively 52 days, 40 days and 68 days. The fourth launching was on October 10, 1972. Observation satellite launchings have now been made routine.

The tasks that can be accomplished by reconnaissance from space, provided atmospheric limb ambiguities prove resolvable, include missile-launch early warning, FOBs deceleration detection and orbital space surveillance through hot and cold body tracking with IR sensors. Suitable orbits include circular ones at 480 or 800 kilometers and some elliptical and geosynchronous ones. A stationary synchronous orbiter could serve as a communications relay for a low circular orbiter to ground stations while the twenty-four hour non-stationary synchronous orbit yields some rather interesting ground tracks over the Eurasian landmass.

One of the major obstacles for satellite reconnaisance is, of course, cloud cover. Certain locations, for example the location of Moscow, are rarely free of clouds. The coordination of observation-satellite launchings with information received from weather satellites can overcome this problem to some extent. Moreover, observation satellites currently in service probably have the capability of changing their orbit to take advantage of breaks in the cloud cover.

Observation satellites contribute to the verification of the SALT agreements in a variety of ways. By photographing missile test sites they help to identify new missile systems, to detect changes in operational procedure that may suggest a change in hardware, and to monitor testing programs. They can watch industrial facilities, including shipyards for the construction of submarines and plants for the assembly of missiles. Intermittent information on critical aspects of transportation networks can be obtained The progress of construction of missile silos, ABM radars or ABM launcher sites can be monitored. Wide area surveys can be made to determine if any activity is underway that violates the agreements.

A new orbital strategic mission, which stems from the steadily increasing strength of the Soviet fleet is the detection and surveillance of ship and submarine traffic. The technical challenge of such surveillance from space lies in the design of radars and their power supplies and in the development of suitable discrimination techniques.

7.1.2.1. *Sensors*

References 12–14

Reconnaissance from space is still limited by sensor resolution and downlink communications channel bandwidths. For these reasons, reconnaissance missions will continue to use recoverable payloads. As is clearly shown by the return-beam vidicon technology of NASA's Earth Resources (ERS) program high-resolution TV from space does not yet approach the spatial resolution needed for military missions.

Reconnaisance from space represents the glamour side of remote sensing. There exists considerable optimism that the comprehensive synoptic coverage of large and vast land and ocean masses obtainable at orbital altitudes will provide important new military capabilities. Some of the sensors being investigated for orbital reconnaisance are listed in Table 7.2, which also indicates possible applications for these sensors. In most of these instrument areas, teams of scientists and engineers are currently cooperating with groups of potential military users to explore applications and to help define additional instrument requirements. In general, the different sensing techniques and applications are being evaluated in test flights over known ground sites to permit evaluation of one or more problems at a time. Known ground sites are a crucial part of any remote sensing program because they permit gathering the characteristic spectral signatures and images against which data gathered on any flight over new territory must be compared in order to know what to look for.

In a typical reconnaisance mission enormous amounts of data are obtained. There is the critical problem therefore of what to do with the data collected. The fact is that the ability to acquire data using satellites far exceeds the ability to interpret and manage the data. There is therefore the question of just how much data is needed for the best interpretation. Gathering more data does not necessarily contribute to gaining more

Table 7.2. *Sensors for orbital reconnaisance*

SENSOR	APPLICATIONS
VISUAL PHOTOGRAPHIC AND TV	GROUND SURVEILLANCE; ICBM sites; troop movements; assess post-attack damage; weather mapping
MULTISPECTRAL PHOTOGRAPHIC AND COLOR TV	GROUND SURVEILLANCE; ICBM sites; troop movements; assess post-attack damage; weather mapping
UV, IR IMAGERY, RADIOMETRY AND SPECTROSCOPY	EARLY EARNING; detects ICBM trail TRACKING; detects launch and possibly reentry points of ICBMs and SLBMs; detects FOBs deceleration.
EM PULSE, IR, OPTICAL, X-RAY, GAMMA RAY AND PARTICLE DETECTORS	SPACE SURVEILLANCE; detects nuclear explosions in space, in the atmosphere and underground
SYNTHETIC APERTURE RADAR IMAGERY AND REFLECTIVITY	GROUND SURVEILLANCE; detects ships and submarines; detects sea state; ground mapping
MICROWAVE IMAGERY AND RADIOMETRY	GROUND MAPPING
ABSORPTION SPECTROSCOPY	REMOTE GEOCHEMICAL SENSING; detection of trace metals; detection of stack gases

information. If one takes note from NASA's ERS program, probably 90 percent of the data gathered to date have not been utilized and, with large multisensor systems now being flown the danger is of ending up prostrate beneath a mountain of utterly useless presentations in real-time viewing or in films, tapes and charts in stored-time operations.

Clearly what is needed in interpreting satellite data is a more efficient means of reducing the enormous amount of obtainable data to a more compact and meaningful form by several orders of magnitude, i.e., programming remote sensors to provide the interpreter with only the data he needs for his particular purpose. In the case of missile early launch warning, the interpreter needs missile data to be emphasized over the natural background of information. In FOBs deceleration detection a similar filtering out of the natural background is required. In general, the useful data must be quickly and readily extracted from the vast amounts of unwanted and therefore redundant data. Even in the relatively sophisticated field of photographic interpretation, automatic discrimination techniques, usually based on recognizing spatial relationships, have not been particularly fruitful in low contrast backgrounds. As we go from optics to other regions of the electromagnetic spectrum, the designer and interpreter uses, first, more basic information on physical properties and spectral characteristics of natural and man-made objects to properly design sensors and interpret the resultant images and, second, needs automatic means of selectively finding the information which is most relevant to his particular problem. This second requirement requires a whole program of data management including communications, storage, processing, retrieval, display and decisions on such questions as hardware, techniques, people and organization—all of which are needed to collect, process, analyze, interpret, present and display the data. Fig. 7.4 shows the various areas in which effort is needed to provide reliable reconnaisance data. In the figure effort is indicated as required for both subsystems and components of a reconnaisance system. In the past much attention has been focused on the more or less straightforward problems of collection hardware in the areas of sensors and communications and considerably less effort has been given to the equally important problems of the less glamorous subsystems and components in the figure. As the various space reconnaisance and surveillance programs mature, this imbalance will undoubtedly even out.

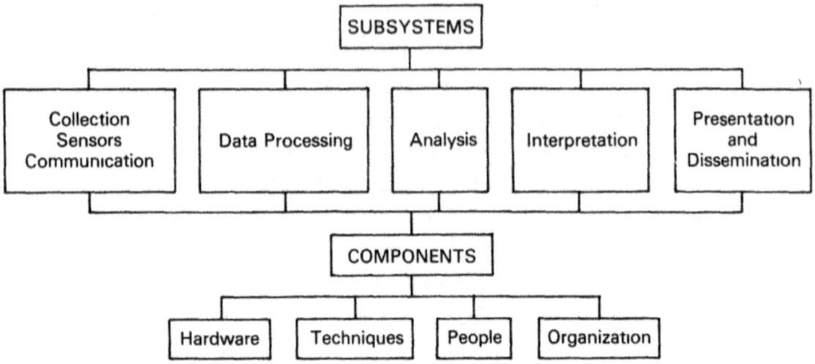

Fig. 7.4. *Data management areas in a reconnaisance system*

In a typical system a number of reconnaisance satellites (RS) are deployed to provide maximum coverage of major portions of the earth. A conceptual scheme for a full coverage real-time reconnaisance system is illustrated in Fig. 7.5. The satellite altitude is primarily a function of the ground resolution which is required and higher resolutions are obtained at the lower altitudes. However, longer satellite lifetimes and longer data-dumping intervals are obtained at the higher altitudes. To eliminate the waiting period between data taking and data dumping a number of communications satellites (CS) are often implemented at somewhat higher altitudes to provide practically instantaneous relaying of reconnaisance data. In this manner,

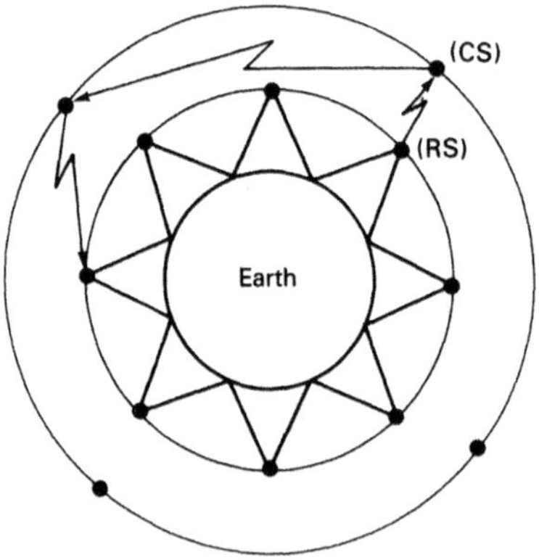

Fig. 7.5. *Full coverage real-time reconnaisance system*

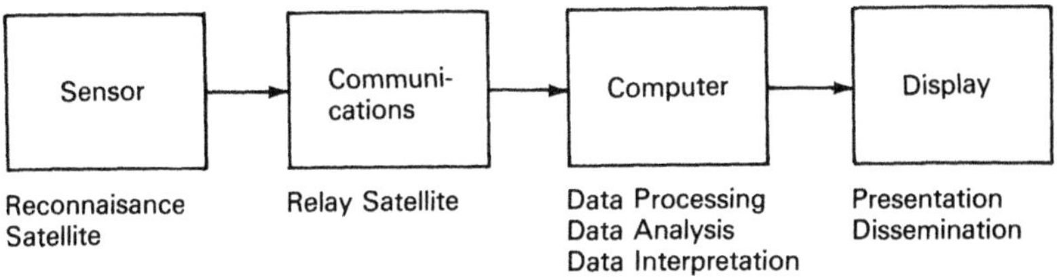

Fig. 7.6. *Major subsystems of a real-time reconnaisance system*

reconnaissance data becomes available to the interpreter in the shortest period following its acquisition by a reconnaissance satellite.

The various subsystems of a real-time reconnaisance system are illustrated in the block diagram of Fig. 7.6 and include sensors, communications, computers and displays. Typical sensors may be photographic, TV, UV, IR scanners and radiometers, synthetic aperture radar and scatterometers. These devices survey the ground and provide their data to the computer via the communication channel. The computer processes the data which is then displayed.

As larger booster rockets have made it possible to put heavier instrument packages into orbit, there has been an increase in the number of sensors and in their resolution and film capacity. At the same time the associated communications, computer and display systems have also become more sophisticated and their data-transmission rates and capacities has increased.

7.1.2.1.1. Spectral Windows. Everything in nature has its own unique distribution of emitted, absorbed or reflected radiation. These spectral characteristics can, along with a knowledge of thermal and quantum noise in instruments, be used to distinguish one thing from another or to obtain information about the size, shape and other physical and chemical properties of objects. Insofar as we know the spectral characteristics of the objects of interest, the general background and the instrument noise, we can pick the appropriate sensor to make the desired measurement at the highest contrast, remembering that for a given collector or aperture diameter we get the greatest spatial resolution where wavelengths are shortest. Of course, detection requires that the energies of radiations or reflections be the greatest. Also, in looking at the earth's surface from a satellite we must deal with losses from scattering, absorption and

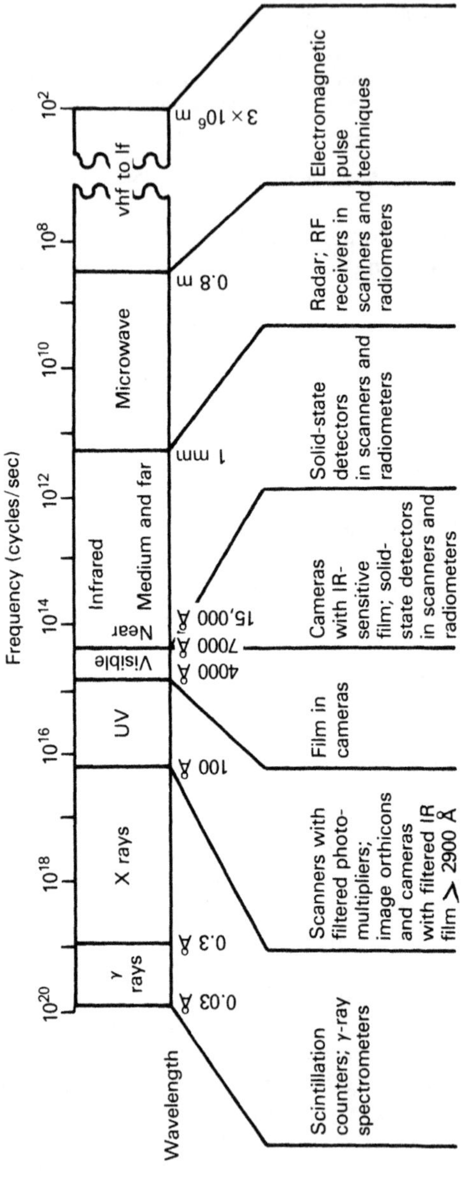

SOURCE: See reference 10.

Fig. 7.7. *Remote sensors used in various portions of the electromagnetic spectrum* reflection by gases, smoke, haze, water vapor and water droplets. These losses are a function of the path length and the number, size and energy of the particles within this path length as well as the wavelength of observation. It should be obvious therefore that we cannot just pick and choose our observational frequencies at random up and down the electromagnetic spectrum. We are forced by nature to channel our attention to those spectral regions where the atmosphere will transmit sufficient radiation to the detector. The most familiar and widely used of these, of course, is the visible region. Here, as well as in adjacent regions of the spectrum, we can use photochemical emulsions to capture light reflected from the sun, moon or some artificial source like a mercury flash lamp. Fig. 7.7 gives a listing of remote sensors used in the various portions of the electromagnetic spectrum.

Because of the fundamental electromagnetic properties of the earth's atmosphere, satellite-borne sensors are restricted to three spectral "windows": the visible-light wavelengths, a broad infra-red band centered on a wavelength of eight microns, and certain radar wavelengths.

7.1.2.1.2. Photography. Satellite photography in the visible region gives the most information on size, shape and relative position of objects of any sensor. Traditionally the pictures have been used in essentially one way: objects are identified on the basis of the ability to recognize them. Pictures of the ground are taken and, in the early satellites and present close-look satellites as well, stored in the satellite for later dumping in recoverable capsules. The later area-surveillance satellites electrically dump the data through appropriate communications channels when the satellite comes into view of a ground station or through communications satellites.

The attempt to distinguish among objects that have the same photographic gray scale and therefore which look the same, for example as do grass, concrete, asphalt and soil runway surfaces on a high altitude panchromatic photograph, has led to the concept of multispectral reconnaisance. Thus multispectral photography is a technique for making pictures simultaneously at several wavelengths and is now increasingly employed in reconnaisance satellites. The theory behind multispectral reconnaisance is that by comparing two or more photographs of the same object made in different regions of the spectrum we learn something about the object that we could not learn by studying the tonal values on just one photograph. When two or more photographs are available for comparisons, an additional degree of freedom has been presumably injected into the analysis.

A typical multispectral camera takes pictures in four spectral bands: 0.5 to 0.6 micrometer, 0.6 to 0.7 micrometer, 0.7 to 0.8 micrometer and 0.8 to 1.1 micrometer. This is the system of a multispectral scanner on board NASA's ERS satellite; the line scanning device employed uses an oscillating mirror to simultaneously record the terrain passing under the spacecraft in the four spectral bands. The electronic signals produced by this television system are then transmitted back to the earth, where they are converted to black-and-white images. The latter in turn can be used to make color composites by registering the images sequentially on color film through the appropriate filters. The ground resolution of the NASA radio-transmission system in no way compares with the ground resolution of the current generation of close-look reconnaisance satellites which require the recovery of film capsules. Midair recovery of a film capsule from a reconnaissance satellite is accomplished by means of specially-equipped aircraft as the capsule floats by parachute through the latter part of its descent. At present large C-130 aircraft are used routinely to recover heavy film capsules dropped by the latest model close-look reconnaisance satellites.

Intriguing as multispectral photography is it should be obvious that, except for simple cases where only a few things are being compared, the addition of so much detail does not suddenly change the photo-interpreter's task to a more happy one. The task at the moment is to eliminate the need for an eye-weary interpreter's comparison of tonal values on a multitude of photos. To accomplish this the information content in the multiple channels must be condensed and not expanded, i.e., the data must be pre-processed and combined electronically to produce an image with enhanced signal-to-background contrast.

Multispectral photography has also been extended into the ultraviolet which is an intriguing region to look at because certain objects like unpainted metal show remarkably greater contrast here than in other regions of the spectrum.

7.1.2.1.3. Television (TV)
Reference 14

Next to photography in its ability to resolve objects is television (TV). Both black-and-white and color TV are based on the same considerations as are visual and multispectral photography. Of course the principle advantage of TV is its potential for real-time operation. To date, however, the resolution afforded by TV does not approach, perhaps by a single order of magnitude, the resolution afforded by photographic sensors. As in photography, objects are identified on the basis of the ability to recognize them, i.e., by correlating patterns.

7.1.2.1.4. Infrared (IR)
References 15–17

In addition to visible-light cameras, infrared (IR) sensors are used in reconnaisance satellites. Since all bodies at terrestrial temperatures radiate energy that is predominantly in the infrared part of the, spectrum, infrared photography is not dependent on the sun for illumination. It is therefore useful for night applications or in polar regions although low-light systems have been developed and may also be in service for these purposes. Perhaps the greatest significance of IR sensors is their ability to detect things such as missile silos that are underground or may be camouflaged. As long as the ground immediately around the object is at a different temperature or has emission characteristics different from those of the surrounding terrain, an underground silo will stand out in an IR picture. For this reason multispectral photography, a technique for making pictures simultaneously at several wavelengths, is increasingly employed in observation satellites. Although longer-wavelength IR photography would require much larger optics to achieve resolutions comparable to those attainable with visible-light systems, such high resolution is probably not needed to accomplish the tasks assigned to the IR sensors.

As stated previously, conventional photography is excellent for resolving recognizable objects but may not tell all we want to know about their physical or chemical states. Moreover, photography is generally restricted to employment during daylight and clear weather. Systems sampling IR radiation from terrestrial objects are free of certain temporal and atmospheric restrictions of visual radiation and at the same time provide unique data about objects. Since IR photography is not dependent on the sun for illumination, it is used in satellites for night applications or in polar regions. IR sensors have outstanding ability to detect missile exhausts.

A common method for obtaining IR imagery is use of an optical-mechanical device in which a rotating mirror scans the terrain in continuous strips transverse to the line of flight. Radiation from each element of the ground scene is reflected from the face of the mirror and focused on a photoelectric detector such as indium antimonide or impurity-doped germanium. The detector output is amplified electronically and may be recorded on magnetic tape or used to intensity-modulate a CRT trace or a light source for recording directly on film. The resulting image is a strip map that superficially resembles a satellite photograph. The vital difference is that the tone in the conventional satellite photograph is a function of light reflectance whereas the tone in the infrared image represents variations in the intensity of the emitted radiation resulting from variations in the product of emissivity and surface temperature. The interpreter of IR images must, therefore, accustom himself to a scene in which the warmer objects appear lighter in tone than the cooler objects.

Absolute measurements of thermal radiation flux are made with a non-scanning device which is simply a radiometer. It looks at one element on the ground and because it is referenced to some known temperature it gives quantitative data that permit calculating the apparent and, in some cases, the actual temperature.

The theoretical resolution of IR devices is given by the Rayleigh resolution criterion. It is not as high as the resolution obtained in the visible portion of the spectrum. Even more important, this resolution is degraded by implementing the components in an actual system. In practice, IR devices achieve two orders less resolution than do optical devices of comparable aperture.

7.1.2.1.5. Radar

References 18–24, 41, 46

Radar permits continuous strip imaging of large areas and has the ability to operate effectively day or night at certain frequencies through all but extremely heavy clouds and rain. Side-looking radars, particularly those which use data-processing techniques to synthesize the effect of an extremely long antenna, give remarkable resolution almost to within one order of magnitude of that which is obtained using photography. However, radar devices are quite heavy and along with their power supplies impose difficult feasibility problems for satellite applications.

Radarscope photos can be displayed free from the foreshortening which is obtained by taking an oblique photograph of a segment of terrain; yet, because the illumination angle is oblique, small textural differences are enhanced by shadows. This makes it easier to detect fractures, faults and large-scale earth-moving projects common to military constructions. In addition it appears possible to penetrate certain vegetation with radar and get a return that is indicative of the underlying surface and its contents.

As with other sensors the identification of objects can be made by analyzing the characteristics of the reflected radar signals. It is possible to see certain things on the basis of long-wave reflectivity alone that might not be seen on a small-scale UV, light or IR photograph, for example, transmission lines and fences. Moreover, small reflectivity variations can be detected, a feature that makes radar an important sensor for studying sea state and for mapping crop types. Also, it might be possible to obtain characteristic signatures for various natural materials on the basis of polarization of the reflected wave. Signatures of water, forest, desert and farmland can be differentiated when obtained by non-imaging radar type systems which measure the terrain-scattering coefficient as a function of the angle of beam incidence. It thus becomes more feasible to differentiate the natural background from man-made targets of military interest using radar than by using other sensors. In addition, the ability to penetrate the ground or sea surface increases as frequencies get lower. At several meters wavelength, objects at or just beneath the surface become visible to the radar.

7.1.2.2. Resolution and Contrast
References 8–11, 25–27

The most useful measure of the quality of a satellite photo-reconnaisance system is ground resolution, a value that is equivalent to the smallest object that can be distinguished on the ground with good contrast. For a camera pointing vertically downward, the expression for ground resolution is:

$$G_0 = \frac{H}{1000\,FR} \tag{7.2}$$

where

G_0 = ground resolution limited by the detector or film; in meters,
H = altitude; in meters,
F = focal length of camera; in meters,
R = joint resolution of the film and optics; in lines per millimeter, and 1,000 is a numerical constant which is characteristic of the units being used for the other quantities.

Equation (7.2) gives the ability of a sensor to resolve between two closely-spaced objects on the ground when the focal length and film resolution are known, as is the case for imaging sensors used in photography. The diffraction limit or Rayleigh resolution criterion of a sensor, on the other hand, is used when the wavelength and aperture of the sensor are known, as is the case for non-imaging sensors and radar. The diffraction limit is given by:

$$G_D = \frac{\lambda H}{D} \tag{7.3}$$

where

G_D = ground resolution limited by diffraction; in meters,
λ = wavelength; in meters,
H = altitude; in meters, and
D = diameter of the aperture; in meters.

Equation (7.3) determines the highest possible resolution of a given sensor. The highest resolution is obtained at the shorter wavelengths.

Figure 7.8 gives the operational potential of various satellite sensors from an altitude of 800 km through the time period of the 1970s and early 1980s. Most significant are the expected improvements of longer focal lengths and better film resolutions of photographic sensors, the improved vidicon bit-rates of TV sensors and the longer synthetic-array lengths of synthetic-aperture radar.

Examples of high-resolution panoramic cameras are those

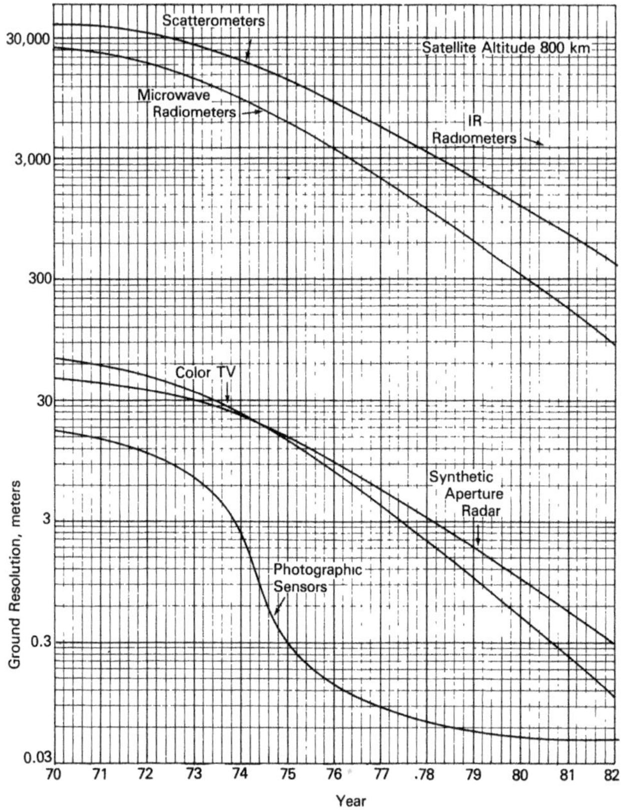

Fig. 7.8. *Ground resolution potential of satellite surveillance sensors*

used in the last three Apollo missions to the moon. These had a focal length of 0.61 meters and photographed the lunar surface from an altitude of about 100 km. The resulting photographs had a ground resolution of about 1 meter. These numbers, inserted in equation (7.2), give an optical resolution for the NASA camera of about 180 lines/per millimeter.

By way of comparison, it has been reported unofficially that the high-resolution cameras used in the newest generation of observation satellites have a focal length of more than 2.44 meters. Taking 2.44 meters as the focal length and 180 lines per millimeter as the optical resolution, one can calculate that the ground resolution of such a camera system from a nominal altitude of 160 km would be 0.366 meters. Actually there is reason to believe that the resolution of the most advanced camera systems now in service may be even better; as early as 1960 experts in the field were talking about prospective U.S. Air Force cameras with focal lengths on the order of 6 meters. Furthermore, films with a resolution considerably better than 180 lines per millimeter are currently available. A compromise must be made, however, between film resolution and film speed, and rather fast film is needed for satellite photography.

Another factor that must be taken into account is the aperture of the camera's optics. Diffraction effects place an upper limit on the optical resolution of any lens system. This limit is never attainable in practice. The limit is given by equation (7.3) so that in view of equation (7.2):

$$R \leq \frac{D}{980\lambda F} \tag{7.4}$$

Thus for a camera with a focal length of 2.445 meters, an aperture of 0.305 meters would result in a diffraction-limited resolution of 180 lines per millimeter, assuming that one is working with visible light at a wavelength of 0.55 micrometer.

In practice a much larger aperture would be employed in order to ensure that the optical system is not limited by diffraction. Therefore a camera with a focal length of 2.445 meters and an aperture of 0.915 meters would provide a ground resolution of 0.448 meters from an altitude of 160 km assuming that it uses film with a resolution of 180 lines per millimeter. With the same film a camera having a focal length of 6.10 meters and an aperture of 1.525 meters would yield a ground resolution of 0.2135 meters. To achieve a ground resolution of 0.1525 meters the same camera would require film with a resolution of about 330 lines per millimeter.

These simple calculations are based on the assumption that the overall resolution of the system is poorer than the poorest contributing factor, which may be either the diffraction-limited resolution of the optics or the line resolution of the film. Moreover, the results do not take into account various operational degradations such as satellite motion, nor do they involve the intrinsic limitation on ground resolution presented by the scattering of light in the atmosphere. The latter limitation amounts to about 5 centimeters for an altitude of 160 km. Nonetheless, it seems clear that the latest U.S. photo-reconnaissance satellites are close to achieving resolutions that are limited only by the atmospheric effects.

Ground resolution is not the only relevant of picture or image quality. It characterizes an image only at the threshold of what can barely be distinguished. The color and light contrast of the objects being photographed strongly affect their distinguishability. The angle of the sun and the intrinsic reflectivity of an object are also important, since these factors influence the object's apparent brightness. In general the task of identifying an object in a photograph is always more difficult than simply locating it. On the other hand, extended objects such as roads or rail lines can often be distinguished even though their dimensions may be far below the calculated ground resolution.

7.1.2.3. Coverage

A string of satellites in polar orbits can provide full coverage of the major portions of the earth of interest. Ground objects, however, are in the field of view of a given satellite in the reconnaissance system for only a few hours. For circular orbits, the orbital period is given by:

$$P = 1.41(1 + H/6400)^{3/2} \tag{7.5}$$

where

P = orbital period; in hours, and
H = altitude; in km.

Thus the 160 km orbit results in an orbital period of about ninety minutes. At this altitude the range to the horizon is about 1,400 km so that a reasonable ground swath for satellite sensors at this altitude, i.e., one which does not require low angles of incidence, would also be at about 1,400 km maximum width or perhaps even slightly less, depending on scanning and data rates needed for the coverage. A series of satellites in a corresponding number of polar orbits could provide satisfactory coverage of much of the ground at any single instant. In this arrangement a ground object would be in full view of at least one sensor in the satellite reconnaissance system almost 100 percent of the time. Thus the addition of a number of satellites in a number of polar orbits increasingly provides full ground coverage. Calculations indeed indicate that between eight to twelve satellites are needed for just such a system.

Two different techniques are currently employed by the USA to retrieve photo-reconnaissance information from space. For the lower-resolution pictures of area-coverage, the film can be developed on the satellite and scanned by a TV camera or a laser system. This information is temporarily stored and then transmitted to the earth when the satellite passes over one of several ground and shipboard stations around the world. Clearly, in order to transmit all the necessary information during the short time in which the satellite is in view of these stations, a communications system with a high data-transmission rate is needed.

For the higher-resolution pictures of close-look coverage the film itself is returned to the ground for processing. This method preserves all the information that is recorded on the film and is therefore employed for very-high-resolution photographs. The exposed film is ejected from the satellite and is returned to the ground in a special reentry package, which is caught in midair by specially-equipped aircraft as it floats by parachute through the latter part of its descent.

Once a film pack is returned to the earth it must be developed to maximum advantage and then interpreted. The transmitted pictures must also be recreated and then interpreted. With a ground resolution measured in small fractions of a meter and with film resolution measured in the hundreds of lines per millimeter, it is clear that by the time enlargements of the photographs are made many man-hours must be expended to interpret them. Indeed, photographic interpretation must be a very large enterprise today both in the USA and the USSR. This is the primary reason for the dual-mode approach by the USA, i.e., using the area-surveillance mode to make photographs at less than maximum resolution in order to identify interesting targets for further photography at a higher resolution using the close-look surveillance mode.

An area-surveillance system similar to the system used in the early U.S. Samos satellites is represented in Fig. 7.9 by a schematic diagram of the interior mechanism of a Lunar Orbiter—the spacecraft used to make a photographic survey of the surface of the moon preparatory to selecting the Apollo landing sites. In both systems the film is exposed and developed by the optical and film-processing elements of the camera subsystem; the developed photographs are then scan-

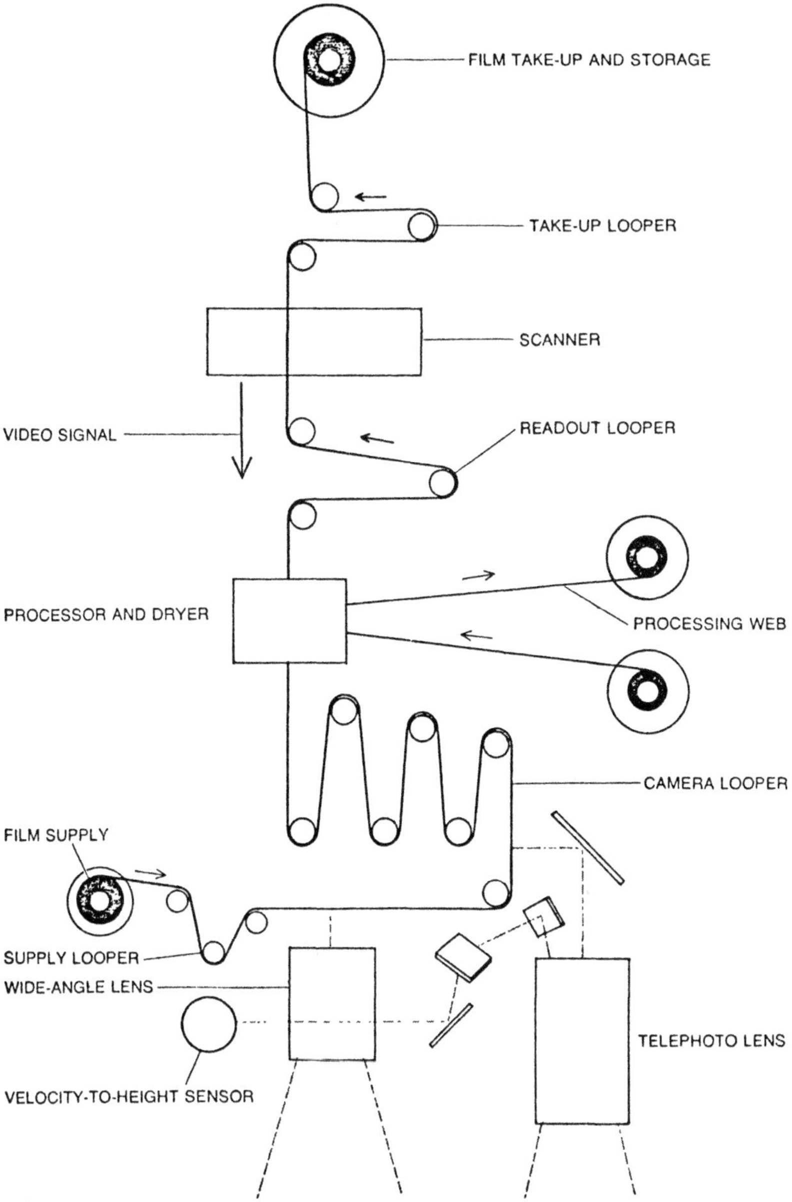

SOURCE: See reference 1.

Fig. 7.9. *Area-surveillance sensor system*

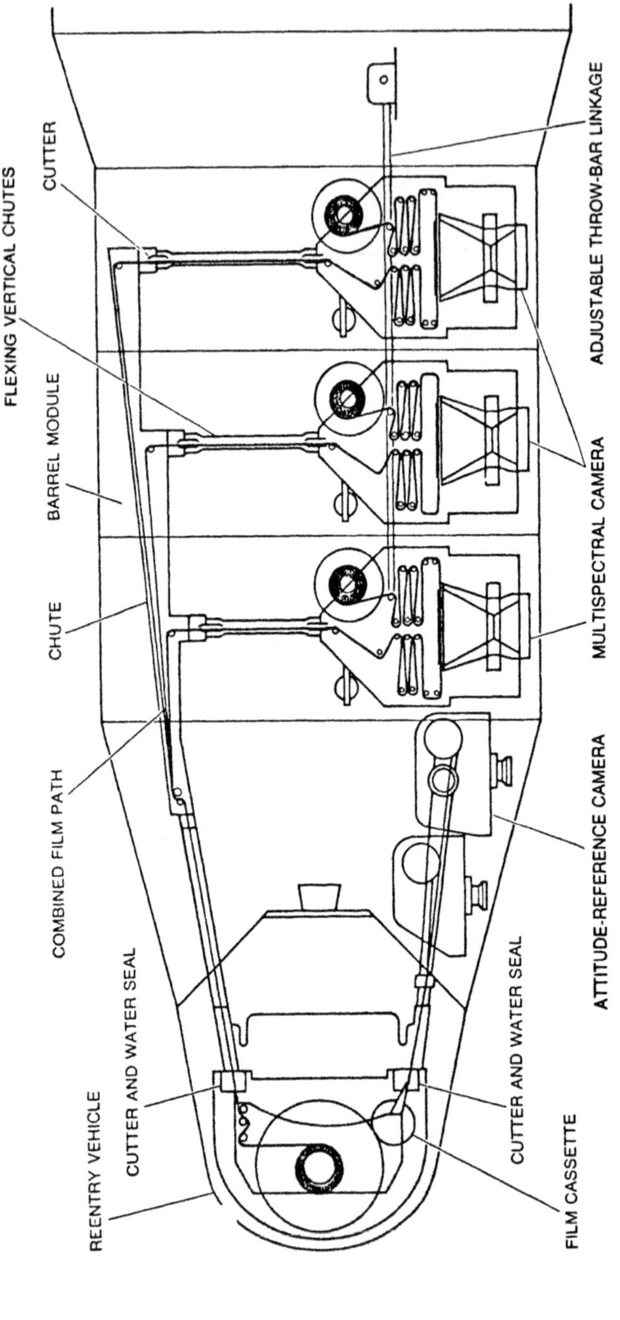

SOURCE: See reference 1.

Fig. 7.10. *Close-look sensor system*

ned by a TV camera, converting the picture into electrical signals for radio transmission back to earth.

A close-look system similar to the system used in the U.S. second-generation Discoverer reconnaissance satellites is represented in Fig. 7.10 by a schematic diagram of the interior mechanism of a proposed non-military satellite. The scheme was considered for use in NASA's earth-resources survey program. In both systems, exposed film is fed from several cameras into a recoverable capsule, which is ejected from the satellite and returned to earth for processing. Since this method preserves all the information on the film, it is preferable for high-resolution photography.

One important aspect of the ground-surveillance systems but not of the early-warning systems is their degradation in performance in the constant presence of clouds. It appears that the chance of using cloud-free photography of any single location over a period of time is only about 25 percent–50 percent and even more in heavy cloud areas. Thus, if preprogrammed and made fully automatic, the reconnaissance system will generate unwanted weather information for this same time duration and true surveillance data only for the balance of this time. Some means therefore are needed to program the taking only of true surveillance data, i.e., data is taken only weather permitting. In this regard, the situation using satellites has not changed from the days of aerial reconnaissance.

7.1.2.4. Sensor Equations

All objects radiate energy by virtue of their temperatures. The fundamental laws governing the emission of radiation can be employed to compute the level of ground signals in the terrain scene. Important sensor equations are those which describe ground signals in the terrain scene, the receiver sensitivity, the background-to-noise ratio and the signal-to-background or contrast ratio. These will now be discussed.

7.1.2.4.1. Ground Signals Spectral Distribution. The law which best describes the emissive power from the terrain scene as a function of both the wavelength and temperature is the well-known Planck distribution law. It describes the emission of thermal radiation within an interval $\Delta\lambda$ at the wavelength λ

from a unit area of terrain surface into the hemisphere above the surface. For a gray body the law can be stated by:

$$J_\lambda = \epsilon_\lambda \frac{C_1}{\lambda^5} \frac{1}{e^{C_2/\lambda T - 1}} \tag{7.6}$$

where

J_λ = the emmissive power; in watts per unit ground area per unit wavelength interval, i.e., watts/cm^3,
$C_1 = 3.732 \times 10^{-12}$ watt-cm^2,
$C_2 = 1.436$ cm-deg K,
λ = wavelength; in cm,
T = absolute temperature; in deg K, and
ϵ_λ = spectral emissivity.

Planck's distribution law applies for all values of the product λT. In the middle- and far-infrared and microwave portions of the electromagnetic spectrum, λT is large so that the ratio $C_2/\lambda T$ is small. For these cases, Planck's equation (7.6) can be simplified to:

$$J_\lambda = \epsilon_\lambda \frac{C_1}{C_2} \frac{T}{\lambda^4} \tag{7.7}$$

This last equation is the so-called Rayleigh-Jeans equation and can be used in the middle-, far-infrared and microwave portions of the electromagnetic spectrum to compute terrain emissions. In this equation it should be noted that both the spectral emissivity ϵ_λ and temperature T are proportional to the emissive power J_λ.

The normal spectral emissivities ϵ_λ of a number of materials are given in Table 7.3 for the case of temperatures ranging between 273–500 deg K. The table illustrates the wide variation in the spectral emissivity between materials and even within a single class of materials.

Any alteration of a surface will change its emissivity, for example, roughening a surface will increase its emissivity while polishing it will decrease the emissivity. The emissivity can therefore only be considered to be a physical property of the material when it is measured for an opaque polished specimen. Composite materials such as paint-films, anodized aluminum or shiny mirrors coated with thin lacquer will possess emissivities which depend on the thickness of film, nature of its organic

Table 7.3. *Spectral emissivity of selected materials at ordinary temperatures*

MATERIAL	EMISSIVITY
MOST PAINT	0.90
SPECIAL ALUMINUM PAINT	0.20
MOST METALS	0.40–0.70
POLISHED METALS	0.10–0.40
HIGHLY POLISHED METALS	0.01–0.10
EARTH, FOLIAGE, WATER, ICE, BRICKS, STONES, CLOTH, GLASS, WOOD, HUMAN SKIN	1.0

composition, etc. Excluding highly-polished metals and restricting attention to ordinary ground temperatures, say 273–500 deg K, many objects are essentially, to within a few dB, black. Thus, the earth, foliage, water, ice, bricks, stones, granular paint films of any color, cloth, glass, wood and the human skin are essentially black insofar as their emissions of radiations at ordinary temperatures are concerned. It must be emphasized that the foregoing remarks related to the spectral emissivities of selected materials apply only within the wavelength region where the materials are opaque and therefore only to temperatures which give rise to negligible intensities of radiation outside those wavelength bands.

Since a pencil beam of width θ intercepts a ground area of $(\theta R)^2/\cos\beta$ at the slant range R and angle of incidence β and since the emitted power from a surface varies as $\cos\beta$ the power which is emitted by the ground in the direction of the sensor is found by multiplying equation (7.7) by the effective intercepted area $(\theta R)^2$ and the wavelength interval B_λ:

$$J = \epsilon_\lambda \frac{C_1}{C_2}\frac{T}{\lambda^4}(\theta R)^2 B_\lambda \text{ watts} \tag{7.8}$$

The power which is received by the sensor can be found by multiplying equation (7.8) by the space factor $1/4\pi R^2$ and noting that $\theta = \lambda/D$ where D is the lens or aperture diameter. The total power at the receiver is therefore given by:

$$P = \frac{\epsilon_\lambda}{4\pi}\frac{C_1}{C_2}\frac{T}{\lambda^2} B_\lambda \text{ watts} \tag{7.9}$$

Thus the power at the receiver is inversely proportional to the wavelength squared λ^2 and directly proportional to the temperature T and wavelength interval B_λ. It can be seen therefore that the power can be expected to increase as the square of the frequency of operation. Thus less emitted power is available at the lower frequencies. An interesting fact is that the power received at the sensor is independent of the lens or aperture size.

7.1.2.4.2. Receiver Sensitivity. In general, the sensitivity limit of a receiver is given by:

$$N_0 = \frac{hf}{e^{hf/kT} - 1} + \frac{hf}{n} \tag{7.10}$$

where

N_0 = noise-power spectral density referred to the receiver input; in watts/Hz,
k = Boltzman constant; 1.38×10^{-23} joules/deg K,
h = Planck constant; 6.625×10^{-34} joule-seconds,
T = absolute temperature; in degrees,
f = frequency of operation; in Hz, and
n = quantum efficiency of the mixer; a dimensionless number less than unity giving the fractional number of carriers generated per incident photon.

For microwave frequencies when $hf \ll kT$ the thermal noise kT dominates and the noise power spectral density N_0 reduces to kT, the familiar expression for the available thermal noise power. The noise drops rapidly as the energy hf of photons approaches the thermal energy kT. At infrared and optical frequencies when $hf \gg kT$ the quantum noise hf dominates and the noise-power spectral density N_0 reduces to hf for the ideal detector ($n = 1$), the familiar expression for the available quantum noise power. Thus at 10.6 microns the ideal receiver has $N_0 = 1.87 \times 10^{-20}$ watts/Hz due entirely to quantum noise effects.

The receiver noise power N in watts is then found by multiplying equation (7.10) by the receiver bandwidth B_f in Hz.

The frequency dependence at the higher frequencies of the noise power spectral density N_0 is plotted in Fig. 7.11, from which it can be seen that an improvement in sensitivity is obtained at the increasingly lower frequencies of operation

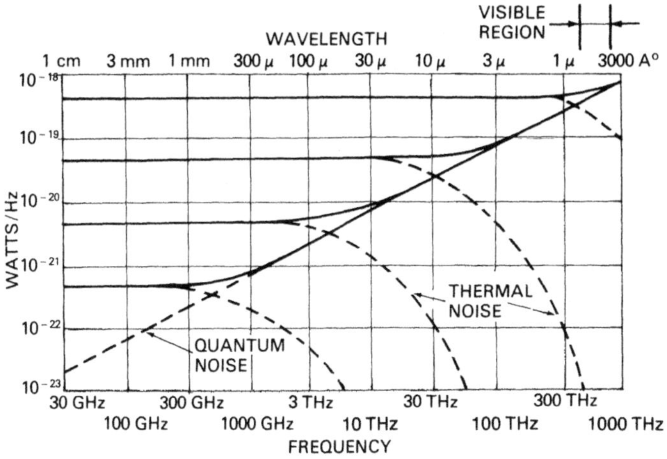

Fig. 7.11. *Frequency dependence of noise power*

below the visible region and especially for ground scenes having increasingly lower temperatures. Similar improvements apply to mechanical tolerances since the performance of optical components is often specified in fractions of the wavelength. Moreover, the longer the wavelength the longer the permissible diameter of the receiver's aperture and the greater the received energy. Thus while optical devices have the greatest potential for resolving objects on the ground they are burdened by the lower sensitivity of their receivers.

7.1.2.4.3. Background and Contrast. The ratio of the power and noise levels at the receiver input can be found by dividing equation (7.9) by the noise power $N = N_0 B_f$. Noting that the wavelength interval-to-bandwidth ratio B_λ / B_f can be expressed in terms of the wavelength and frequency as $(\lambda/f)(1+(B_f/2f)^2)$, the power-to-noise ratio can be written as:

$$\frac{P}{N} = \epsilon_\lambda \frac{C_1}{4\pi c C_2 h} \frac{T}{f}(1+(B_f/2f)^2) \tag{7.11}$$

Thus to a first order the power-to-noise ratio at the receiver is inversely proportional to the frequency of operation f and proportional to the temperature of the ground T which is being sampled by the beam. It should be noted that the spectral emissivity ϵ_λ and the temperature T are both functions of the pointing direction of the beam, i.e., of the angles of incidence and train of the beam pointing direction. It should also be noted that the power-to-noise ratio is independent of the slant range R and is dependent only on the ground spectral emissivity ϵ_λ, temperature T and the frequency of operation f.

In a typical terrain scene, a sensor beam samples the terrain at different angles of incidence and train with each beam direction intercepting a different resolution cell on the ground. Provided that variations exist in the emissivity-temperature product $\epsilon_\lambda T$ from cell-to-cell, the power-to-noise ratio will produce tonal variations in the picture or strip chart which is being recorded. In general a single cell contains two types of information: unwanted background information from naturally-occurring objects such as forests, water, mountains, deserts, etc., and the wanted-signal information from the man-made objects of interest such as missile sites, military installations, ships, aircraft, missiles during movement or launch, etc. Both the background and signal information in a single resolution cell contribute to the power which is received at the receiver. We can define the various interrelationships as follows:

$$\frac{P}{N} = \frac{B}{N}\left(1 + \frac{S}{B}\right) \tag{7.12}$$

where

P/N = power-to-noise ratio,
B/N = background-to-noise ratio, and
S/N = signal-to-noise ratio.

Inspection of equation (7.12) indicates that the absence of a signal in a given resolution cell ($S = 0$) defines the power-to-noise ratio as being the background-to-noise ratio while the presence of a signal ($S \neq 0$) defines the power-to-noise ratio in the broader sense of including the signal as well. This latter definition is of prime interest in reconnaissance work since it is desired that the contrast ratio be as high as possible, i.e., $S \gg B$. When this happens equation (7.12) defines the power-to-noise ratio as being the signal-to-noise ratio with the background having been completely suppressed by the sensor. We now briefly inquire into those factors which tend to increase the contrast in a reconnaissance sensor.

In general, if the man-made target projects an effective area A_s in the direction of the sensor, then it can be easily shown that the temperature T in equation (7.11) can be replaced by an effective temperature T':

$$T' = \left(1 + \left(\frac{\epsilon_t T_t}{\epsilon_b T_b} - 1\right)\frac{A}{(\theta R)^2}\right)T_b \tag{7.13}$$

where

T', T_b = ground temperatures in the presence and absence of a target; $T_b = T$ in equation (7.11),
ϵ_t, T_t = emmisivity and temperature of the target,
ϵ_b, T_b = emissivity and temperature of the background,
$A, (\theta R)^2$ = projected areas of the target and resolution cell in the direction of the sensor.

Equation (7.13) indicates those factors which tend to provide a contrast between the target and its background. The expression within the inner brackets is determined solely by the target and background emissivities and temperatures. The ratio $A/(\theta R)^2$ on the other hand is determined solely by the resolving power of the sensor. In order that the presence of a target of interest can be remotely sensed with a high contrast it is only necessary to have, in view of equation (7.13), the following condition:

$$\left(\frac{\epsilon_t T_t}{\epsilon_b T_b} - 1\right)\frac{A}{(\theta R)^2} \neq 0 \tag{7.14}$$

Thus for a given ratio of the emissivity-temperature product of the target and background, inequality (7.14) requires the highest possible resolving power to obtain contrast between targets of interest and the background. Furthermore the higher the resolving power of a given sensor the higher will be the contrast ratio of its recording. It should be emphasized however that increasing the sensor resolution will increase the contrast between the target and the background only when the emissivity-temperature product of the target is different from that of the background, i.e., when $\epsilon_t T_t \neq \epsilon_b T_b$. The degree by which increasing the sensor resolution improves the contrast depends exclusively on the deviation of the emissivity-temperature product between the target and the background. Since higher resolutions are obtained at the higher frequencies of operation it is therefore to be expected that a better contrast between signals and the background is obtained at these frequencies.

Contrasting basically requires the determination of the emissivity-temperature products of the various targets of military interest (man-made) and of the background (naturally-) occurring at all frequencies of the spectrum and at all possible target aspects. The selection of sensors is made at those frequencies which provide the highest possible resolving power consistent with target-background emissivity-temperature performance.

7.1.2.5. *Communications*

References 30–35, 43, 45, 53

In effect a multitude of communications channels exist between the ground area being covered by the satellite system and the data interpreter. Each channel begins at some resolvable point on the ground scene and traces through the reconnaissance satellite making the observation, through the appropriate satellite-ground communications link and finally, following processing and analysis, to the data interpreter. Of interest here are the capacity of the communications system, the actual data rates and the communication system efficiency. These will now be discussed.

7.1.2.5.1. Capacity. The information capacity of a single communications channel is given by the well-known Shannon formula:

$$C_i = W_i \ln\left(\frac{P_i}{N_i} + 1\right) \tag{7.15}$$

where

C_i = ith channel capacity; in bits/sec,
W_i = ith channel bandwidth; in Hz, and
P_i, N_i = received power and noise in ith channel.

The total information capacity of the communication system, i.e., of all i channels, is found by adding together the sum of individual channels:

$$C = \sum_i W_i \ln\left(\frac{P_i}{N_i} + 1\right) \tag{7.16}$$

where C is the communications system capacity and the summation extends over all the simultaneous channels in the system, i.e., between all the resolvable points in the ground scene and the data interpreter. For the case of a reconnaissance system with a 1 meter ground resolution which is mapping the Eurasian ground scene—an area approximately equal to 10^{13} m^2—the number of channels in the communication system will be about 10^{13}.

To illustrate, if it can be assumed that the ground scene is relatively uniform from point-to-point, i.e., the emissivity-temperature product varies uniformly over the ground scene, then the power-to-noise ratio in equation (7.16) becomes a constant for which case:

$$C = W \ln \frac{P}{N} \tag{7.17}$$

where $W = \sum_i W_i$ and the power-to-noise ratio is considered to be much greater than unity.

Equation (7.17) better indicates the fact that the communications system capacity is determined by the system bandwidth and the power-to-noise ratio. In this regard it should be noted that the system capacity will increase with the carrier frequency, since the resolution of ground objects is higher at the higher frequencies. Since more bandwidth is available at the higher frequencies, higher resolutions, to be useful, must be paid for by having a higher system capacity. This means that more data becomes available and, to be useful, must all be processed and analyzed.

7.1.2.5.2. Actual Data Rate. If a ground scene comprises 10^{13} communications channels—the case for the Eurasian landmass under surveillance by a reconnaissance system with 1 meter resolution—and if all elements or resolution cells of sensors in the ground scene can be sampled every 100 seconds then the actual data rate is 10^{11} bits/sec. This is a rather enormous rate but is quite consistent with the task of an early-warning missile launch, which is perhaps the most demanding task of the satellite reconnaissance system. Of course, lower sampling rates which may be required for normal aerial and close-look surveillance will result in correspondingly lower rates for the amount of data required in these tasks. The point is that early-warning tasks require high sampling rates.

In general, the actual data rate R for samplings of a ground scene is given by:

$$R = \frac{A_g}{r^2 t_s n} \tag{7.18}$$

where

R = actual data rate; in bits/sec,
A_g = area of the ground scene; in m^2,
r = sensor resolution; in m,
t_s = time between samplings (1/sampling rate); in sec., and
n = number of satellites on-scene

It can be seen from equation (7.18) that in a given ground scene the actual data rate will increase proportionally as the sampling rate and inversely as the sensor resolution squared and the number of satellites on-scene. If all satellites on-scene simultaneously sample the scene at the sampling rate $1/t_s$ then $n = 1$. If on the other hand satellites on-scene can be synchronized to sample in-sequence so that the entire ground scene is covered in the sampling interval t_s, then $n > 1$, thereby reducing the actual data rate per satellite by the number of satellites n. In retrospect, increasing the sensor resolution increases the system contrast but is paid for by increasing both the system capacity and actual data rates. This means that more data becomes available in less time and to be useful this must all be processed and analyzed.

7.1.2.5.3. Efficiency. Shannon showed that whenever the information rate R is less than C, then there exists some coding modulation scheme with as low a decoded error probability as may be desired, whereas if $R > C$, the error probability cannot approach zero and more coding generally makes things worse. Thus given the information rate R the first step is to design the communications system to pass the actual data rate with acceptable error. Now whether coding is needed or not depends on the expected power-to-noise ratio. If P/N is greater than about 10 then no coding is needed, while if P/N is less than 10 coding of the information is required to maintain acceptable errors. In any case, the actual data rate must satisfy the inequality:

$$R \leq 0.1 f \ln\left(\frac{P}{N}+1\right) \tag{7.19}$$

where f is the carrier frequency and it has been assumed that the bandwidth is 0.1 times the carrier frequency, a reasonable assumption for bandwidth-limited sensors.

Thus an actual data rate of 10^{11} bits/sec would require a minimum sensor operating frequency of 435 GHz, at which level the system error would be negligible when $P/N = 10$. No coding of the information would be required for this case. On the other hand the system would require a minimum carrier frequency 2.3 times as great for $P/N = 1$, at which level the system error would be considerable. The system carrier frequency must be increased even further if coding is used to reduce the system error for this last case. Even further trades between the system carrier frequency and bandwidth, i.e., the frequency and bandwidth of operation of sensors and the P/N ratio are possible. Whichever the case, the actual data rate must satisfy inequality (7.19).

If the efficiency of a communication system is defined to be the ratio of the actual data rate and system capacity R/C then operation at full system capacity with acceptable error indicates 100 percent efficient operation. In general, coding a transmission reduces the system efficiency since additional bandwidth is utilized to transmit (and receive) the code, such bandwidth being in excess of that which is required to pass the data alone (with increased error).

Communications requires that the actual data rate of ground information will not exceed the system capacity. Furthermore, the communication system is power-limited at large P/N ratios and bandwidth-limited at low P/N ratios. Coding of transmissions may be employed at the cost of adding bandwidth to maintain acceptable errors in the communication system.

The communication system portion of a satellite reconnaissance system passes a vast amount of unwanted data and relatively, in comparison, a very small amount of wanted data. Since it may be possible to provide a high contrast between these respective classes of data the possibility exists that the communication system can be designed as a filter passing only the wanted data. This could be accomplished by thresholding the data at some convenient point in the communication system.

7.1.2.6. *Data Processing*
References 36–38

What is the best way to process reconnaissance data and, once processed, how good is the data? The basic problem is to filter the naturally-occurring unwanted data and to pass the man-made signal data. What is desired therefore is to maximize the signal-to-background or contrast ratio S_t/B_t in each channel of the reconnaissance system. This particular problem requires, first, the *a priori* characterization of both the signal and background statistics and, second, the design of the optimized processor to handle these signals. Since the background power B_t is not a universally random quantity characterized by a mean and square deviation but varies locally from point-to-point in the ground scene, it is to be expected that B_t statistics will be different for land and ocean data, i.e., deserts, mountains, forests, water, etc. Similarly since S_i varies for each type of target it is to be expected that S_i statistics will differ from target to target, i.e., missiles during launch or reentry, missile sites, ships, etc. The first problem in data processing is therefore to determine *a priori* the signal and background statistics. Once this is accomplished the processor can be designed for optimal processing of the data.

7.1.2.6.1. Contrast Statistics. The contrast ratio can be found using equation (7.13) and is given as:

$$\frac{S_t}{B_t} = 1 + \left(\frac{\epsilon_t T_t}{\epsilon_b T_b} - 1\right)\frac{A}{(\theta R)^2} \tag{7.20}$$

It should be recalled that the emissivity and temperature of both the target and the background vary from point-to-point in the ground scene and that these quantities as well as the projected target area are functions of the angles of incidence and train of the direction of the beam. Thus the contrast ratio S_t/B_t is expected to fluctuate from sample-to-sample as do the signal S_t and background B_t.

The statistical nature of the target signal and background is illustrated in Fig. 7.12, which gives the probability $P(P)$ of a given power level for the case where the mean signal power S_t

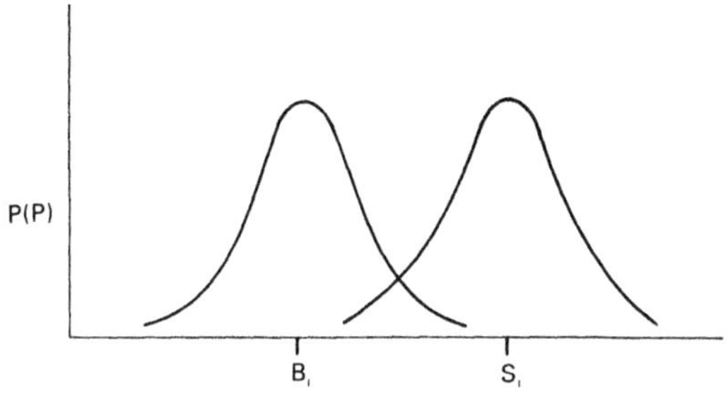

Fig. 7.12. *Signal and background statistics*

exceeds the mean background power B_i in a given channel, thereby giving a positive contrast which can be quantitatively computed using equation (7.20). It should be noted that a negative contrast is also possible and would occur when the emissivity-temperature product of the background exceeds that of the signal. The importance of contrast statistics cannot be underestimated since knowledge of it permits the design of optimized data-processing techniques.

7.1.2.6.2. Computer Transfer Function. The processing is optimized, i.e., the contrast ratio is maximized at the computer's output when its transfer function equals, except for a constant, the complex conjugate of the frequency spectrum of the input signal divided by the input noise power density spectrum, i.e.:

$$H_i(f) = \frac{S_i^*(f)}{B_i(f)} \tag{7.21}$$

where

$H_i(f)$ = ith channel computer transfer function,
$S_i^*(f)$ = ith channel complex conjugate signal frequency spectrum,
$B_i(f)$ = ith channel background power density spectrum.

The input signal-to-background or contrast ratio is:

$$\frac{S_i}{B_i} = \frac{|s_i(f)|^2}{B_i} = \int \frac{|s_i(f)|^2}{B_i(f)}\, df \tag{7.22}$$

The signal-to-background contrast ratio at the output of the processor is:

$$\left(\frac{S_i}{B_i}\right)_{\text{out}} = \frac{\left|\int s_i(f) H_i(f)\, df\right|^2}{\int B_i(f) |H_i(f)|^2\, df} \tag{7.23}$$

The maximum contrast at the output occurs when equation (7.21) holds, i.e.:

$$\left(\frac{S_i}{B_i}\right)_{\text{out}} = \int \frac{|s_i(f)|^2}{B_i(f)}\, df = \frac{S_i}{B_i} \tag{7.24}$$

Equation (7.24) indicates that the best processing of the reconnaissance data just preserves the contrast which is available at the input to the processor. The i-channel processor, therefore, whose transfer function for each information channel is the complex conjugate of the signal frequency spectrum divided by the background power density spectrum, is the best way to process the data, i.e., to maximize the contrast ratio at the processor output. Once processed the data is only as good as the available input contrast ratio which is essentially determined by the target and background characteristics and the sensor resolution, assuming of course that this all has been preserved in the communication system.

The data processor essentially functions as a bank of matched filters in parallel—one for each information channel—and acts to detect man-made targets in a natural background. The output of the processor is therefore the i-channel ground scene under surveillance, with each channel or cell representing targets and background with the system contrast. Thresholding logic may be employed to emphasize the contrasting for displays.

To date tonal contrast in reconnaissance data varies perhaps from as low as 2:1 to as high as 10:1, especially when using non-photographic sensors. Since no amount of processing of the data can be expected to improve the existing instrument contrast, it becomes mandatory that the appropriate transfer function be established for the processing of data lest information be lost in the process.

Once a reconnaissance picture has been obtained by the sensor and preserved by the communications and data-processing channels it becomes available for data analysis and, following this, for data interpretation. Analysis and interpretation tasks differ mainly in their objectives. Data analysis emphasizes what the data represents , i.e., identification of objects, while data interpretation emphasizes what the data means, i.e., threat potential of objects. Both tasks are less amenable to automation than is the straightforward task of data processing and are accomplished primarily by skilled analysts and interpreters.

Data processing requires that the man-made target signals and the natural background statistics must be determined beforehand. Once determined, these statistics can be used to implement the processing of reconnaissance data for maximum contrast availability. It can be shown that the maximum contrast which can be made available to the data analyst and data interpreter can never exceed the contrast provided by the sensor as modified by the communications and data-processing channels. The analysis and interpretation tasks are not as readily amenable to automation as is the processing for maximum contrast. The analysis and interpretation of reconnaissance data therefore must be accomplished by skilled analysts and interpreters.

7.1.2.7. Data Display

The monitoring of vast amounts of reconnaissance data is a tedious task which must be accomplished by human operators. The presentation and dissemination of data in terms of coverage and resolution must therefore be accomplished in a manner which assists operators in detecting objects of interest when such objects are being registered in the system.

Basically, each "picture" of the ground scene which is taken by the reconnaissance system must be available to the data analyst and data interpreter both in real- or stored-time operations. Picture availability is therefore the end product of a vast reconnaissance system involving numerous satellites in orbit and including subsystems of data collection, processing, analysis, presentation and dissemination, with each particular subsystem in turn involving hardware, techniques, people and organizations. The effort of this vast complex of subsystems and components results in "pictures" the great majority of which have no strategic interest and become supplementary data to be used for training and research purposes. Thus the reconnaissance system must be capable of performing the normal routine peacetime surveillance tasks but at the same time must also be capable of detecting non-routine missile launches. The former tasks, possibly, can be relegated to the more leisurely tasks of stored-time operations of analysis and interpretation whereas the latter task of early warning must be accomplished in real time. The system therefore involves a dual response characteristic. Accordingly system displays must be of two general classes; early warning and monitoring.

Unfortunately, most effort in implementing satellite reconnaissance systems to data, as measured by priority, interest and dollars, has gone into the collection system hardware. The successive subsystems (data processing, analysis and interpretation, presentation and dissemination) and components (hardware, techniques, people and organization) have received the least support. As a consequence therefore the presentation and dissemination of reconnaissance data in the areas of hardware, techniques, people and organization have received the least amount of development effort. These no doubt will become the most profitable areas for concern as the reconnaissance program matures and implements the required system response in both real- and stored-time operations.

In terms of coverage and resolution, the presentation and dissemination subsystem components must provide both real- and stored-time displays which monitor the entire ground scene under surveillance. The magnitude of the required effort to provide the desired end product on a timely basis cannot be underestimated. Multisensor imagery data must be collected, filed, plotted, duplicated, enlarged, reduced and combined so that mosaics can be made, maps produced, both true and color prints made and specialized atlases and encyclopedias made, stored and published. The various displays in either real or stored time will need imagery analysts and interpreters, such as specialists in photo, radar, and IR analysis, and interpretation for all seasons, wavelengths and sensors as well. Some will be area specialists while others will be functional specialists. For example, one man may specialize in the Eurasian terrain, another in missile detection, another in background interpretation, and so forth. Their interests intersect but clearly are different. There are specialized viewers and projection apparatus, measuring instruments of various sorts, automatic isodensitometers, recording microdensitometers, dark rooms, photo labs and mass printing and transparency production facilities for both black-and-white and color film end-products; library and filing facilities with retrieval of various types of data; and computers and their human assistants in quantity. So much for stored-time operations in which manual and semiautomatic processing, analysis, interpretation and dissemination is the general routine. Even more importantly, the real-time system response must be implemented so that all these and many more functions can be accomplished almost simultaneously.

Data display requires that the presentation and dissemination of reconnaissance data be accomplished in real- and stored-time operations and in terms of the ground scene coverage and resolution. The emphasis is for the timely availability of early-warning data and the high quality monitoring of routine data. Each task involves hardware, techniques, people and organizations of presentation and dissemination subsystems.

7.1.3. Satellite Early Warning

References 1, 42, 45

Parallel to the observation-satellite developments of the late 1950s and early 1960s there was an expensive and unsuccessful program to develop a satellite early-warning system. For years this program was plagued by unreliabilities in its hardware and by the inability of its IR sensors to distinguish between rocket-exhaust plumes and sunlight reflected from high clouds. By 1963 an interim capability had been achieved, but not until recently did a high-confidence system exist. The first two operational vehicles of the satellite early-warning network were launched on May 5, 1971, and were placed in near synchronous orbits over the Indian Ocean at about 65 degrees east longitude. Their orbits have a 10 degree inclination to provide more northern coverage than a truly synchronous orbit would allow. A slight ellipticity of orbit increases the time they spend over the Northern Hemisphere. Two satellites are required to provide continuous coverage. Although the primary mission of these satellites and others deployed at different longitudes is to provide early warning of an ICBM attack, they also provide a capability to monitor Russian missile tests.

7.1.4. Satellite Inspection

Reference 39

The interception of a satellite is required only when its hostile intent is known. It is therefore necessary to implement inspection capability using another satellite or, as an alternative to this, to implement inspection of the satellite target from the ground.

While surprisingly detailed information can be obtained about orbiting vehicles from the ground these vehicles can be optically or electronically equipped to deceive the ground-based observations. For this and related reasons the requirement exists for manned and unmanned satellite-inspection capability. As a consequence, rendezvous techniques must be developed.

7.1.4.1. Rendezvous Techniques

A rendezvous system must guide the inspector satellite so as to make its velocity vector equal to the velocity vector of the satellite target. Assume, for example, that a potentially hostile satellite is orbiting the earth in a 400 km circular orbit at an inclination with respect to the equator of 50 degrees. If the launch site for the rendezvous and inspection is Cape Kennedy (latitude 28.5 deg N) the rotation of the earth will bring the launch site through the plane of the satellite target's orbit twice each sidereal day (23 hr, 56 min, 4 sec); in turn the satellite target will cross the 28.5 deg N parallel twice during each 92 minutes, 32 seconds of its orbit. The four simplest possibilities for rendezvous with the satellite target are identified as using minimum thrust, catch-up maneuver, shortest time and rendezvous turn. Each possibility will now be described briefly.

7.1.4.1.1. Minimum Thrust. The minimum expenditure of thrust occurs when the inspector satellite can be launched in the plane of the target satellite, i.e., both orbital planes coincide. Theoretically, the in-plane launch is the most efficient technique but it calls for extremely precise timing. Because the timing requirement is so stringent the launch window, defined as the time period which is available to launch the inspector vehicle within accepted range safety limits, is very narrow and a launch delay measured in seconds could mean a delay of a number of days before another launch window became available, even though the orbital planes coincide twice daily.

7.1.4.1.2. Catch-Up Maneuvering. The launch window constraint can be avoided by launching the inspector satellite into an elliptical orbit with apogee near the launch site, the apogee occurring near the target satellite and only a short distance behind the target, and then circularizing the inspector-satellite orbit with appropriate maneuvers. For example, assuming that the target is 12 deg ahead in circular orbit in the same plane, a velocity increment of 7.6 km/sec will cause the two vehicles to meet after two orbits. Adding a second velocity increment of 0.045 km/sec at this point makes the two orbits identical so that inspection or docking can take place.

7.1.4.1.3. Shortest Time. The shortest-time technique involves making a direct non-coplanar launch from the launch site to an intercept point in the target satellite's orbit—as predicted from the target's ephemeris—and then turning the inspector satellite to the required velocity and direction of the target. This method, however, requires the most thrust for in-space maneuvering. For example, an inspector satellite fueled with a liquid propellant offering the theoretical maximum of specific impulse ($I_{sp} = 400$ sec) would expend propellant almost half its weight to make a 20° turn while in a 400 km circular orbit.

7.1.4.1.4. Rendezvous Turn. If the interceptor satellite is launched in a direction which meets the target over a point on earth 90 spherical degrees (10,000 km) downrange from the launch site, the required angle of turn is always the least. The required turn angle decreases even further when the launch site is as close as possible to the target orbit plane.

7.1.4.2. *Propulsion Requirements*

In-space propulsion is required to make three basic corrections: altitude change, plane change, and speed-up change.

Normally an altitude change will be made by a minimum energy or Hohmann transfer maneuver in which the inspector vehicle follows a semi-elliptical path to either a higher or lower orbital altitude, accomplishing the transfer by application of an initial and a terminal velocity impulse. The incremental velocity ΔV required to ascend or descend through altitudes from 160 to 1,000 km is on the order of one hundred m/sec.

A change in plane without a corresponding change in a satellite's velocity is accomplished by giving the satellite a speed increment in a direction normal to the plane which bisects the initial and final orbital planes. The magnitude of the velocity increment is given by:

$$\Delta V = 2V \sin\frac{\delta}{2} \qquad (7.25)$$

where

ΔV = increment of speed needed to rotate the orbit plane through an angle of δ degrees,
V = velocity of satellite; remains unchanged during plane change, and
δ = angle of plane change; in degrees.

It can be seen from equation (7.25) that a plane change of 60° requires a velocity increment which equals the satellite velocity. Even for low values of plane change or rotation the increment in speed for the change is in the order of 1,000 m/sec. Thus a 6° plane change requires a speed increment of 140 m/sec for a typical near-earth altitude of 500 km in which the satellite velocity is 9,000 m/sec.

The increment in velocity required for a plane change cannot be obtained in excess of the ideal velocity equation given by:

$$\Delta V = I_{sp} g \ln \frac{w_i}{w_b} \qquad (7.26)$$

where

ΔV = velocity increment,
I_{sp} = specific impulse of propellant; in sec.,
g = acceleration of gravity, and
w_i/w_b = mass ratio; w_i is the initial weight of the satellite prior to plane change, w_b is the final weight of satellite after propellant has been burned and satellite has made the plane change.

Inspection of equation (7.26) indicates that for a maximum available specific impulse using a liquid propellant with $I_{sp} = 100$ sec and a satellite design with a mass ratio of 2, the maximum available incremental velocity is about 3,300 m/sec which, in view of equation (7.25), corresponds to a maximum possible plane-change angle of about 15° in typical maneuvers. It is therefore concluded that plane-change maneuvers must be held to a minimum.

Finally, speed-up changes can be accomplished with reasonable velocity increments on the order of about 100 m/sec provided that strict launch window conditions are met.

In summary, propulsion requirements for rendezvous operations are quite stringent but can be accomplished within available constraints. Additional flexibility for satellite inspection operations requires the development of high-performance satellite engines and high mass fractions.

7.1.4.3. Launch Window Requirements

Because of range safety considerations, launch azimuths at the U.S. Eastern Test Range are limited to 44 to 110 deg and at the Western Test Range to 170 to 301 deg. Figure 7.13 illustrates

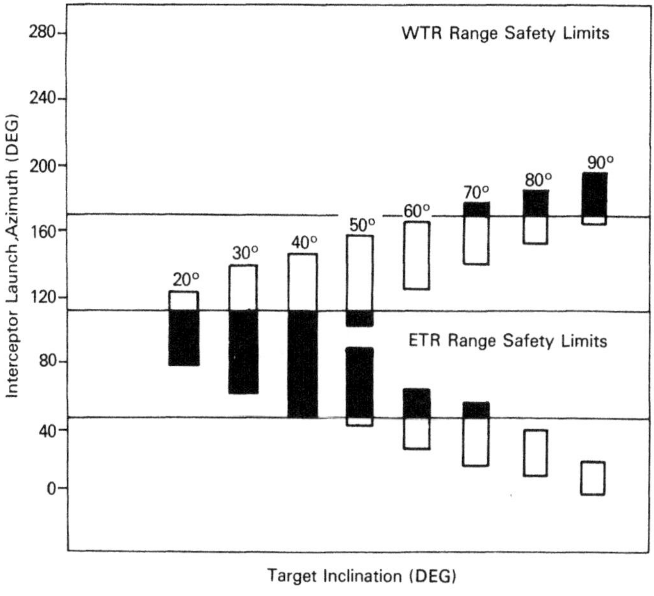

SOURCE: See reference 39.

Fig. 7.13. ***Range safety limits in U.S. rendezvous operations*** the effect of safety limits on satellite interceptor operations as a function of the target satellite inclination.

In the figure, the vertical bars show the azimuth sweeps over which injection into typical planes can be achieved and how some particular orbits are placed off-limits for range safety considerations. Thus a target inclination of 60° is a plane for which WTR rendezvous operations are severely restricted. Of course, range safety limits can be always lifted in an emergency.

7.2. Treaty Verification

References 1, 4, 40, 45, 47–52

National or unilateral, as opposed to on-site or cooperative, verification techniques include economic analysis, diplomacy, content analysis of documents and speeches, interviewing travelers and participants in scientific meetings, and espionage. Although the utility of these information sources cannot be denied, they do have the disadvantage of relying on inference. Reconnaissance and surveillance, on the other hand, are dependent primarily on the physical properties of electromagnetic sensors and therefore provide less ambiguous information and broader coverage than other techniques. Specifically omitted from the present discussion because of the lack of sufficient information are the ships, aircraft, satellites and land-based receivers used for electronic and communications intelligence. The information gathered by these devices, such as radar transmission frequencies and the nature of the Russian and Chinese communications networks, may be of greater relevance to military planning than the other techniques. To the extent that communications can be intercepted and decoded, however, the intelligence community may be able to go beyond its assessment of technical capabilites to clues about political intentions.

Both the OTH radars and the early-warning satellites can be used to help verify the testing restrictions included in the SALT I and II agreements. Beyond that they can be used in conjunction with land-based line-of-sight radars to indicate when and where missile tests are taking place and to provide some information about the type of test. These systems would therefore be used to monitor any agreement that imposed numerical or geographical restrictions on missile-testing. As long as a missile test area is within view of a line-of-sight radar its point of ballistic impact can be determined. Some characteristics of the reentry vehicle, such as size and shape, can also be determined from radar observation. The precision of such determinations would depend on the detailed characteristics of both the radar and the reentry vehicle.

The most useful observations for monitoring long-range missile tests are those made from ships and aircraft in the region of impact of the RV. From terminal radar and photographic observations detailed information about the reentry system can be derived. If one compares the calculated impact point with the observed impact point and knows the local weather conditions at the point of impact, one can make estimates of the mode of reentry through the atmosphere. For example, one can estimate whether the RV has been designed to glide past its ballistic-impact point or to fall short of it. From the close-range radar pictures of the RV and its observed speed of passage through the atmosphere, estimates can be made of its weight and ballistic coefficient (a constant that represents the efficiency of the RV in overcoming air resistance). If a powered terminal maneuver is attempted, observing radar and IR sensors should be able to detect it. Multiple-warhead tests can be easily detected near the impact site, if they are not detected earlier by other techniques.

Since 1961 U.S. ships have monitored Russian missile shots into the Pacific. These observations not only have allowed the intelligence community to keep abreast of qualitative improvements in Russian missile technology but also have provided data on radar "signatures" of Russian missiles, supplying an important input for the design of ABM radars.

The verification of any qualitative restrictions on ICBMs that might emerge from future SALT agreements would presumably rely on these land-based line-of-sight radars, OTH radars, early-warning satellites and shipboard sensors. An examination of their possibilities and limitations is therefore important to any realistic assessment of what may be possible and desirable in such agreements.

One approach that has been suggested in order to inhibit qualitative improvements in offensive missiles is to restrict the number of missile tests that each side is allowed every year. It is argued that if the restricted number were small enough, the competition between the testing of existing hardware and the development of new hardware would eliminate most or all new developments. Such limitations of tests could be verified.

Counting missile firings is just what the fixed land-based radars, OTH radars and early-warning satellites do best. If the purpose of such an agreement is to restrain qualitative improvements, however, a useful extension would be to insist that all sanctioned tests be conducted along designated flight corridors and at preannounced times. Such an extension would facilitate the task of ensuring that qualitative improvements have not been made in spite of the numerical limitations.

Where qualitative restrictions themselves are concerned, the verification problem becomes more difficult. Thus a prohibition against new types of boosters, multiple warheads and guidance systems could probably be verified with high confidence, particularly if it were imposed on top of the previously suggested restrictions limiting the number of tests. Since the various identifying signatures of current systems are well known, any new hardware would be distinguishable by an unrecognized signature.

Other less restrictive prohibitions can also be considered. For example, since terminal maneuvering can probably be detected, either directly with IR and radar sensors or directly by comparing the calculated and the observed trajectories, such maneuvering might be prohibitable. A prohibition of this type would automatically rule out terminal guidance, which would serve no purpose without corrective terminal maneuvering.

Another possibility, although it would be more difficult to verify, would be to restrict improvements in accuracy. Although an observer can tell where an RV lands and probably can tell whether the vehicle went through deliberate terminal maneuvers or exploited an aerodynamic shape in order to arrive at its target, he cannot tell from his own observations alone where it was supposed to land. Therefore information on missile accuracy must come from second-order inference. One way to estimate the accuracy of the guidance systems, but not the reentry techniques, is to monitor the apparent accuracy of the inertial guidance on space missions. Presumably the technology for space shots is also available for ICBMs. The USA has a very capable worldwide network of radars in the Space Detection and Tracking System, and these systems can be used for making such inferences. Nonetheless, trying to limit accuracy by constraining inertial-guidance systems does not seem to be a fruitful approach. Instead, prohibitions against giving reentry vehicles terminal-maneuvering capabilities and high ballistic coefficients might be used to indirectly inhibit improvements in accuracy.

The presence of several warheads can be detected by terminal radar, shipboard radar and possibly by other systems. Could a multiple-warhead system be tested with only one warhead at a time, however, thereby avoiding detection? The full answer must depend on both the detailed structure of the hardware and the confidence required in such a system. A partial answer can be formulated by recognizing that in a world characterized by mutual deterence the major concern of each side is that the opponent not achieve a capability that can be used for a preemptive, counterforce attack. It seems unlikely that such a capability could be developed and brought to a status of high reliability without a large number of full-system tests. There is a problem, however, in translating the likelihood that the USSR could not create such a high-confidence system while it was constrained by an arms-control treaty into a certainty on the part of the USA that such a system has not been developed. In this connection the distinction between MRVs (multiple RVs) and MIRVs (multiple independently-targetable RVs) may be quite inconsequential, depending on the nature of the systems. For a small separation of the RVs, systems with either mechanical separation or independent guidance could be conceived. To distinguish one such system from another would be difficult and of little value. For a wide separation, however, independent guidance would be required in order to prevent a degradation of accuracy. Presumably a system that can provide wide separation could also be programmed for small separation and could perhaps be tested without being recognized for what it was. It therefore seems difficult if not impossible to impose verifiable restrictions on multiple-warhead systems except by a total prohibition or a freezing of present systems.

References

1. T. Greenwood, "Reconnaissance and Arms Control", *Scientific American*, February, 1973. Reconnaissance satellites are the chief means relied on by the USA and the USSR to verify each other's compliance with the SALT I accords. What bearing will they and related systems have on SALT II?

2. A. Katz, "Observation Satellites: Problems and Prospects", *Astronautics*, April, June, July, August, September, October, 1960.
3. P. Klass, *Secret Sentries in Space*, Random House, 1971.
4. T. Greenwood, "Reconnaisance, Surveillance and Arms Control", Adelphi Paper No. 88, The International Institute for Strategic Studies, 1972.
5. Eds., "Backscatter Radar on 2 Coasts to Detect Planes over Horizon", *Electronic Design*, July 6, 1972.
6. Eds., "New OTH-B radar to be CW bistatic", *Electronic Design*, 12 April, 1975.
7. J. Headrick and M. Skolnik, "Over-the-Horizon Radar in the HF Band", *Proc. IEEE*, June, 1974, pp. 664–673. Over-the-horizon (OTH) HF radar using sky-wave propagation via refraction by the ionosphere is capable of detecting targets at distances an order of magnitude greater than conventional microwave radar limited by the line of sight. Some of the characteristics, capabilities, and limitations of OTH radar based on the experience of the MADRE radar as developed by the Naval Research Laboratory are described. Also discussed is the application of OTH radar to air-traffic control and to the remote sensing of sea conditions.
8. K. Stehling, "Remote Sensing of the Oceans", *Astronautics and Aeronautics*, May, 1969. Ocean-scanning satellites with all known useful sensors no doubt lie ahead, but a hardheaded look to the immediate future finds a polar orbiter with video and microwave radiometer going up first.
9. G. Manella, "Aerospace Sensor Systems", *Astronautics and Aeronautics*, December, 1968.
10. D. Parker and M. Wolff, "Remote Sensing", *International Science and Technology*, July, 1965. No object can remain completely invisible if we but learn to put highly sensitive detectors to the task of capturing its subtlest emanations.
11. Special Issue, "Remote Environmental Sensing", *Proc. IEEE*, April, 1969. A general reference. Attention is directed to the articles on remote electromagnetic sensing of the earth's surface.
12. J. Horan, "Landsat: Multispectral Eye in the Sky", *IEEE Spectrum*, March, 1978. Data collected by the Landsat satellites' remote sensors are exploited in mapping, meteorology, and resource identification.
13. P. Pryor, "The Performance of Imaging Sensors Aloft", *Astronautics and Aeronautics*, September, 1971. An expert provides the background needed to understand the debates over film versus TV and other scanners, and airplanes versus satellites for remote sensing.
14. S. Mende, "A Low Light Level Slow Scan TV Camera for Satellite Application", *Proceedings IEEE*, March, 1969.
15. Advertisement, "High-Sensitivity Infrared (10.6 micron) Hetero-

dyne Receivers with Gigahertz IF Capability", *IEEE Spectrum*, June, 1968.
16. J. Horan, "Spacecraft infrared imaging", *IEEE Spectrum*, June, July, 1968. In two parts: principles and applications and systems engineering aspects.
17. H. Beauchamp, "Infrared Aerial Photography", *Countermeasures*, March, 1976. Near IR terrain illumination equipment for night aerial photo-reconnaissance is becoming more common as technological improvements push the state-of-the-art of electro-optics. The Swedish Defense Material Administration has developed a new pod-mounted IR aerial reconnaissance system which will be used on Sweden's new all-weather attack aircraft, the SAAB 37 Viggen.
18. K. Tomiyasu, "Remote Sensing of the Earth by Microwaves", *Proc. IEEE*, January, 1974.
19. News Brief, *Electronic Design*, 24 May, 1974. The Soviet Union, according to the Pentagon, has put radar carrying satellites into orbit to monitor surface-ship traffic on the high seas. The radar's high-energy requirements are believed to be supplied by radio isotope sources.
20. S. Kazel, "Airborne mapping of terrain using narrow pulsewidths; high resolution even with wide beam antennas", *Proc. IEEE*, April, 1972.
21. "Active Microwave Systems for Space Applications", Session 4, WESCON, Los Angeles, September 14–17, 1976.
22. "No Place to Hide", *Countermeasures*, December/January, 1976. A new wide-angle integrated multifrequency synthetic aperture radar, mounted on airborne platforms, can see through dense foliage to seek out tactical targets.
23. "Problems Corrected on Ocean Surveillance Satellite", *Microwave Systems News*, February, 1978.
24. "U.S. seeks to advance reconnaissance lead", *Electronic Warfare*, April, 1978. Reports on aircraft reconnaissance.
25. A. Katz, "Let Aircraft Make Earth-Resource Surveys", *Astronautics and Aeronautics*, June, August, October, 1969.
26. Ed., "ERS Resolution: Old Numbers Out", *Astronautics and Aeronautics*, April, 1970.
27. A. Katz, "ERS Resolution: Old Numbers In Again?", *Astronautics and Aeronautics*, August, September, 1970.
28. "U.S. Satellite to Keep Eye on Red Missiles", UPI, 1971. News item on launching of a satellite equipped for early warning of a missile or orbital bomb attack.
29. "Satellite Navigation", *Countermeasures*, December/January, 1976. In the mid-1980s, a new satellite navigational concept known as NAVSTAR global positioning system will circle the earth, providing highly accurate three-dimensional location and speed

determinations for use anywhere on earth at any time, and in any kind of weather.
30. Special Issue, "Satellite Communications", *Proc. IEEE,* February, 1971.
31. Special Issue, "Satellite Communications", *Proc. IEEE,* March, 1977.
32. Communications Satellite Systems", Session 32, WESCON, San Fransisco, September 16–19, 1975.
33. H. van Trees et al., "Communications Satellites: looking into the 1980s", *IEEE Spectrum,* December, 1977.
34. "Satellite Technology Evolves", *Microwave Systems News,* December, 1977. Design of antennas and such subsystems as multiplexers is pushing forward to meet increasing demands of satellite communications.
35. "Navy Develops Satcom Terminal", *Electronic Warfare,* February, 1978.
36. G. Nagy, "Digital image processing activities in remote sensing for earth resources", *Proc. IEEE,* October, 1972.
37. "Satellite Computer Developments", *Countermeasures,* December/January, 1976.
38. E. Teicholz, "Processing Satellite Data", *Datamation,* June, 1978.
39. J. Fusca, "Catching a Satellite", *Space/Aeronautics,* September, 1965. Describes the problem of rendezvous and inspection of a satellite by another satellite.
40. R. Hartman, "War in Space", *EW/DE,* October, 1978. Increasing U.S. and Soviet military dependence on satellite based systems ensures that battlefields of the future will extend to outer space.
41. R. Kallas, "Synthetic Apertures Studied for Satellite Radar", *MSN,* September, 1978.
42. Ed., "Strategic Warfare", *Space/Aeronautics,* January, 1969. See Reference 1.16.
43. "SCDRS: New Life for Military Satcom?", *Microwave Systems News,* September, 1979.
44. "Want a Satellite Picture of New England?", AP News item, 28 January, 1971.
45. D. Capuano, "Satellite-ology Soviet Style", *Telecommunications,* October, 1979. From the launch of the first Sputnik until the present, Soviet space scientists have displayed ingenuity in developing practical applications of satellite technology. Gives launch sites, tracking and ground support, communications satellites, remote sensing, geodesy and mapping, navigation, military command and control and reconnaisance applications.
46. "Prospects for Radar Satellites Hinge on Emerging Technologies", *Microwave Systems News,* April 1979. Space based radar program promises ability to track targets ranging from ships to cruise missiles and MIRVs. Advances in antenna structures and monolithic GaAs are crucial.

47. L. Aspin, "The Verification of the SALT II Agreement", *Scientific American,* February, 1979. The U.S. has at its disposal ample "national technical means" of surveillance to detect any attempt by the U.S.S.R. to gain a significant military advantage by violating a new arms pact.
48. H. Scoville, "Iranian Listening Posts Aren't Needed to Verify SALT" *Los Angeles Times,* 11 March, 1979.
49. J. Mason, "SALT II: a state-of-the-art treaty" *IEEE Spectrum,* May, 1979. Compliance must be verifiable by technical means, so technology determines the treaty's leeway and its restrictions.
50. K. Bacon, "Can SALT II be Policed", *Wall Street Journal,* 9 July, 1979. The outcome of the verification debate could well depend on trust—the amount of trust Senators place in the nation's intelligence agencies and in the administration's willingness to challenge the Soviets over any suspected violations.
51. F. Moncrief, "SALT Verification", *Microwaves,* September, 1979. How we monitor the Soviet arsenal with elint radar and spacecraft.
52. "The New Military Race in Space", *Business Week,* 4 June, 1979.
53. "Defense Satellite Communications in the 1980s", *Countermeasures,* December/January 1976. See Reference 1.29.

8

Command Control and Communications

The quadruplet of warning, command-control, communications and weapons systems forms a closely-knit overall system which supports the national command authorities, the President and the Secretary of Defense. While the early warning (EW), command-and-control (C^2), communications (C) and weapons (W) systems were developed along traditional vertical military departmental lines, with limited horizontal capabilities, the pressures have mounted since World War II for a common management of assets across military lines. Since 1971, the intertying of systems has evolved from a consolidation of individual service systems into an integrated broadly-based overall system of EW, C^2, C and W elements. Fig. 8.1 shows the major systems in their interrelationships and in their relationship to the threat (T).

The threat (T) represents the opponent's weapon (W) element, which is part of his command-and-control loop similar to Fig. 8.1. Threat T is characterized by having signals and communications, weapons capabilities and vulnerabilities. With the exception of the threat (T), all other blocks in Fig. 8.1 are under the jurisdiction and, preferably, control of the National Command. The early warning (EW) system comprise early-warning radars, over-the-horizon radars, satellite reconnaissance systems and a variety of signal intelligence systems. These systems all receive input from the threat (T) and then pass important information to the command (C^2) through a communications link (C). In turn, the control portion of the command (C^2) will pass the most urgent threat or target information to the weapon (W) element. Weapons (W) receive commands from the command-and-control (C^2) function through a communication link (C). Weapons (W) include the triad of offense forces, the ABM and bomber forces, and satellite forces, all of which have been subjects of this book. As

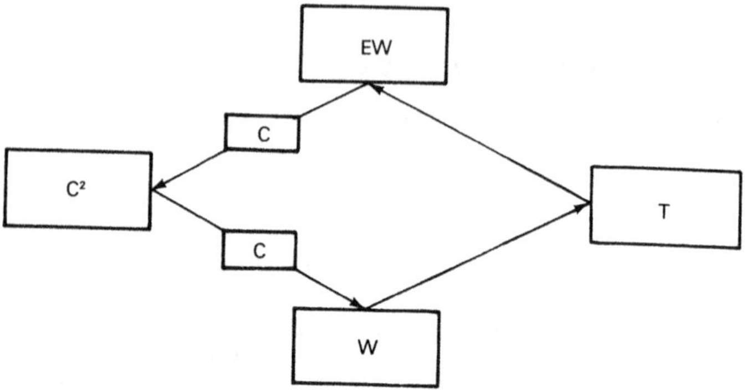

Fig. 8.1. *Command and control loop*

a consequence, weapons (W) will be discussed only within the broader concept of the command-and-control loop. The application of the broad system of Fig. 8.1 is universal; it applies to tactical as well as to the strategic problems of command and control, the difference being mainly the level of the threat (T). For this reason, the strategic problem can be understood as an escalation of the tactical problem of command and control.

In Fig. 8.1, it can be seen that the command-and-control loop is basically a servomechanism. As such, it is possible to apply the standard theory of servomechanisms in order to optimize the system operation, performance and response. To accomplish this requires that the various blocks of the system can be specified in terms of their "bandwidths" and "response times". While this is easier said than done it is possible, at least in principle. Of course, it will be realized that the simple block diagram of Fig. 8.1, at the strategic level, is a plurality of interrelating loops, each with a portion of the threat (T). In any case, the early-warning (EW) block will provide a first threshold function to the signals it receives from the threat (T) and will pass along only the more important signals. The command (C^2) will provide a second threshold function to the signals it receives from the early warning (EW) and will pass along only the most urgent signals. Thus, not only a servomechanism is to be postulated but one that also operates in a noise environment. No doubt much of the noise will be provided by the threat (T) as it attempts to jam key elements of the loop. However, an equally important part of the noise is inherent to any large and complex non-automatic system.

8.1. Warning

References 1–8

Early-warning radar has been discussed previously in the context of bomber defense, early-warning and over-the-horizon radar. Such radar of course complements the reconnaissance satellites and both radars and satellites are complemented by a variety of signal intelligence systems. The triad of radars, satellites and signal intelligence systems forms the EW element of the command-and-control loop in Fig. 8.1.

8.1.1. Signals Intelligence

References 9–14, 71

Signals intelligence (SIGINT) is a vital component of EW. It includes both electronic intelligence (ELINT) and communications intelligence (COMINT). ELINT is associated with the detection, identification and location of radars and similar emitters of weapons systems while COMINT is associated with communications, telemetry and other information links. The exercise of SIGINT reveals who is where, doing what and provides important information as EW to the commander or diplomat. More particularly, ELINT reveals where the present enemy forces are and what they are doing, while COMINT reveals not only the present enemy presence but his thinking; COMINT includes the art of breaking codes.

In the microwave ELINT spectrum, pulsed radar signals are used extensively for navigation, search, detection, tracking, guidance and other military applications. It is therefore desirable to classify radar signals by their unique characteristics or signatures in order to determine their application and owner. The parameters of interest are the pulse repetition interval, width, amplitude, frequency and other less obvious items. The pulse characteristics can be stable on each recurring pulse of an emitter's pulse train, or they can be changing with each pulse emitted. The emitter may also be emitting for a small fraction of a second, or it may be one of several emitters transmitting simultaneously in the area of surveillance. The ELINT task may be further complicated by atmospheric noise, dropout of pulses, and reflections of pulses. To recognize a radar signal under such actual conditions requires the capability of a computer-directed system that can work in real time to simultaneously process and identify pulse characteristics of multiple signals. Typically such a system is designed to monitor signals known to exist in the area of surveillance and to notify the operator if any unexpected or unknown emitter is present.

8.1.2. Airborne and Shipborne Radar Warning Systems

References 15–29, 37–39, 51–58

The emerging generation of EW systems are characterized by the use of digital computers, and stored program logic to perform the data analysis and system control functions. The application of computers to EW systems has permitted the incorporation of far more comprehensive data-analysis techniques than is practical with the analog processing techniques used in the past. As a result, more effective EW systems that are capable of operating in an increasingly, complex signal environment are becoming available.

A typical modern military aircraft radar warning system is shown in Fig. 8.2 and comprises an antenna, detector/amplifier, processor analyzer, controller indicator and display units. The antenna is used to detect enemy radar signals that are present in the area of surveillance. It is located in the nose and tail of the aircraft and sometimes in the wingtips to determine the direction of radar pulses. The detector/amplifier filters out unwanted signals and amplifies relevant signals. It is located near the antennas. The processor/analyzer analyzes radar signals to determine the type of radar, location and operating status. It sends video signals to the display and control unit and audio warnings to the operator's earphones. It is located in the main equipment bay of the aircraft. The controller/indicator provides an indication of the type of threat detected through lights on a control panel in the cockpit. The operator can control the whole warning system with this unit. Finally, the display shows the pertinent information relative to the location and types of radars being detected by the warning system.

The airborne EW concept evolved from the need for better surveillance, command and control of forces. An airborne platform with a long-range radar, computer, displays and communications equipment offers an optimum solution. An aircraft can search more area by virtue of its altitude or distance from the command post center and can provide over-

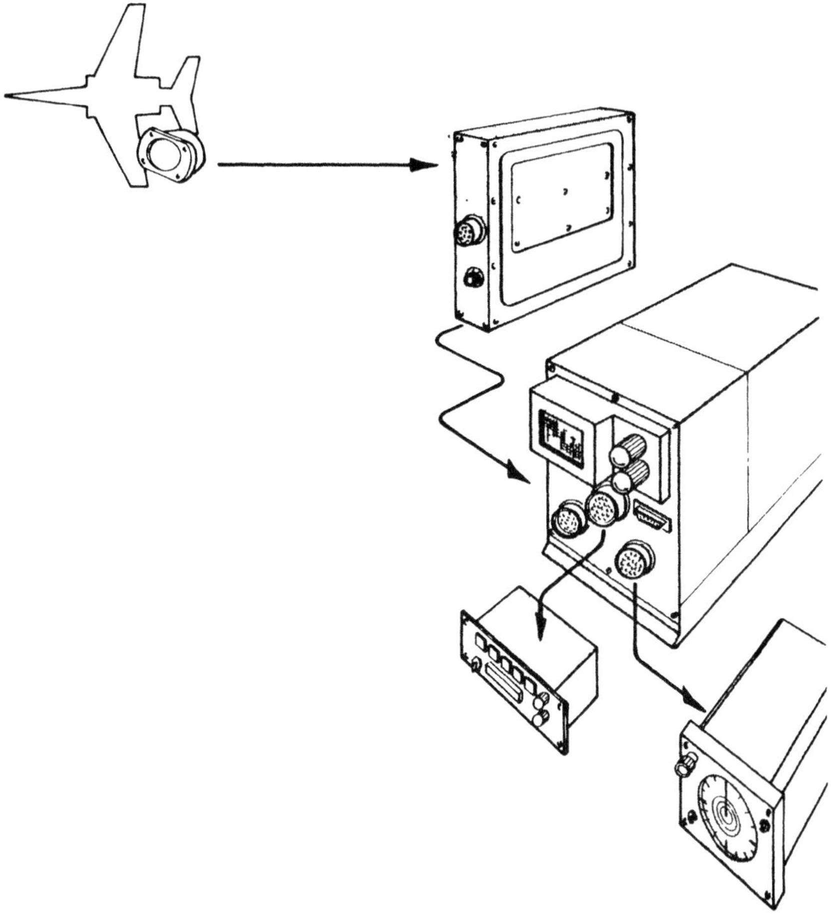

SOURCE: See reference 20.

Fig. 8.2. *Military aircraft radar warning system*

the-horizon viewing of an impending attack by low-flying aircraft and cruise missiles or fast-moving surface ships. The airborne platform can also serve as an advanced command post to vector interceptors against surface and airborne threats and direct attack aircraft and missiles against targets. The EW aircraft selected by the U.S. Air Force is the E-3 AWACS and by the U.S. Navy the E-2 Hawkeye. At 10,000 m such radar-equipped aircraft can spot small targets at ranges of 500 km and direct fighter aircraft and missiles against these targets. This capability extends over both tactical and strategic engagements. Equally important, these aircraft are equipped with sophisticated EW receivers which can report enemy emitters at extremely long range.

While shipborne EW has a lesser strategic value than does airborne EW, by virtue of the ship's surface location, it too has evolved along similar lines and for the same basic reasons, namely surface and subsurface early threat warning and surveillance. The signal environment of surface ships includes signals from a variety of aircraft, cruise missiles, anti-ship missiles, enemy task forces, as well as the normal noise and interference from friendly forces. The rapid direction finding (DF) capability of EW is needed to facilitate the pointing of weapons and correlation with search radars. Thus, the surveillance function of the surface-ship system requires passive detection for providing early alert against long-range threats, and data collection for evaluation of current, tactical and, to a lesser degree, strategic, situations.

8.2. Command and Control

References 30–36, 66, 67

As shown in Table 8.1, the command-and-control (C^2) center in Fig. 8.1 receives important information from the early-warning (EW) element and issues urgent information to the weapons (W) element. The command-and-control system organizes the important (incoming) information and, like a threshold, draws some kind of a conclusion.

Table 8.1. *Information transfers by the command-and-control* (C^2) *center*

INFORMATION	
IMPORTANT (INCOMING)	URGENT (OUTGOING)
WARNING STATION	WEAPONS STATION/TYPE
WARNING COVERAGE	WEAPONS CAPABILITY
NUMBER AND LOCATION OF TARGETS	WEAPONS ASSIGNMENT TO TARGETS
TARGET ACTIVITY	BATTLE ENGAGEMENT

The important information made available to the command system by the warning element includes the number and location of targets and the target activity. If the conclusion warrants, the command-and-control system (C^2) will issue urgent information to the weapon (W) element. The urgent information made available to the weapon (W) system is the assignment of weapons to targets and the order for battle engagement.

From the need for centralizing command-and-control functions has evolved the worldwide military Command-and-Control System (WWMCCS), whose communications counterpart is the Defense Communications System (DCS).

The tactical command-and-control system allows a commander to direct and control his forces and make maximum use of weapons systems assigned to him in the theater. He needs to know the readiness of his forces and, in a very timely manner, have the capability to command and control them. He must counter every action and assure and maintain the initiative by employment of his forces. He must support his superior command and also support parallel units of command. The strategic command-and-control system is quite similar, but at a higher level. The changeover point between tactical and strategic command and control is not clearly demarcated but appears to be just below the major command headquarters of the major service commands. As such, both tactical and strategic command and control systems have and are being used interchangeably, for example, the E-3 AWACS and E-2 Hawkeye airborne command posts. In either case, the command-and-control system consists of a variety of ground and/or airborne equipment, computers and procedural systems and is the classical man-machine interface.

8.2.1. Airborne Command and Control

References 19, 24, 38–40

A heavy reliance in tactical air war has been placed on the E-3A Airborne Warning and Control System (AWACS). However, while airborne command-and-control systems can also be used for strategic war, these have different requirements. These requirements have to do mainly with the level of command authority by each type of system. In general, airborne command-and-control system configurations vary in accordance with mission requirements. A typical tactical system can include between twenty and thirty operation and command stations. Strategic systems, however, can be configured to accommodate in excess of ninety operating personnel. The configurations for both types of systems include facilities for one or more radio operators, teletype operators, and command staff personnel in addition to the aircraft- and equipment-operating personnel. Since command-and-control missions usually involve long-duration flights, facilities for crew comfort and convenience, such as galleys, lavoratories, bunks, rest seats and tables, and air conditioning, also are integral parts of the system.

A purely strategic airborne command-and-control system is the E-4 airborne command post. The E-4 system is designed to provide the National Command Authority (NCA) and Strategic Air Command with survivable command, control and communications (C^3) capabilities under severe conditions of a general war including nuclear or hostile electromagnetic environments. The E-4 can accommodate a larger battle staff, accomplish a broader scope and complexity of C^3 functions and provide greater nuclear survivability than the existing tactical oriented systems. The E-4 aircraft will have thirteen separate communications subsystems covering the frequency spectrum from very low frequency (VLF) through super high frequency (SHF) communications. It can be airborne for periods up to seventy-two hours and can accommodate ninety-four operating personnel.

8.2.2. Measures of Performance

References 39–41

The command-and-control system must have the ability to transmit and receive information with other elements of the command-control loop. It must have the means for rapidly processing vast amounts of information from many sources, to filter the information and to provide the essential information to the decision-maker. The degree of its degradation by concerted enemy action must be minimum through improved security, reliability, redundancy and hardening against both nuclear and jamming effects. Finally it must provide orders that assure the suppression or acceptable rate of killing the threat.

A major issue of control is how to use available weapons.

Once fired, costly missiles cannot be recalled and the user's capability is systematically reduced through his expenditure of relatively expensive weapons, whether they hit or miss their targets. In contrast, electronic warfare (also EW) weapons can be used as often as desired without attrition of the main force assets. Some day, with proper equipment and training, it is possible, at least in principle, that electronic combat will determine to a great extent the results of a battle engagement prior to a launch of principle weapons by both adversaries. For example, the destruction of key communications links in Fig. 8.1 could prevent the command center from ordering his battle engagement plan. In the meantime, it is the commander who must decide whether to hold off his expendable forces.

8.3. Communications

References 42–50, 59–70

The need for centralizing communications has led to the creation of the Defense Communications System (DCS) whose command-and-control counterpart is the Worldwide Military Command and Control System (WWMCCS). About 60 percent of the DCS is in the continental United States (CONUS), and for the most part is leased, the remaining 40 percent is overseas, being primarily government-owned and operated. The DCS utilizes all transmission media ranging from cables to commercial and military satellites and has various types of circuits going to over 3,000 locations in about seventy-five countries throughout the world. Most international transmission service is provided over leased facilities. The principle switched networks included are the Automatic Voice Network (AUTOVON), Automatic Digital Network (AUTODIN), and Automatic Secure Voice Communications Network (AUTOSEVCON). The communications network spans the globe over ground, air and satellite links.

8.3.1. Defense Satellite Communication System (DSCS)

References 43, 44

Along with the vast ground network portion of the DSC there has evolved the Defense Satellite Communication System (DSCS), which provides long-haul wideband digital service for secure voice of high quality, imagery and other special digital services.

The DSCS-II currently in use provides the military command with a high-capacity, secure, strategic communications via a super-high frequency (SHF) satellite network. DSCS-II became operational in February 1974 with the deployment and successful operation of two satellites. A subsequent attempt to put two more satellites in orbit failed in May 1975 because of a launch vehicle malfunction. As a result, the launch of six replenishment vehicles was delayed until late 1976. With the deployment of all six replenishment satellites, the DSCS-II system will serve as a primary strategic communications network beginning in 1979. The DSCS accommodates a wide variety of user terminal equipment, ranging from small mobile terminals for ground deployment and those carried on board ships and aircraft to large fixed ground stations. By 1980, the number of terminals will exceed 400.

8.3.2. Data Processing

References 44–46

The program to supply an intercomputer network for WWMCCS was begun in 1971 as a means of finding how computer internetting could benefit the command-and-control functions, and as a method for applying the technology of computer internetting to the particular problem of WWMCCS.

The effect of processing messages by network control programs in the host computer through communications links is to make a terminal served by the local computer appear as if it were local to a remote computer.

The WWMCCS Automatic Data Processing (ADP) internetting program objective is to provide direct computer-to-computer exchange of information via the DCS network. Internetting makes the information flow easier and more rapidly improves the system capability to maintain operation under conditions of partial failure; it provides the capability to share information processing between several locations, and a potential for more efficient information storage and retrieval. The packet-switching technology which had been developed to interconnect computers in the Defense Advanced Research Projects Agency (DARPA) network was chosen as the basic system of the WWMCCS intercomputer network. This tech-

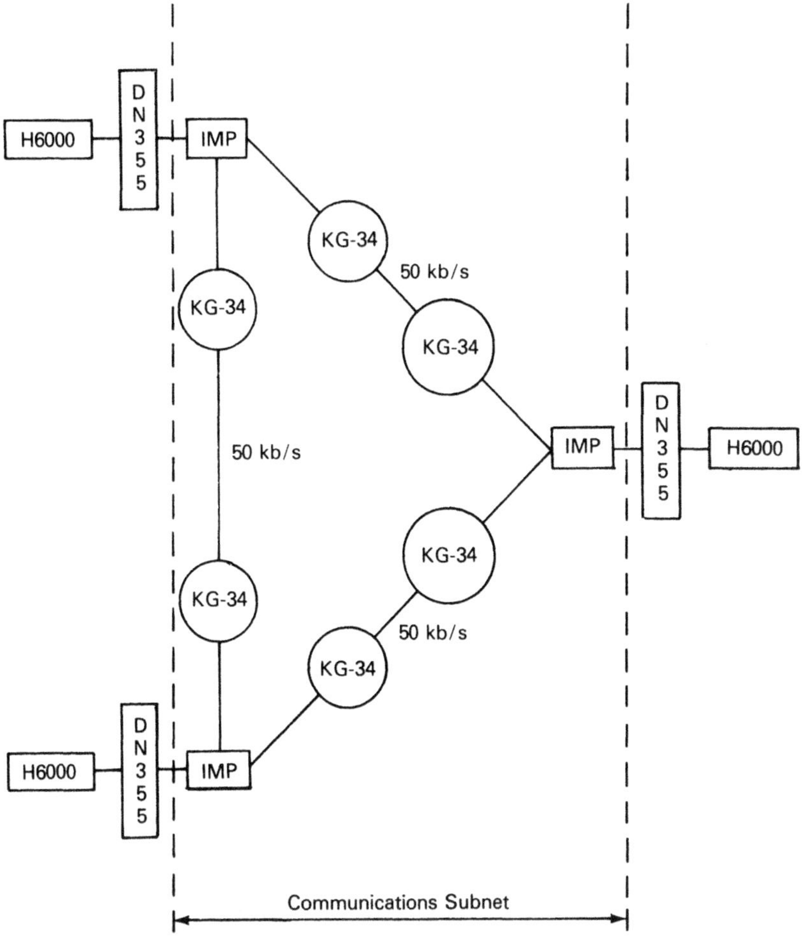

SOURCE: See reference 45.

Fig. 8.3. *Initial* (3-*mode*) WWMCCS *network*

nology, however, had to be modified to include security and operational features required by WWMCCS. The intercomputer network now includes communications links to pass messages between host computers and a network control program to provide an interface between users and tasks being performed at host computers at remote locations. The communications subnet is a direct copy of the packet-switching network developed by DARPA. In order to accomodate WWMCCS it has been modified to permit link encryption of the communications lines.

Figure 8.3 depicts the initial WWMCCS network. At each site is a H6000, host processor, and its associated front-end processor, the Datanet 355. Messages processed from this local site are forwarded to the Interface Message Processor (IMP) which is an H716 minicomputer. The IMP breaks the message down into 1,000 bit packets, and passes these packets to the transmission lines via a cryptographic device and mode (KG-34). The communications subnet consists of 50 kilobit transmission lines that interconnect the IMPs which perform the packet-switching function. When a message is received from a remote computer, the local IMP collects the data packets and forms them into a message to its host computer.

8.3.3. Security

References 36, 37, 47–49

Recent local wars have directed emphasis to the problem of securing the operation of vital tactical communications links in the command-and-control loop. Recognizing this need for a tactical communications capability for multi-service command and control, a Joint Tactical Information Distribution System (JTIDS) was initiated in 1975 to improve the flow of information at the multi-service level. The system is a further evolution of the Mystic Link program and provides direct and instantaneous communications between the Air Force, Army, Navy and Marine Corps units of an integrated command. JTIDS, when fully operational by 1980, will handle up to 33,000 individual users (subscribers). Tactical information will be sent to specific subscriber terminals using time division multiple access (TDMA) to permit the transmission of more data. Techniques such as spread spectrum, data interleaving and frequency hopping are used to prevent eavesdropping and give more jamming protection than is possible with the more conventional communications networks. The first TDMA terminal was developed for the E-3A AWACS and provides a jam-resistant radio terminal that enables the aircraft to exchange secure, real-time information over a single network on a time-shared basis. The TDMA terminals are designed to provide continuous communications exchange that result in an information pool that is being continually updated by the individual subscribers, and is made available to all network members. The ARC-181 TDMA terminal includes a communications processor, signal processor, transmitter, receiver, high-power amplifier and control, and display panel.

Strategic C^3 is not too far removed from tactical C^3 and jam-resistant communications systems are needed at both levels. In conventional communications, a bandwidth is used that is just wide enough to transmit the information involved. However, using systems that spread their transmissions over a much wider bandwidth has the military advantage of jam resistance since the jammer must also spread his signal over the wider bandwidth. If the signal spectrum is spread sufficiently wide so that its magnitude is below a receiver's noise level, then it is virtually impossible to detect the signal if the receiver is hostile and does not have the signal code. There are many modulation techniques which can be used to obtain this result and these include using pseudo-random noise (PRN), frequency hopping, time hopping and chirping. The PRN-coded systems include systems which are uniquely suited to function in a jamming environment and TDMA burst-type code modulation systems which are in use by aircraft.

The first implementation of spread spectrum and TDMA is in the E-3A AWACS and the first terminal will be installed in the nineteenth E-3A, produced in 1979; subsequently TDMA terminals will be retrofitted into the first eighteen planes. This particular TDMA system is intended for use by NATO to cope with the dense signal environment and intensive jamming anticipated in Western Europe.

The effective security of the strategic DSCS network requires protecting both the communications and satellite control links against both friendly and hostile signal environments. Spread-spectrum techniques are to be used for this purpose and additional protection may be provided by the use of satellite multibeam antennas which spatially discriminate between the user and interfering signal sources.

References

1. B. Miller, "Rising Costs Stimulate the Search for Better Ground Surveillance Radar", *Electronic Warfare*, February 1978. In an effort to cut soaring maintenance expenses, defense planners

everywhere are reevaluating their surveillance radar systems with an eye to refinement.
2. "Radar Requirements on the Upswing", *Microwave Systems News*, March 1978. Radars for test ranges, early-warning air defense, intelligence and shipboard radars are being planned. PAVE PAWS, the new SLBM system, is to be erected at Otis Air Force Base, Mass.
3. "Air Force Alaskan Radars to Go Modern", *Countermeasures*, May 1978. The manned bomber watch was initiated more than twenty years ago. Strung along Alaska's western coast is a seven-site linkup. Other network radars are located more inland. This is the SEEK IGLOO system.
4. "Alaskan Radar to be Updated", *Microwave Systems News*, July 1978.
5. "Young Firm Captures Radar LNA Contract", *Microwave Systems News*, May 1977.
6. Eds., "Backscatter Radar on 2 Coasts to Detect Planes over Horizon", *Electronics Design*, July 6, 1972.
7. Eds., "New OTH-B radar to be CW bistatic", *Electronic Design*, 12 April, 1975.
8. J. Headrick and M. Skolnik, "Over-the-Horizon Radar in the HF Band", *Proceedings IEEE*, June, 1974, pp. 664–673. Over-the-horizon (OTH) HF radar using sky-wave propagation via refraction by the ionosphere is capable of detecting targets at distances an order of magnitude greater than conventional microwave radar limited by the line of sight. Some of the characteristics, capabilities, and limitations of OTH radar based on the experience of the MADRE radar as developed by the Naval Research Laboratory are described. Also discussed is the application of OTH radar to air-traffic control and to the remote sensing of sea conditions.
9. S. Saperstein and J. Campbell, "Signal Recognition in a Complex Radar Environment", *Countermeasures*, March 1977.
10. "System identifies radar fingerprints", *Microwaves*, April 1977.
11. K. Arkind, "Realistic ELINT Design Considers Antenna Pattern Trade-Offs", *Microwave Systems News*, October/November 1976. Proper design can yield an optimum balance between factors like sensitivity and signal density, narrow beams and coverage, or gain and beamwidth.
12. T. Bibbens and J. Lake, "COMINT/SIGINT: Our First Line of Defense", *Countermeasures*, March 1977.
13. S. Lipsky, "Understanding DF Accuracy of Radar Warning Receivers", *Electronic Warfare*, April 1978. Here are some considerations affecting the DF accuracy of broadband intercept and warning receivers. Is it possible to compute the expected DF accuracy before the design?
14. "ERADCOMs Elements Detailed", *Electronic Warfare*, May 1978. The U.S. Army's Electronics Research and Development Com-

mand (ERADCOM) has sole responsibility for all the services intelligence, surveillance and target material resources.
15. E. Ebersol, "Rush is on to complete first AWACS radars", *Microwaves,* September 1973. Westinghouse is building three pre-production units for early flight testing.
16. News item, "AWACS flies through first trials", *Microwaves,* February 1975.
17. R. Levine, "Boeing Co. Wins Air Force Award of $247.6 Million", *Wall Street Journal,* 29 April, 1975. Six production models set in midst of hard sell for controversial radar jet.
18. J. Rhea, "Takeoff Clearance for AWACS", *Southern California Electronic News,* 19 May, 1975.
19. B. Walsh, "An Eagle in the Sky", *Countermeasures,* July 1976. The E-3 Airborne Warning and Control System (AWACS) provides the "big picture vision and understanding for battle management through a unique all-altitude surveillance, command, control and communications capability". The AWACS consists of special avionics and a large surveillance radar. The U.S. Air Force is developing it to fulfill tactical and air defense force's needs.
20. "Threat Simulation for Radar Warning Systems", *Countermeasures,* August 1975. The latest generation of military radar warning systems is being controlled by computer-aided electronics to provide faster, more accurate information. When strike aircraft are threatened by anti-aircraft fire or surface-to-air missiles, the airmen's lives and the success of their mission often depend on the proper operation of the equipment to permit the best countermeasures response. The effective validation of the warning system can be assured through simulation—that is, through cycling the equipment through all the scenarios which it may encounter.
21. L. Schwartz, "The EW system evaluation challenge", *Countermeasures,* August 1975. Describes new generation of radar warning and jammer management systems design, system testing and evaluation challenges.
22. J. Montgomery, "Digital Processors Expand Radar Warning Capabilities", *Countermeasures,* August 1975.
23. T. Harper, "New Trends in EW Receivers", *Countermeasures,* December/January 1976.
24. "Airborne Navonics", *Countermeasures,* February 1976. Airborne early warning, a concept which began during World War II, has matured into an essential element of fleet defense and control of air operations because of the added capabilities afforded by electronics. The E-2C aircraft, a veritable storehouse of computers, sensors, displays, and communications equipment, incorporates the latest airborne Navonics (naval electronics) technology to complete its mission as an airborne early warning and control platform for U.S. Navy fleet operations throughout the world.
25. "Modern Shipboard EW", *Countermeasures,* February 1976. In

modern warfare, a complex electromagnetic environment poses many threats to the survivability of naval vessels in a tactical theater of operation. New sophisticated electronic equipments provide threat early-warning protection for the U.S. Navy's fleet of surface and subsurface platforms.

26. J. Lake "Air EW—They have forgotten again", *Countermeasures*, November 1976.
27. "Navy to Equip Fleet with Multibeam EW Systems", *Microwave Systems News*, July 1977.
28. T. Paff, "Digital Control Impacts Design", *Microwave Systems News*", August 1977. Smart microwave design is producing equipment with simpler circuits and less rigorous demands on components.
29. "TASES Replacements Considered", *Electronic Warfare*, May 1978. A replacement for the Tactical Airborne Signal Exploitation System (TASES) scuttled by Congress because of cost.
30. B. Bell, "C^3 for Tactical RPVs", *Countermeasures*, May 1976.
31. J. Lake, "C^3 Software Myopia", *Countermeasures*, December 1976.
32. B. Walsh, "C^3 The Big Picture", *Countermeasures*, March 1977.
33. "Army Plans BETA System Test", *Electronic Warfare*, February 1978.
34. R. Hartman, "C^3 Prologue", *Countermeasures*, April 1978.
35. "Tactical C^3 in the Infantry", *Countermeasures*, April 1978.
36. J. Fawcette, "C^3: Key Challenges Face Military Planners", *Electronic Warfare*, June 1978. Spiraling costs, technological complexities loom as major obstacles at both the strategic and tactical levels.
37. J. Lightfoot et al., "An Experimental TDMA Network for Airborne Warning and Control Systems", *Microwave Systems News*, December/January 1976. With its TDMA (Time Division Multiple Access) communication system AWACS demonstrated improved operational capability in such areas as surveillance, control, detection of penetrating aircraft, distribution of radar tracking information to surface C^2 centers, and control of friendly interceptors.
38. W. Shockley, "The E-4 Airborne Command Post", *Countermeasures*, July 1976. Airborne command-and-control systems are used for both strategic and tactical applications. In strategic situations, these systems are used by top-level command personnel to monitor worldwide military operations and allow integrated command of all friendly forces. In tactical applications, airborne command-and-control systems are used by local area commanders to monitor battlefield operations and control the ground and airborne forces under their command.
39. C. Burdick et al., "EW–C^3 Interrelationships", *Countermeasures*, July 1977, August 1977.
40. A. Gallotta, "C^3 and EW in the Navy", *Countermeasures*, April 1978.

41. "Air and Space C³", *Countermeasures*, April 1978. AF tactical command and control concepts were obtained during an interview with Brig. Gen. J. Creedon, U.S. Air Force, Deputy Director for Tactical Information, Directorate of Operations and Readiness. General Creedon's responsibilities include both tactical and strategic command-and-control. Like the strategic command-and-control system, with its Worldwide Military Command and Control System (WWMCCS), the tactical command-and-control system consists of a blend of newly emerging hardware, software, and procedural systems—the classical man-machine interface.
42. H. Inose, "Communications Networks", *Scientific American*, September 1972. A channel with many sources and many destinations forms a network. Effective networks call for good switching and resourceful design to minimize the number of branches and to maximize their capacity. The ARPA network, operated by the Advanced Research Projects Agency, is one of several emerging computer networks connecting various education and research institutions over the country.
43. "Communications Satellite Systems", Session 32, WESCON, San Francisco, September 16–19, 1975. The paper by H. Wynne and D. Kendall gives the background for the Defense Satellite Communications system, currently in its second phase (DSCS-II), in the 1980s.
44. "Defense Satellite Communications in the 1980s", *Countermeasures*, December/January 1976. The defense satellite communications system (DSCS) is a worldwide military command-and-control communications network in use by the U.S. Department of Defense. Because of an unsuccessful attempt in May 1975 to deploy two additional satellites, new direction and urgency have taken over the program. Other improvements in the launch weight capability of the Titan IIIC launch vehicles, redesign of the satellite payload, improvements in the capability and effectiveness of the electronics and the advent of the space shuttle as a primary launch vehicle have added new dimensions to the defense satellite communications systems.
45. L. Paschall, "Command, Control and Technology", *Countermeasures*, July 1976. Lt. Gen. L. Paschall, U.S. Air Force is Director, Defense Communications Agency (DCA). As Director, General Paschall is responsible for management and direction of the worldwide Defense Communications System. He is also responsible for system engineering and technical support to the National Military Command System, and for provision of technical support to the worldwide military command-and-control standard automatic data-processing systems. In his capacity as Manager, National Communications System, he is responsible for providing effective direction to the worldwide National Communications System, which includes the communications facilities of the various

federal agencies. The director, DCA, is also Chairman, Military Communications-Electronics Board, providing a liaison point for joint and international communications matters.
46. I. Green, "DARPA: Investing in the Future", *Countermeasures*, August 1977.
47. T. Campobasso, "Telecommunications Impact on Strategy", *Countermeasures*, June 1977.
48. R. Hartman, "TRI-TAC: A Planned Revolution", *Countermeasures*, June 1977.
49. L. Cuccia, "Spread Spectrum Techniques are Revolutionizing Communications", Microwave Systems News, September 1977. Programs like NAVSTAR GPS, SEEK TALK, JTIDS, and the TDRSS satellite all rely on spreading transmissions over wide bandwidths to provide high resistance to jamming, security, and multiple access. Technology for correlators and code processing has rapidly evolved to make these systems possible.
50. J. Fawcette, "Mystic Link Revealed", *Microwave Systems News*, September 1977. The Joint Tactical Information Distribution System (JTIDS) is intended as the major tactical communications net from the 1980s on. As a hybrid spread spectrum and TDMA link, it uses advanced signal processing that may be based on digital LSI, CCDs, or SAWs.
51. G. Houser "Lamps MK III" ME/C October 1978. Helicopter protects fleet from hostile ships and submarines.
52. "The silent war", *Countermeasures*, March, 1975. See Reference 1.26.
53. "PAVE PAWS Tests Scrutinized", *Microwave Systems News*, June, 1978. PAVE PAWS is a detection and early-warning system to guard against a sea-launched ballistic missile (SLBM) attack on the continental U.S. It will also track satellites in orbit.
54. P. Nahin, "Can Land-Based Strategic Bombers Survive an SLBM Attack?", *IEEE Transactions on Aerospace and Electronic Systems*, March, 1977. See Reference 1.51.
55. G. Clancey, "New Breakthroughs in ASW Signal Processing", *Countermeasures*, October/November, 1975.
56. "Airborne Counter Threat to Submarines", *Countermeasures*, September, 1975. An overview of the long-range patrol aircraft whose mission is to locate and track enemy submarines.
57. "PAVE PAWS hounded by another lawsuit", *Microwaves*, September, 1979.
58. "USAF's SEEK IGLOO Ready Soon", *Microwave System News*, November, 1979.
59. Special Issue, "Satellite Communications", *Proc. IEEE*, February, 1971.
60. Special Issue, "Satellite Communications", *Proc. IEEE*, March, 1977.

61. H. van Trees et al., "Communications Satellites: looking into the 1980s", *IEEE Spectrum*, December, 1977.
62. "Satellite Technology Evolves", *Microwave Systems News*, December, 1977. Design of antennas and such subsystems as multiplexers is pushing forward to meet increasing demands of satellite communications.
63. "Navy Develops Satcom Terminal", *Electronic Warfare*, February, 1978.
64. "SCDRS: New Life for Military Satcom?", *Microwave Systems News*, September, 1979.
65. D. Capuano, "Satellite-ology Soviet Style", *Telecommunications*, October, 1979. See Reference 7.45.
66. K. Black, "Command, Control, & Communication", *Telecommunications*, June, 1979. The US Joint Chiefs of Staff have identified five high priority areas, including Command, Control, & Communication, where readiness and force effectiveness must be improved.
67. D. Solomon, "Defense Telecommunications, Command and Control", *Telecommunications*, June, 1979.
68. S. Moskowitz, "Military communications systems hop on digital bandwagon", *Microwaves*, May, 1979.
69. "Improvements Badly Needed in Military Communications", *MSN*, May, 1979. Emphasis is shifting from satellites to less-vulnerable, more-flexible radio networks, and large balloon-borne transponders are being resurrected.
70. J. Boning, "Satellite Networks Meet Government Needs", *Telecommunications*, June, 1979.
71. R. Jones, *The Wizaed War: British Scientific Intelligence 1939–1945*, Coward, McCann & Geoghegan, Inc., 1980.

Electromagnetic Propagation Effects,

A1. Attenuation of RF Waves by Absorption
Reference 1

This appendix provides a series of graphs which determine the attenuation of RF waves by the atmosphere through absorption.

Blake has integrated the attenuation rate along ray paths through a standard atmosphere to find the total absorptions in decibels, which may be expected as functions of radar range *R* and angle of arrival ϵ. The calculations cover the frequency region 100-10,000 mc. Some of Blake's results are shown in Figs. A1.1(a–d). It is emphasized that this is a theoretical description of the absorption process, and has not been verified by experiment.

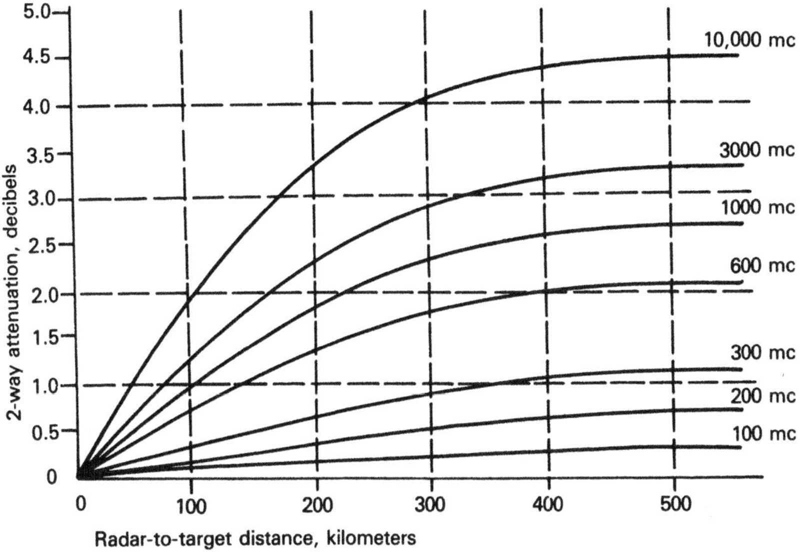

A1.1(a). *Radar atmospheric attenuation—0° ray elevation angle*

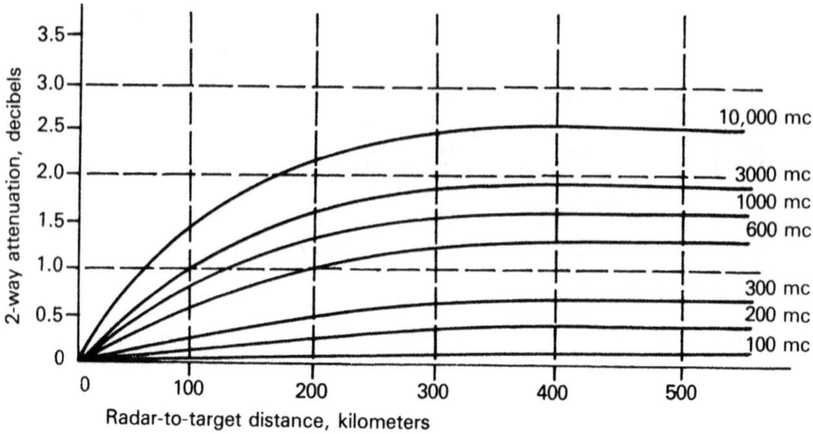

Fig. A1.1(b) *Radar atmospheric attenuation—1° ray elevation angle*

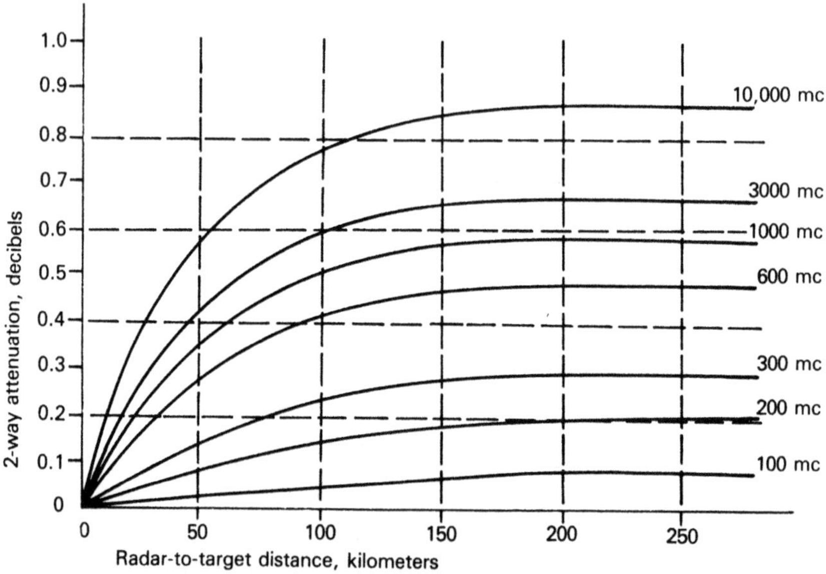

Fig. A1.1(c). *Radar atmospheric attenuation—5° ray elevation angle*

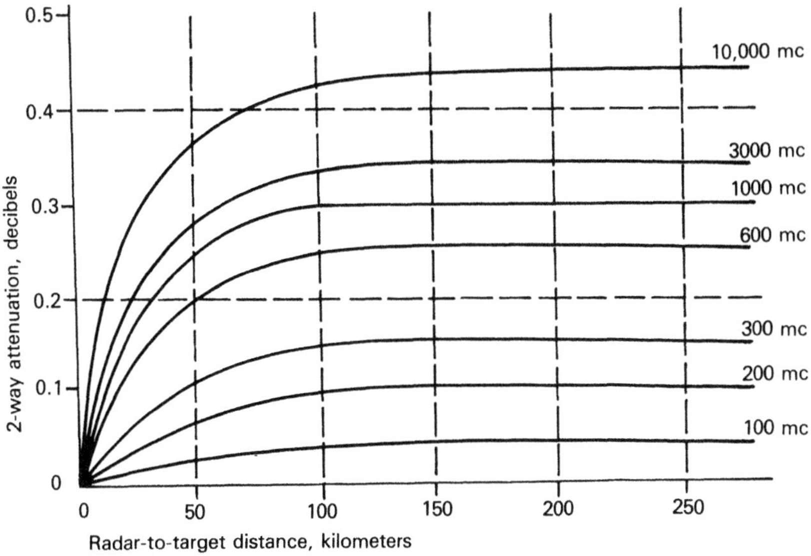

Fig. A1.1(d). *Radar atmospheric attenuation—10° ray elevation angle*

A2. Attenuation of RF Waves by Precipitation

References 2, 3

This appendix provides a graph which determines the attenuation of RF waves by the atmosphere through precipitation.

Radar meteorologists have studied the attenuation of microwaves for rainfall. Figure A2.1 compares the attenuation versus wavelengths for various rainfall rates. The results predict intense absorption at the high frequencies. Experimental evidence confirms the theoretical predictions in many cases. Generally speaking, for frequencies between 2–10 GHz, only heavy downpour will provide serious attenuation. However, radar attenuation by rain usually exceeds attenuation caused by absorption.

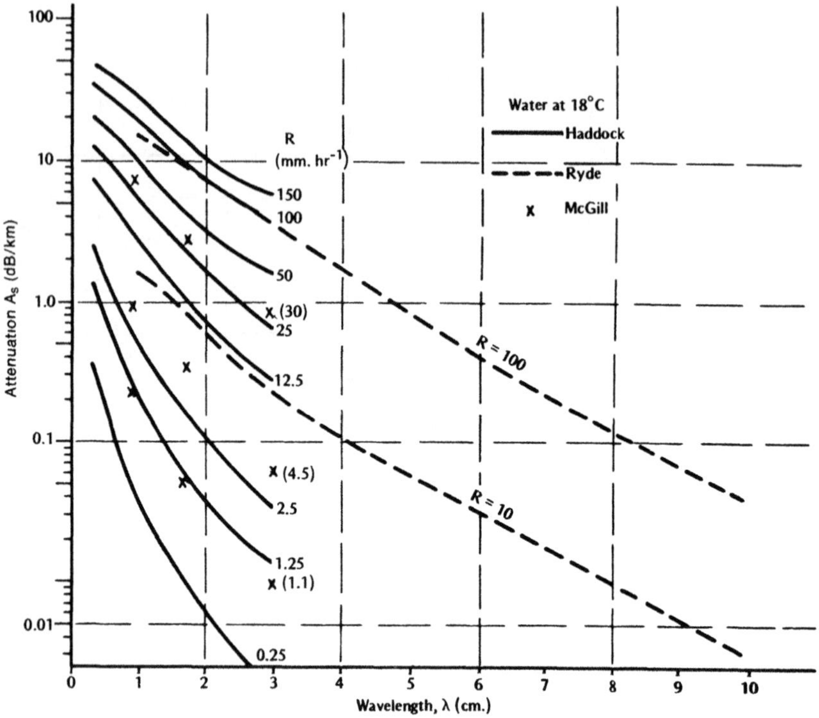

Fig. A2.1. *Attenuation versus wavelength for rainfall of various rates*

A3. Refraction of RF Waves by the Ionosphere
References 2–14

This appendix provides a series of graphs which determine the refraction of RF waves by the ionosphere.

P. Lister and T. J. Kenesha have computed the bending of radar signals by the ionosphere. The results are shown in Figs. A3.1 through A3.10.

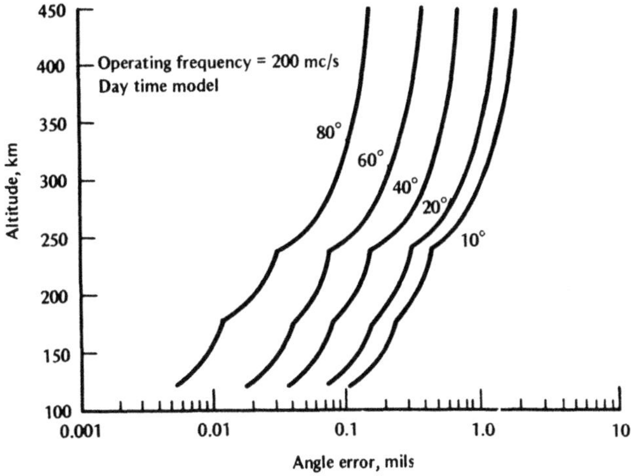

Fig. A3.1(a). *Sighting errors for various beam angles*

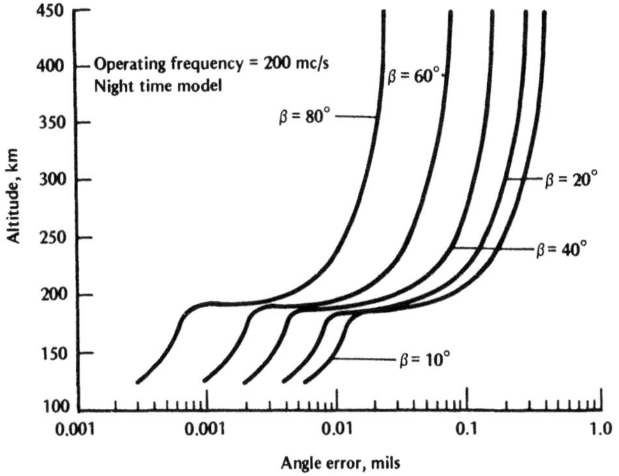

Fig. A3.1(b). *Sighting errors for various beam angles*

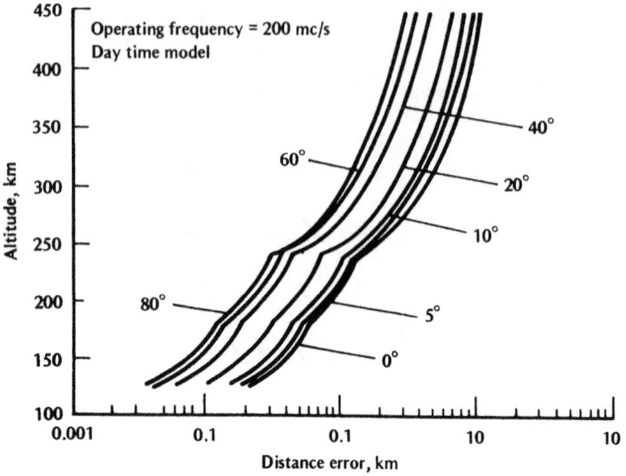

Fig. A3.1(c). *Sighting errors for various beam angles*

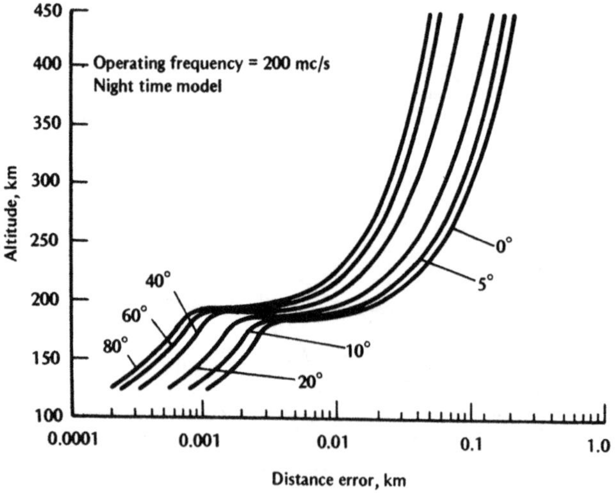

Fig. A3.1(d). *Sighting errors for various beam angles*

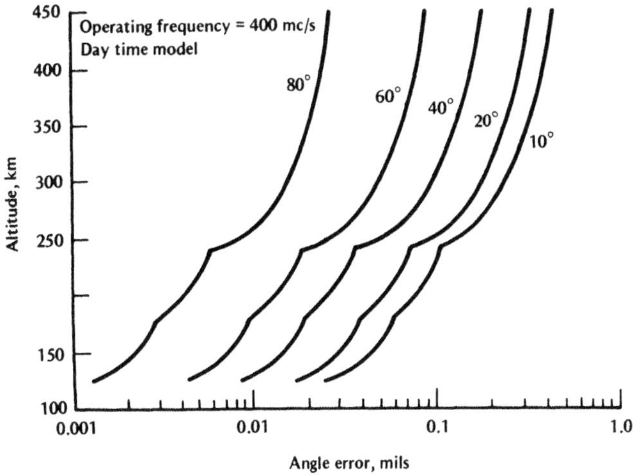

Fig. A3.2(a). *Sighting errors for various beam angles*

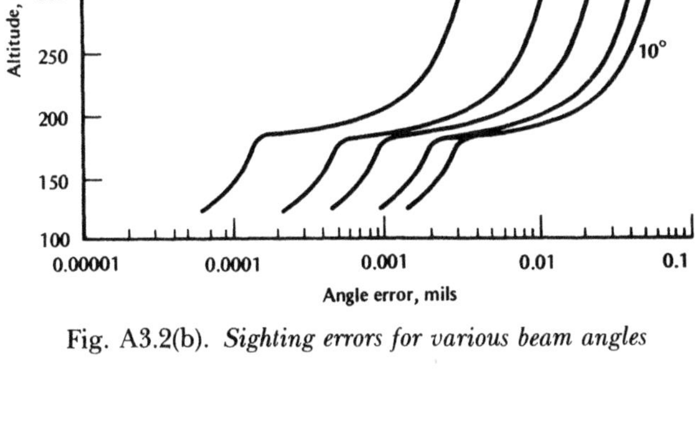

Fig. A3.2(b). *Sighting errors for various beam angles*

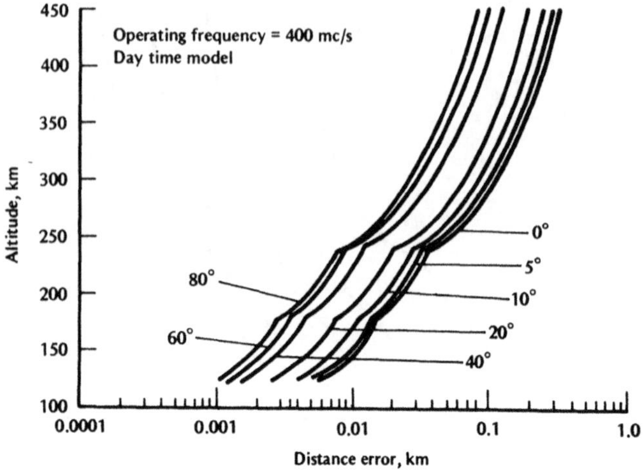

Fig. A3.2(c). *Sighting errors for various beam angles*

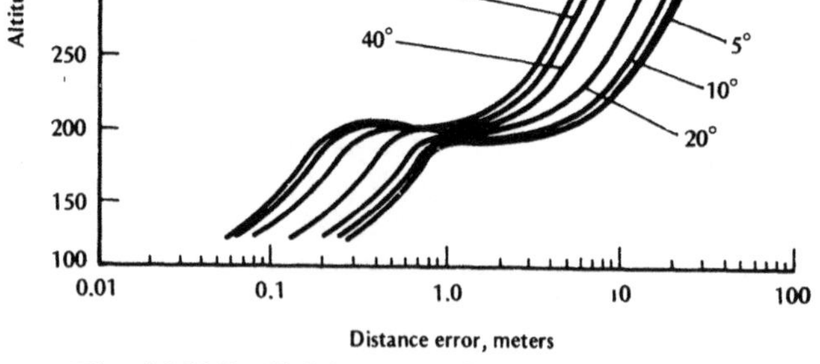

Fig. A3.2(d). *Sighting errors for various beam angles*

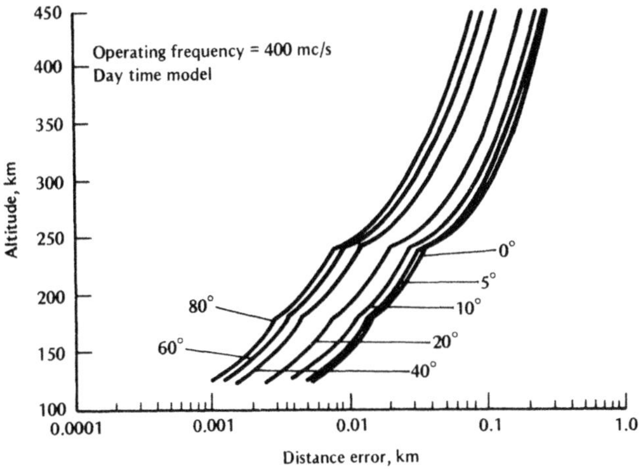

Fig. A3.2(e). *Sighting errors for various beam angles*

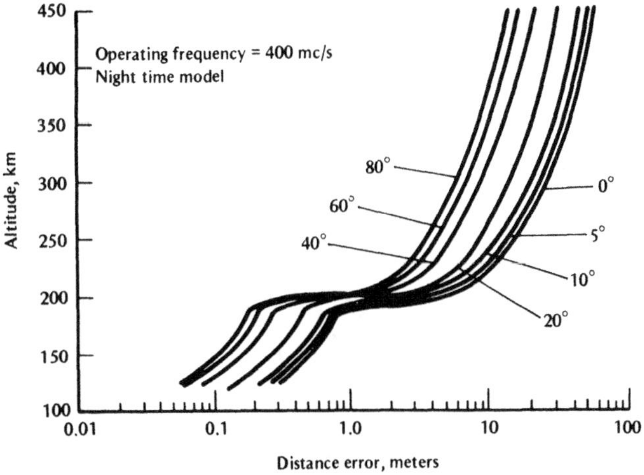

Fig. A3.2(f). *Sighting errors for various beam angles*

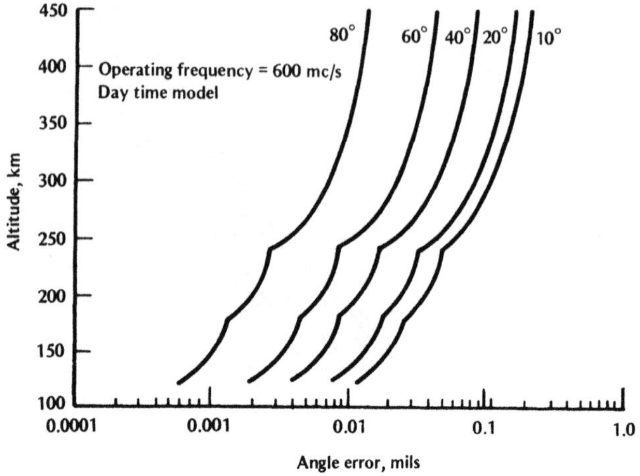

Fig. A3.3(a). *Sighting errors for various beam angles*

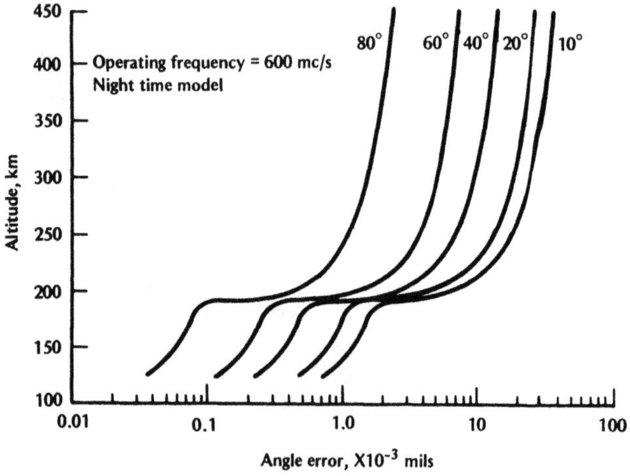

Fig. A3.3(b). *Sighting errors for various beam angles*

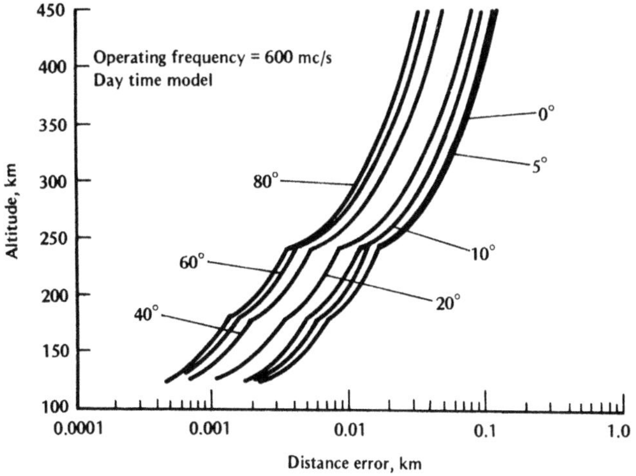

Fig. A3.3(c). *Sighting errors for various beam angles*

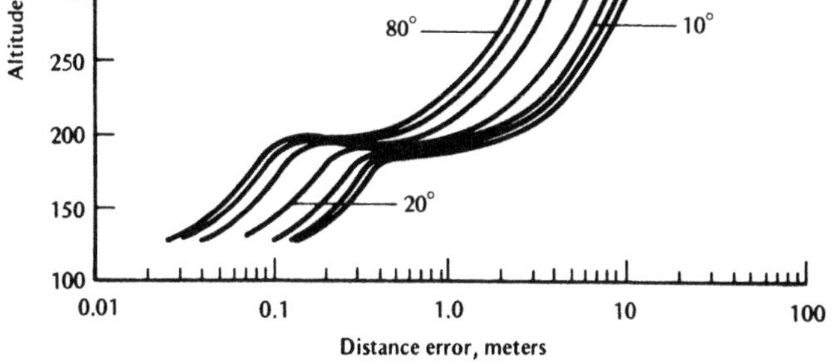

Fig. A3.3(d). *Sighting errors for various beam angles*

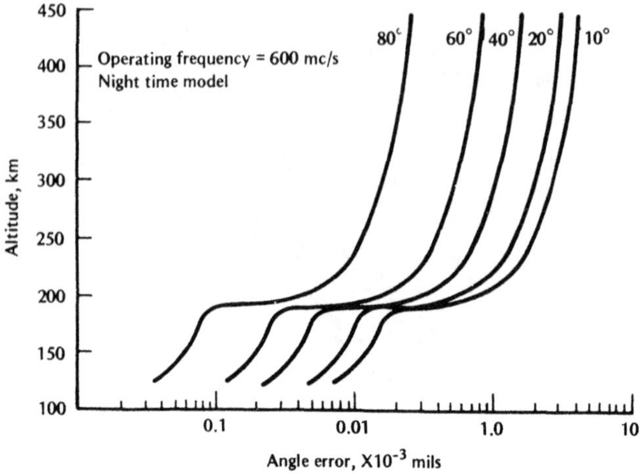

Fig. A3.3(e). *Sighting errors for various beam angles*

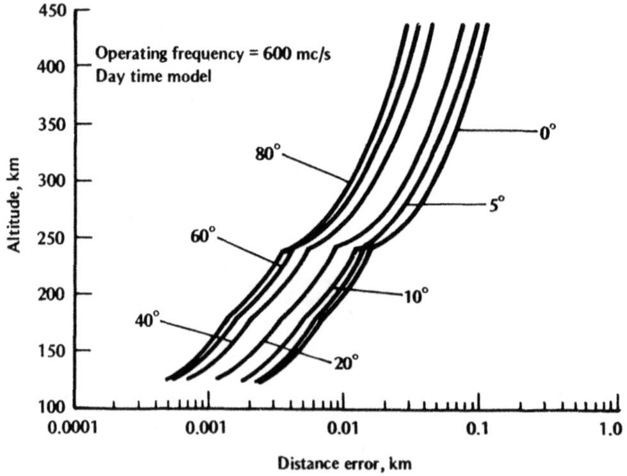

Fig. A3.3(f). *Sighting errors for various beam angles*

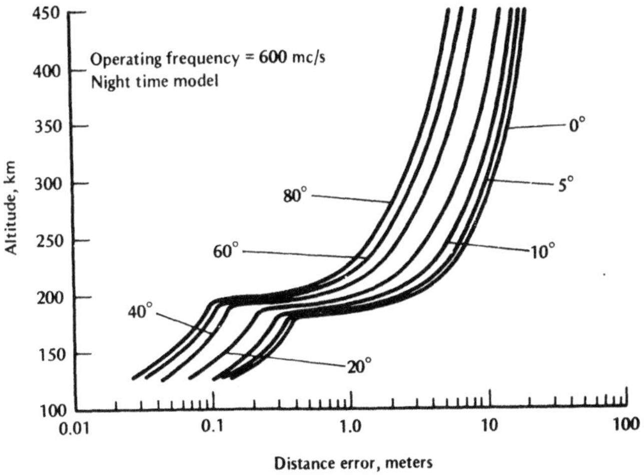

Fig. A3.3(g). *Sighting errors for various beam angles*

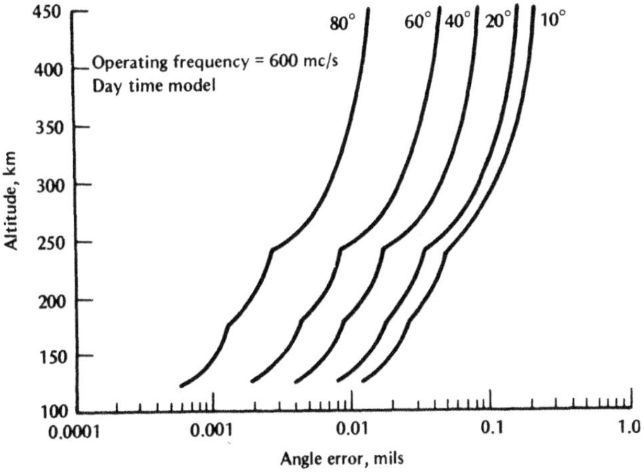

Fig. A3.3(h). *Sighting errors for various beam angles*

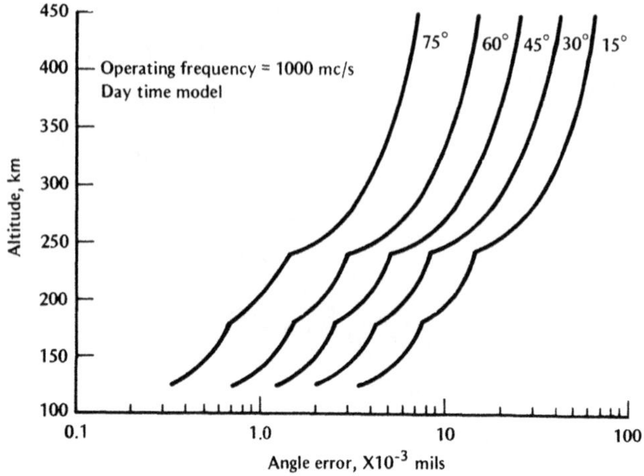

Fig. A3.4(a). *Sighting errors for various beam angles*

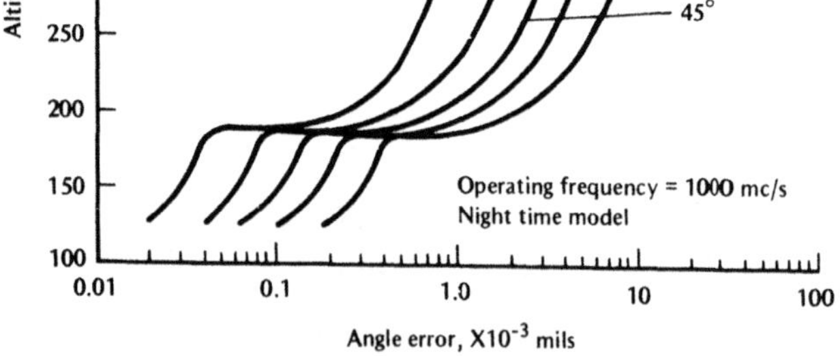

Fig. A3.4(b). *Sighting errors for various beam angles*

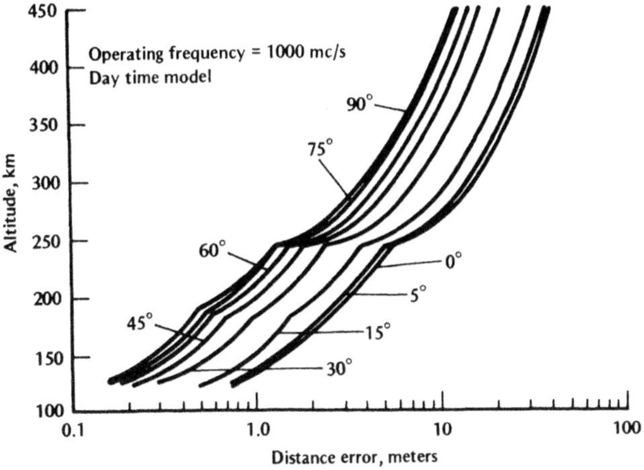

Fig. A3.4(c). *Sighting errors for various beam angles*

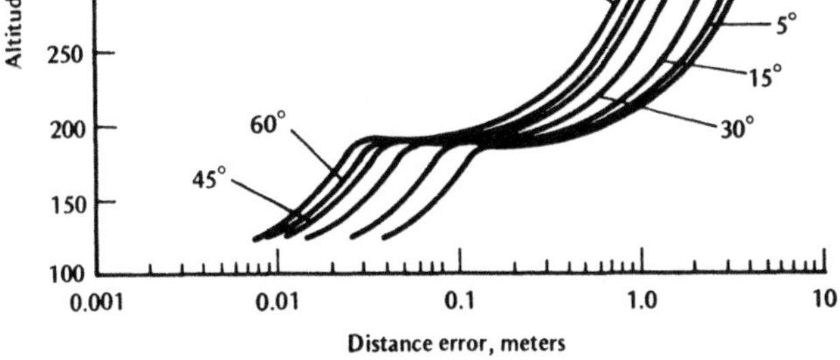

Fig. A3.4(d). *Sighting errors for various beam angles*

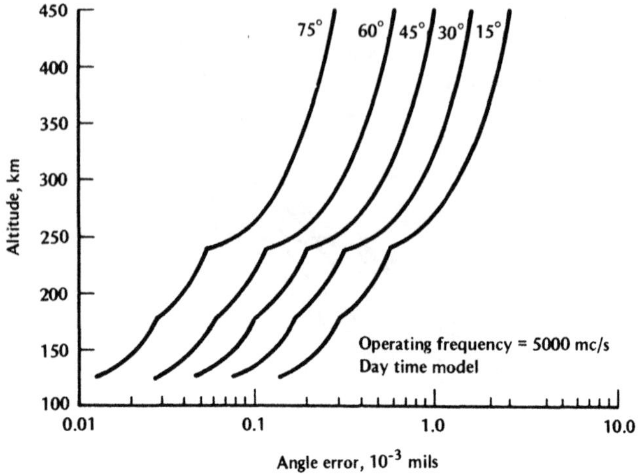

Fig. A3.5(a). *Sighting errors for various beam angles*

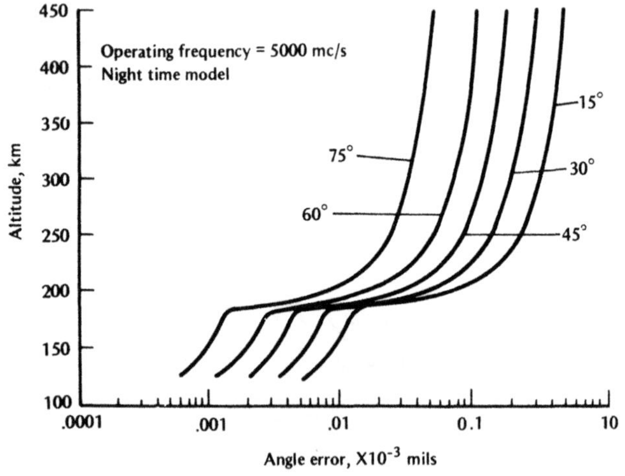

Fig. A3.5(b). *Sighting errors for various beam angles*

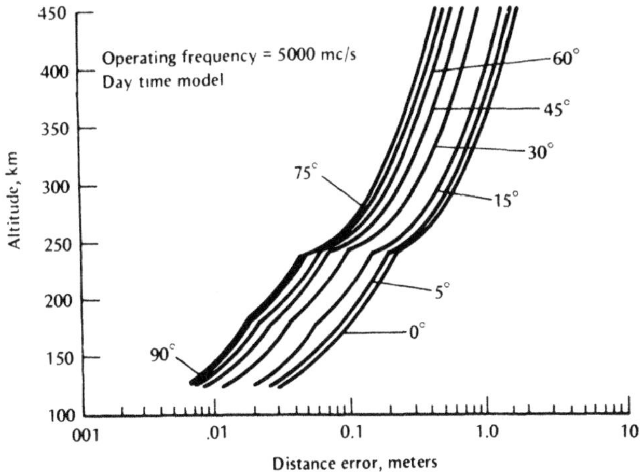

Fig. A3.5(c). *Sighting errors for various beam angles*

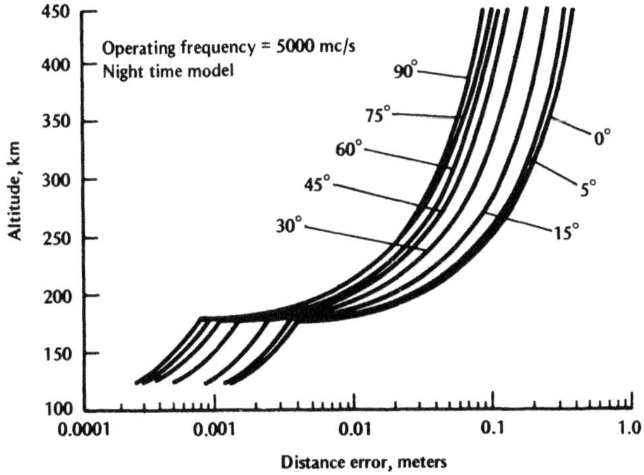

Fig. A3.5(d). *Sighting errors for various beam angles*

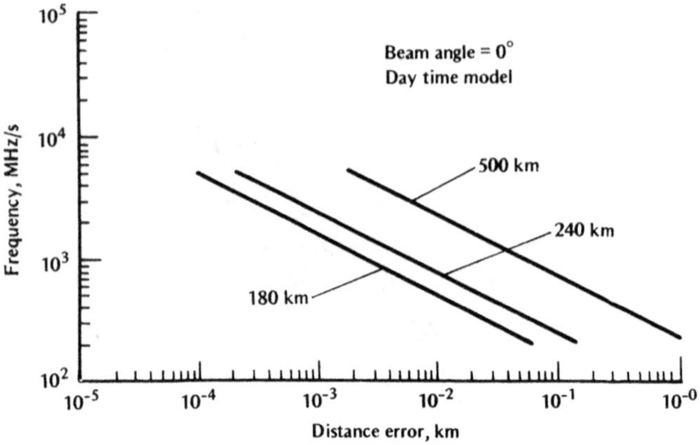

Fig. A3.6(a). *Sighting errors for various altitudes*

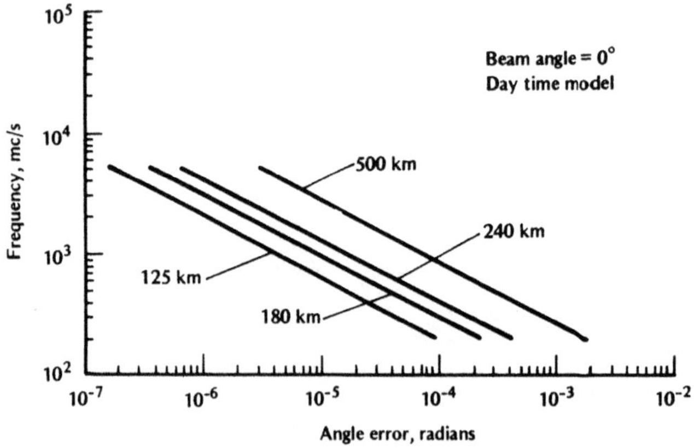

Fig. A3.6(b). *Sighting errors for various altitudes*

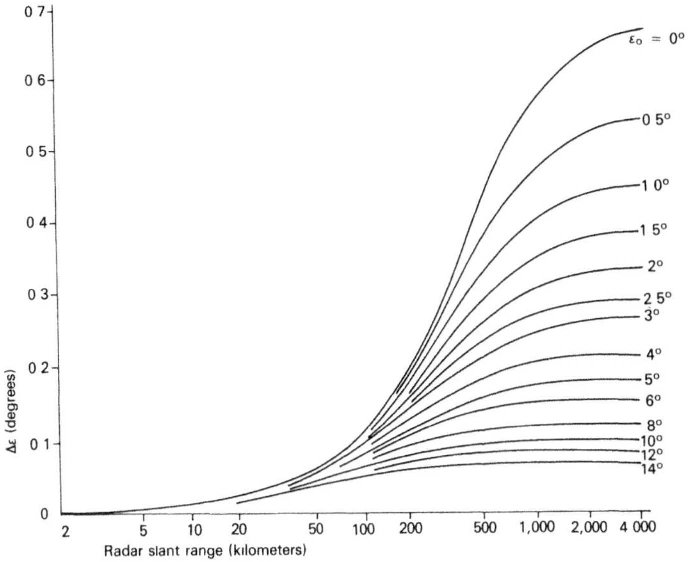

Fig. A3.7(a). *Elevation angle of arrival corrections versus range and elevation angle of arrival (0° – 14°)—April exponential model atmosphere*

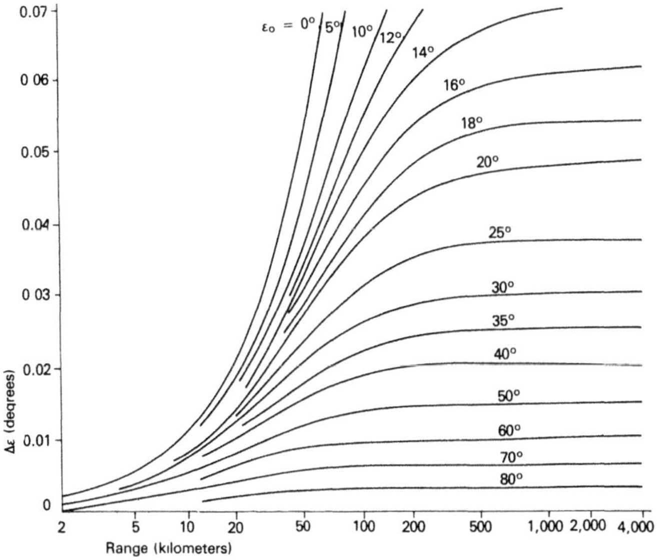

Fig. A3.7(b). *Elevation angle of arrival corrections versus range and elevation angle of arrival (0°–90°)—April exponential model atmosphere*

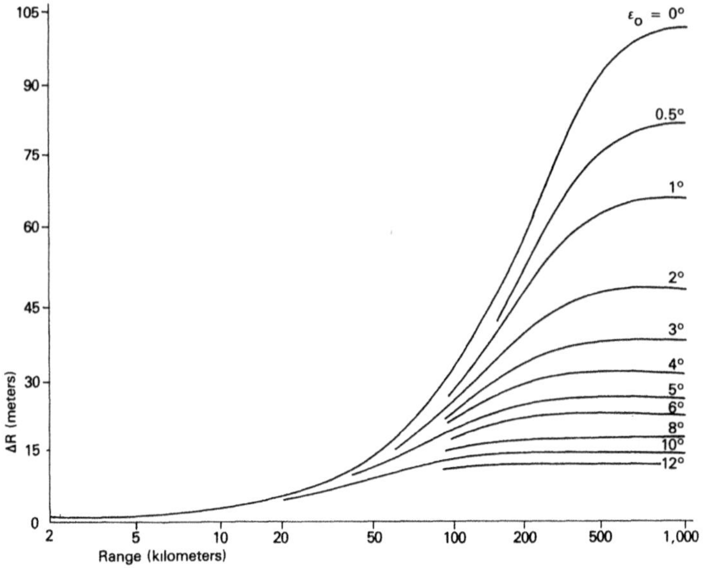

Fig. A3.8(a). ***Range corrections versus range and elevation angle of arrival (0°–12°)—April exponential model atmosphere***

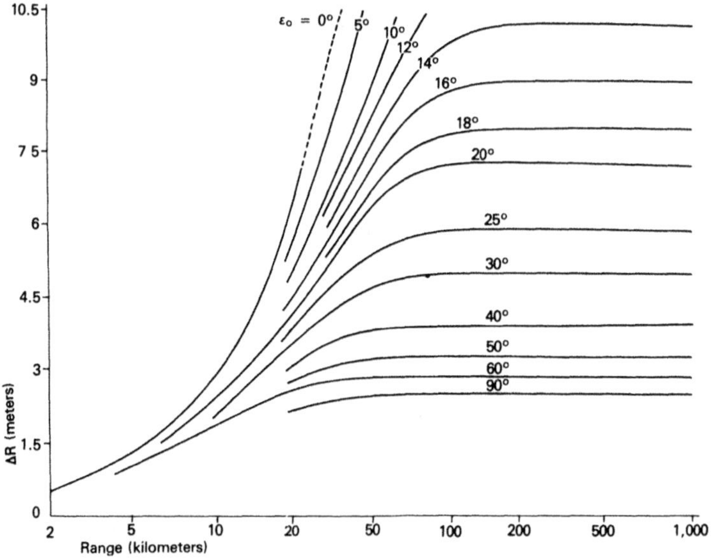

Fig. A3.8(b). ***Range corrections versus range and elevation angle of arrival (0° – 90°)—April exponential model atmosphere***

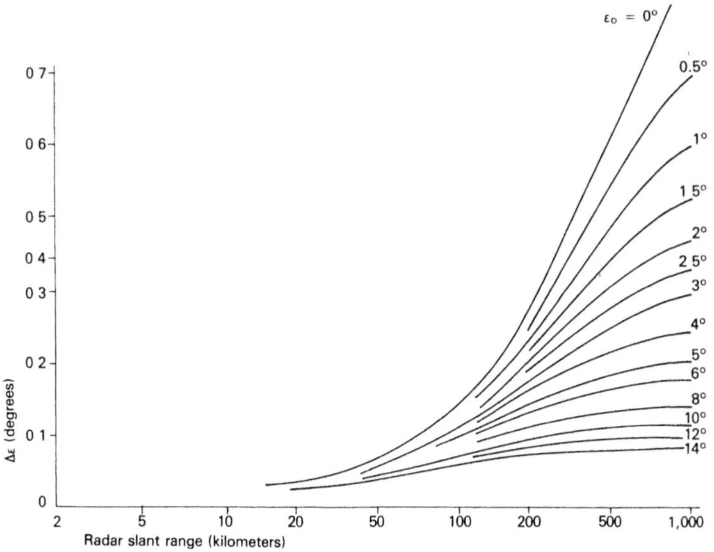

Fig. A3.9(a). *Elevation angle of arrival corrections versus range and elevation angle of arrival (0°–14°)—July exponential model atmosphere*

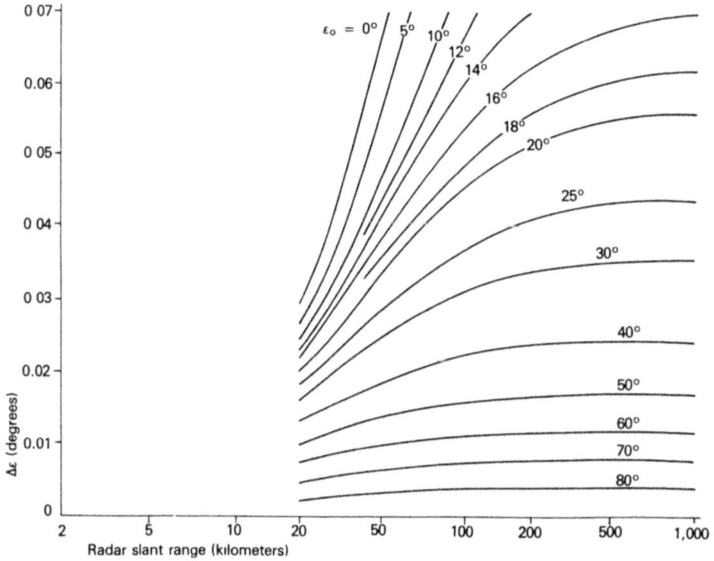

Fig. A3.9(b). *Elevation angle of arrival corrections versus range and elevation angle of arrival (0° – 90°)—July exponential model atmosphere*

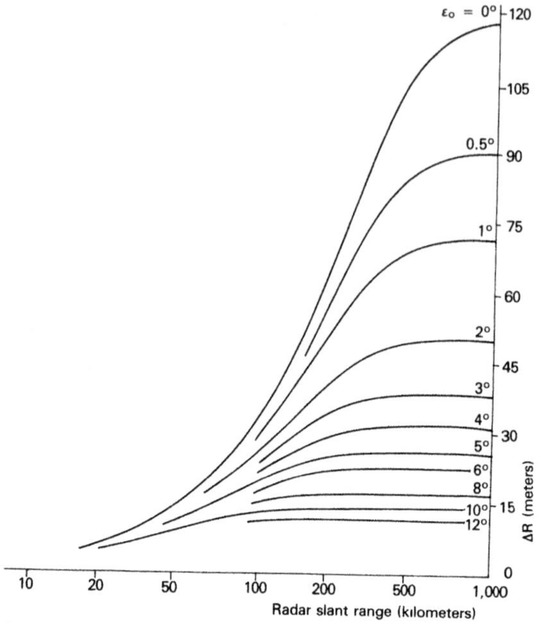

Fig. A3.10(a). *Range corrections versus range and elevation angle of arrival (0°–12°)—July exponential model atmosphere*

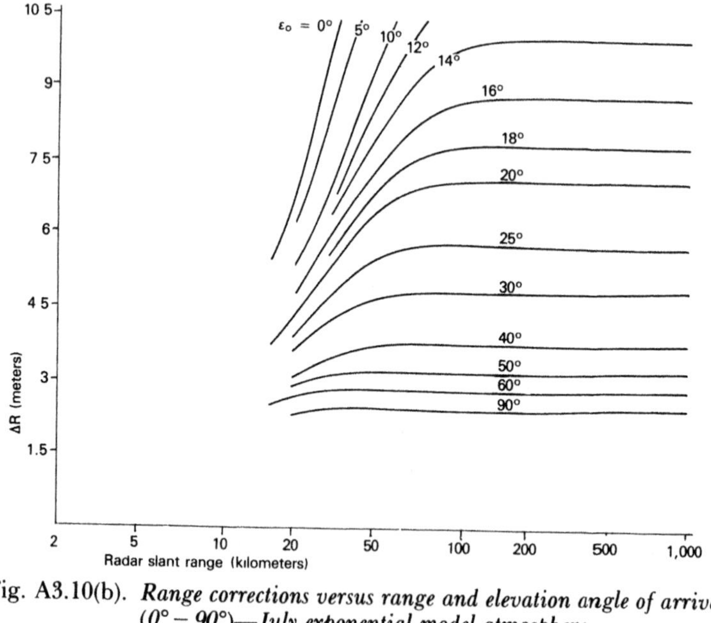

Fig. A3.10(b). *Range corrections versus range and elevation angle of arrival (0° – 90°)—July exponential model atmosphere*

References

1. L. V. Blake, *Curves of Atmospheric Absorption Loss for Use in Radar Range Calculations*, USNRL, Washington, D.C., NRL Report 5601, March 23, 1961.
2. *Handbook of Geophysics*, USAF, Macmillan, 1960.
3. D. Atlas et al., *AF Surveys in Geophysics*, No. 23, Geophysics Research Directorate, Cambridge, Mass., 1952.
4. Von Handel and Hoehndorf, "High Accuracy Tracking of Space Vehicles", *IRE Transactions on Military Electronics*, October 1955.
5. *Standard Atmosphere—Tables and Data for Altitudes to 65,000 Feet*, NACA Report No. 1235.
6. R. A. Minzner and W. S. Ripley, *The ARDC Model Atmosphere*, Air Force Cambridge Research Center (AFCRC), TN-56-204, December 1956.
7. K. S. W. Champion and R. A. Minzner, *Atmosphere Densities from Satellite and Rocket Observations*, AFCRC, May 1959.
8. *Cospar International Reference Atmosphere*, Interscience, 1961.
9. E. A. Guillemin, *Communication Networks*, Vol. II, Wiley, 1935.
10. C. M. Crain, "Survey of Airborne Microwave Refractometer Measurements", *Proceedings of the IRE*, October 1955.
11. Mason, Bauer and E. Wilson, *Radio Refraction in a Cool Exponential Atmosphere*, Lincoln Laboratories, M.I.T., Cambridge, Mass., TR-186, August 27, 1958.
12. G. H. Millman, "Atmospheric Effects in UHF and VHF Propagation", *Proceedings of the IRE*, August 1958.
13. M. I. Skolnik, *Introduction to Radar Systems*, McGraw-Hill, 1962. Particular attention is directed to Chapter 11.
14. G. R. Burrows and S. S. Attwood, *Radiowave Propagation*, Academic Press, 1949.

Index

A

ABM 11
— analysis 184
— costs 82
— display 120
— electronic equipment 185
— facility 184
— features 189
— selection 180
— testing 119
absorption, RF waves App. A1
accuracy 106, 194
— data 144
— measurement 158
advanced, aircraft 14, 19
— strategic missiles 18
— surveillance 326
adverse environment, operation 159
aided detection, operator 303
airborne, C^2 395
— radar 9
— warning 392
Airborne Warning and Control System (AWACS) 315
aircraft, advanced 14, 19
— interceptor 315
analysis
CBR&S effects 248
— ECM 215
— error 162
— nuclear effects 227
— performance 189
— systems 43
angle tracking, passive 302
— receiver 279
antenna, ECM characteristics 210
— gain 283
— jammer 296
— sidelobe suppression 284
anti-aircraft, systems 15
Anti-Ballistic Missile System (ABM) 11
antisatellite, systems 13, 15
array, advantages 160
— characteristics 147
— disadvantages 160
atmosphere, reflections 330
attenuation, RF waves App. A1, App. A2
automatic gain control (AGC) 272
avoidance, noise 267
AWACS 8, 315

B

background, sensor 363
balanced mixers 281
beacon, operations 93
Big Bird, satellite 336
biological, survivability 236
— weapons 236
bomber, costs 318
— defense 313
— Russian 318
— strategic 22
bonus point, target 228

C

cancellation, noise 269
capacity, communications channel 366
— target 202
CBR&S, effects analysis 248

— international agreements 250
— threat 236
— vulnerability 246
C^2, systems 389, 394
C^3, systems 394
CFAR 271
chaff, cloud volume 298
— dipole cross section 298
— environment 95
— glint 98
characteristics, array 147
— target 145
Chemical, Biological, Radiological Systems & Sabotage (CBR&S) 236
chemical weapons 240
collision, course 128
Command Control and Communications (C^2, C^3), systems 389, 394
command, guidance 131
communications 389, 397
— efficiency 368
— channel capacity 366
— satellite 113
— security 400
components, error 163
compression, time 289
— pulse 276
computer, operations 110
— optimization 288
— transfer function 371
computation time 172
confusion ECM 212, 259
constant false alarm rate (CFAR) 271
contrast, sensor 351, 363
— statistics 370
control, radar 117
conventional, warheads 133
costs, ABM 82
— bomber 318
— systems 48
course, pursuit 127
coverage, sensor 355
cruise missiles 22
cumulative, kill probability 37
curvefitting, error 171

D

data, accuracy 144
— C^3 398
— satellite 370
— smoothing 166, 199, 293
Data Processing System (DPS) 80
deception ECM 213, 217, 260
Defense Satellite Communications System (DSCS) 397
defense, systems 10
design, analysis 184
— features 189
— procedure 179
— radar 154, 179
— selection 180
detection, operator aided 303
— probability 88
— radar 155
— strategic 6
— target 91
development, systems 45
dipole cross section 298
direct point, target 229
Discoverer, satellites 335
discrimination, altitude 265
— signal 279
displays, ABM 120
diversity, range 268
— spatial 268
doppler, technique 279
DPS 80
DSCS 397
dynamic range, receiver 301
dynamics, target 146

E

early warning, radar 314
Early Warning, satellite 336, 375
ECCM, fast time constant 274
— FM pulse 277
— forward biasing 274

— frequency agility 267
— frequency diversity 268
— techniques 266
ECM, analysis 215
— antenna characteristics 210
— confusion 212, 259
— deception 213, 217, 260
— design 260, 262
— discrimination altitude 265
— frequency stability 211
— location 210
— objectives 261
— power 211
— signal-to-noise ratio 262
— survivability 209
— techniques 211
— threat 210
— vulnerabilities 212
effects, jamming 259
— weapons 223
efficiency communications 368
electronic equipment, ABM 185
enhancement, signal 276
engineering, systems 43
environment, chaff 95
— nuclear 99
equations, sensor 359
error, analysis 162
— components 163
— curvefitting 171
— flexure 176
— inputs 173
— model 172
— parametric curves 177
— platform 176
— radar 176
— stabilization system 173
— target 165, 173
evolution, systems 41

F

facility, ABM 184
false alarm, probability 88
fast time constant, ECCM 274
FB-111 17
filter, general equation 168
— midpoint 169
— matched 288
flexure, error 176
forward biasing, ECCM 274
frequency, agility 267
— diversity 268
— ECM 211
— FM pulse 277
— stability 211
functions, radar 179

G

gain, antenna 283
game theory geometric technique 29
— Monte Carlo technique 31
— strategy 38
general, filter equation 168
— radar equation 84
geometric technique 29
geometry, target 146
glint, chaff 98
ground, radar 5
guidance 128
— interceptor 118
— preset 129
— terminal 132

H

hard limiting 271
Hibex 12
human operator 120

I

ICBM, Alarm satellite 7, 334
— systems 16
incapacitating weapons, biological 238
— chemical 243
infrared, sensors 349
inputs, error 173
inspection, satellite 375
integration, predetection 277
— post-detection 278

intelligence, signals 391
— systems 5
interceptor, aircraft 315
— guidance 118
— operations 121
— Spartan 78
— Sprint 79
interference, suppression 275
international agreements, CBR&S 250
ionosphere, refraction App. A3

K

kill, mechanisms 133
— probability 33, 37

J

jammer, antenna gain 297
— power 296
jamming, effects 259
— environment 94, 203
— range 297
— suppression 296

L

launch window, satellite 279
lead, collision course 128
— pursuit course 127
lethal, area approach 34
— biological weapons 236
— chemical weapons 240
line, target 231
location, ECM 210
— subsystem 147
logarithmic, receivers 273
loss, ohmic 285
— propagation 286

M

maneuvering, satellite 376
matched filter, computer 288
measurement, accuracy 158
mechanism, kill 133
memory extension 289
midpoint filter 169
minimum thrust, satellite 376
Minuteman 17
Missile Site Radar (MSR) 75
mission objectives 179
— requirements 42
model, error 172
— value 49
Monte Carlo, technique 31
MSR 75
multipurpose, satellite 7, 335
— weapons 320

N

navigational triangle 124
noise, avoidance 267
— cancellation 269
— figure 287
— jamming 203
— reduction 168
— silencing 270
— suppression 269
non-catastrophic, operation 300
nuclear, effects analysis 227
— environment 99
— survivability 222
— threat 222
— vulnerability 226
— warheads 134

O

objectives, ECM 261
— mission 179
observation, time 284
ocean surveillance, satellite 8, 335
offense, systems 16
ohmic losses 285
operations, adverse environment 159
— beacon 93
— computer 110
— interceptor 121
— non-catastrophic 300

— radar 83
— telemetry 93
— tracking 92, 192
operator, aided detection 303
— over-ride 304
— trained 302
optimization, computer 288
— radar 282
OTH 329
over-ride, operation 304
over-the-Horizon (OTH) radar 329

P

PAR 72
parabolic, trajectory 122
parametric, error curves 177
passive, angle tracking 302
performance, analysis 189
— C^2 396
— radar 155
— system 45
Perimeter Acquisition Radar (PAR) 72
photography 347
platform, error 176
Polaris 17
polarization 281
policy, strategic 41
Poseidon 17
position, uncertainty 177
post detection, integration 278
power, ECM 211
— jammer 296
— transmitter 282
precipitation, attenuation App. A2
predetection, integration 277
preset, guidance 129
probabilistic, technique 33
probability, detection 88
— false alarm 88
— kill 37
— survival 214
processing, signal 118
propagation, loss 286
propulsion, satellite 377
pulse, compression 276
— repetition frequency (PRF) 280
— width 280
pursuit, course 127

R

radar, airborne 9
— control 117
— design 154, 179
— early warning 314
— error 176
— functions 179
— general equation 84
— ground 5
— MSR 75
— operations 83
— optimization 282
— OTH 329
— PAR 72
— performance 155
— sensors 350
— waveforms 101
radiological, survivability 236
— weapons 244
range, bandwidth 279
— diversity 268
— jamming 297
reacquisition 301
receiver, angle tracking 279
— dynamic range 301
— logarithmic 273
— noise figure 287
— range 279
— sensitivity 362
reconnaissance, satellite 332
reduction, noise 168
reflection, atmosphere 330
refraction, ionosphere App. A3
reliability 159
rendex-vous, satellite 376
repetition frequency, pulse 280
replicated-dispersed, target 231
requirements, mission 42
— system 143

resolution 103, 156, 194, 351
RF waves, attenuation App. A1, App. A2
— refraction App. A3
Russian, bombers 318

S

SABMIS 13
sabotage 236, 245
Safeguard 71
SAMs 317
SAMOS, satellites 6, 334
satellite, Big Bird 336
— communications 113
— contrast statistics 370
— data processing 367, 370
— Discoverer 335
— early warning 336, 375
— ICBM Alarm 7,334
— inspection 375
— launch window 279
— maneuvering 376
— minimum thrust 376
— multipurpose 7, 335
— ocean surveillance 8, 335
— propulsion requirements 377
— reconnaissance 9, 332
— surveillance 8, 335
search 115, 190
security, communications 400
selection, target 116
sensitivity, assessment model 56
— receiver 362
— time constant (STC) 275
sensor, equations 359
sensors 341
— background 363
— contrast 351, 363
— coverage 355
— IR 349
— radar 350
— TV 349
Sentinel 12
shipborne, warning 392
sidelobe, suppression 284
signal, discrimination 279
— enhancement 276
— intelligence 391
— processing 118
signal-to-noise ratio, ECM 262
silencing, noise 270
single shot, kill probability 33
siting 286
SLBM 20
smoothing 199
— position data 166
— velocity data 166
spatial, diversity 268
Spartan 78
spectral, windows 345
Sprint 79
SRAM 18
stabilization system, error 173
strategic, advanced missiles 18
— bombers 22
— policy 41
— systems 1
— warning and detection 6, 391
systems, analysis 43
— anti-aircraft 15
— anti-satellite 13, 15
— C^2 389, 394
— cost 48
— defense 10
— development 45
— engineering 43
— evaluation 41
— ICBM 16
— intelligence 5
— offense 16
— performance 45
— requirements 143
— Safeguard 71
— strategic 1
subsystem, location 147
Surface-to-Air Missiles (SAMs) 317
survivability, chemical 236
— biological 236
— ECM 209
— nuclear 222

— radiological 236
— sabotage 236
surveillance, advanced 326
— satellite 8, 335
suppression, interference 275
— jamming 296
— noise 269
survival, probability 214
synchronization and blanking 276

T

target, bonus point 228
— capacity 202
— characteristics 145
— combinations 233
— detection 91
— direct point 293
— dynamics 146
— error 165, 173
— line 231
— replicated-dispersed 231
— selection 116
— tracking 116
— trajectory 146
targeting,weapon 226
technique, doppler 279
— ECCM 266
— ECM 211
— Monte Carlo 31
— probabilistic 33
telemetry 93
terminal, guidance 132
testing, ABM 119
threat, CBR&S 236
— ECM 210
— nuclear 222
time, compression 289
— computation 172
— observation 284
tracking, geometry 146
— operation 92, 192
— target 116
trained, operator 302
trajectory, parabolic 122
— target 146
transmitter, power 282
treaty, verification 380
triangle, navigational 124
triangulation 302

U

uncertainty, position 177
— velocity 177

V

value model 49
— sensitivity assessment 56
VELA 7, 334
velocity, uncertainty 177
verification, treaty 380
vulnerability, CBR&S 246
— ECM 212
— nuclear 226
vulnerable area approach 36

W

war games 24
warhead, conventional 133
— nuclear 134
warning 5, 391
— airborne 392
— shipborne 392
— strategic 6
waveforms, radar 101
weapon, effects 223
— multipurpose 320
— targeting 226
weapons, biological 238
— chemical 243
— nuclear 134
— radiological 244
width, pulse 280
windows, satellite launch 279
— spectral 345

ABOUT THE AUTHOR

From 1950 to 1966 James N. Constant was employed in the U.S. Defense Industry by such firms as Hughes Aircraft, General Dynamics and Aerospace Corporations. During this period he worked on the development of air-to-air fire control, ground-to-air point defense, tactical and strategic area defense weapons systems, and the ECM and ECCM systems. Between 1966 and 1968 he was a strategic weapons and countermeasures consultant of the Department of Defense for defining Minuteman and Polaris MIRV and countermeasures systems.

Since 1968 he is an independent consultant whose clients include major U.S. and foreign corporations and several foreign governments.

Mr Constant is the author of *Introduction to Defense Radar Systems* and *Gravitational Action*, as well as numerous articles. He holds thirty patents for radar and optical identification and coding, real time synthetic radar, collision avoidance, reconnaissance radiometer, chaff system, digital and optical computers, digital matched filters and correlators, feedforward filters, gravitational mass detector, etc.

FIGURE CREDIT LIST

FUNDAMENTALS OF WEAPONS SYSTEMS

Part One (Offense Systems)

1.12, 1,14 From "The Dynamics of the Arms Race", G. Rathjens. *Scientific American,* April 1969. Copyright © 1969 by Scientific American. All rights reserved.

2.3 From "Military Technology and National Security", H. York. *Scientific American,* August 1969. Copyright © 1969 by Scientific American. All rights reserved.

5.2 From "Multiple-Warhead Missiles", H. York. *Scientific American,* November 1973. Copyright © 1973 by Scientific American. All rights reserved.

6.9 From "Missile Submarines and National Security", H. Scoville, *Scientific American,* June 1979. Copyright © 1972 by Scientific American. All rights reserved.

6.26 From "Antisubmarine Warfare and National Security", R.L. Garwin. *Scientific American,* July 1972. Copyright © 1972 by Scientific American. All rights reserved.

7.1 From "Defense Against Bomber Attack", R. English and D. Bolef. *Scientific American,* August 1973. Copyright © 1973 by Scientific American. All rights reserved.

1.11 Reprinted from "Strategic Warfare", D. Fink. *Science and Technology,* October 1968.

6.2, 6.3 Reprinted from "The Ocean", S. Telson. *International Science and Technology,* February 1966.

Tables 1.1, 2.5, 6.6, 6.8, 7.1, 7.2 From Alton H. Quanbeck and Barry M. Blechman, *Strategic Forces: Issues for the Mid-Seventies,* Copyright © 1973 by the Brookings Institution.

6.7, 6.8, 6.10-6.13 From "Trends in Marine Technology", O.H. Oakley (Naval Skip Engineering Center, NAVSEC). *Astronautics and Aeronautics,* April 1966. A publication of the American Institute of Aeronautics and Astronautics.

3.8, 3.9 From "Empirical Formulas for Ballistic Reentry Trajectories", R. Bate and R. Johnson. *ARS Journal,* December 1962.

4.2 From "Air Ionization in the Hypersonic Laminar Wake of Sharp Cones", F.L. Fernandez and E.S. Levinsky. *AIAA Journal*, October 1964. A publication of the American Institute of Aeronautics and Astronautics.

4.3 From "Hypersonic Wakes and Trails", L. Lees. *AIAA Journal*, March 1964. A publication of the American Institute of Aeronautics and Astronautics.

1.21 From "The Strategic Survey", International Institute for Strategic Studies, London.

6.6 From "Oceanography in Perspective", V.L. Giannini. IEEE 9th Annual Lecture Series, April 24, 1969. Copyright © 1969 by the Institute of Electrical and Electronics Engineers.

1.18 From "Ballistic Missile Defense Radars", G.M. Johnson. *IEEE Spectrum*, March 1970. Copyright © 1970 by Institute of Electrical and Electronics Engineers.

6.39 From "Acoustic Communication is Better than None", V.C. Anderson. *IEEE Spectrum*, October 1970. Copyright © 1970 by the Institute of Electrical and Electronics Engineers.

6.51 From "Radio Communication in the Sea", R.K. Moore. *IEEE Spectrum,* November 1967. Copyright © 1967 by the Institute of Electrical and Electronics Engineers.

6.38, 6.40, 6.42-6.46 From "Sonics in the Sea", L. Batchelder, IEEE Proceedings vol. 53, October 1965. Copyright © 1965 by IEEE.

Table 2, Appendix 1 From *Rocket Propulsion Elements*, G.P. Sutton, 2nd Edition, John Wiley and Sons, N.Y., 1956.

6.16 From "Soviet Naval Deployment", *Los Angeles Times*, November 8, 1970. Copyright © 1979 by Los Angeles Times.

4.1 From *Introduction to Radar Systems,* M.I. Skolnik, McGraw-Hill, N.Y., 1962.

2.1, 2.4, 2.5, 2.7, Table 2.1 From "The Next ICBM's", *Space/Aeronautics Magazine*, March 1966, Ziff-Davis N.Y.

2.2 From "Strategic Warfare", *Space/Aeronautics Magazine*, January 1969, Ziff-Davis N.Y.

2.13, 6.23, 6.24 From "ULMS: Strategic Emphasis Shifts Seaward", F. Leary, *Space/Aeronautics Magazine,* June 1970, Ziff-Davis, N.Y.

6.15, 6.17, 6.18, 6.25, 6.50 From "Who's Winning the Undersea War?" F. Leary, *Space/Aeronautics Magazine,* February 1969, Ziff-Davis, N.Y.

6.30, 6.34, 6.49 From "Search for Subs", F. Leary, *Space/Aeronautics Magazine,* September 1965, Ziff-Davis, N.Y.

Table 6.3 From "Centralized Direction Helps Navy's ASW Program", *Space/Aeronautics Magazine,* September 1965, Ziff-Davis, N.Y.

6.36, 6.37, Table 6.7 From "VSX-Sub Hunting from the Air", M.P. London, *Space/Aeronautics Magazine,* April 1969, Ziff-Davis, N.Y.

Part Two (Defense Systems)

6.1 Courtesy of Boeing Aerospace Company.

2.17 From "Anti-Ballistic Missile Systems", R.L. Garwin and H.A. Bethe. *Scientific American*, March 1969. Copyright © 1969 by Scientific American. All rights reserved.

7.1, 7.9, 7.10 From "Reconnaissance and Surveillance Systems", T. Greenwood. *Scientific American,* February 1973. Copyright © 1973 by Scientific American. All rights reserved.

7.7 Reprinted from "Remote Sensing", D. Parker and M. Wolff. *International Science and Technology*, July 1965.

Table 6.1 From Alton H. Quanbeck and Barry M. Blechman, *Strategic Forces : Issues for the Mid-Seventies,* Copyright © 1973 by the Brookings Institution.

2.1-2.4, Table 2.1 From "Ballistic Missile Defense Radars", G.M. Johnson. *IEEE Spectrum,* March 1970. Copyright © 1970 by Institute of Electrical and Electronics Engineers.

2.11 From "Design of Electronic Scanning Radar Systems", P.J. Kahrilas. IEEE Proceedings, vol. 56, no. 11, 1968. Copyright © 1968 by the IEEE.

2.12, Tables 3.3-3.5 From "The AN/FPS-85 Radar System", by J. Emory Reed, IEEE Proceedings, vol. 57, no. 3, March 1969. Copyright © 1969 by IEEE.

4.4, 4.6, 4.8-4.12, Table 4.1 From "Aircom Survivability Considerations", O.G. Becker and H.E. Struck, 2nd Winter Convention on Military Electronics IRE, Los Angeles, February 1961. Copyright ©1961 by IEEE.

3.4, 3.8 From "Tracking Error Analysis of a Range Ship", K.E. Peress and A. Schwartz. *Sperry Engineering Review*, Fall 1962. Reprinted by permission of Sperry Rand Corporation.

8.2 From "Threat Simulation for Radar Warning Systems", *Countermeasures,* August 1975. Reprinted by permission of MILITARY ELECTRONICS/COUNTERMEASURES, 2065 Martin Ave., Suite 104, Santa Clara, CA 95050, USA.

8.3 From "Command, Control and Technology", L. Paschall, *Countermeasures*, July 1976. Reprinted by permission of MILITARY ELECTRONICS/COUNTERMEASURES, 2065 Martin Ave., Suite 104, Santa Clara, CA 95050, USA.

Tables 1.1-1.3 From "Strategic Warfare", *Space/Aeronautics Magazine*, January 1969, Ziff-Davis N.Y.

7.13 From "Catching a Satellite", J. Fasca, *Space/Aeronautics Magazine,* September 1965. Ziff-Davis N.Y.

2.6, 2.7 From "General Radar Equation", *R&D Handbook 1962-1963,* W.M. Hall, *Space/Aeronautics Magazine*, Ziff-Davis, N.Y.

COLOPHON

letter: Baskerville 10/11 and 9/10
setter: European Printing Corporation
cover-design: Bram de Blécourt

GPSR Compliance

The European Union's (EU) General Product Safety Regulation (GPSR) is a set of rules that requires consumer products to be safe and our obligations to ensure this.

If you have any concerns about our products, you can contact us on ProductSafety@springernature.com

In case Publisher is established outside the EU, the EU authorized representative is:

Springer Nature Customer Service Center GmbH
Europaplatz 3
69115 Heidelberg, Germany

Batch number: 10830935

Printed by Printforce, the Netherlands